Linear Control Systems

THE UNIVERSITY SERIES IN MATHEMATICS

Series Editors: Sylvain E. Cappell, *New York University*
Joseph J. Kohn, *Princeton University*

Recent volumes in the series:

COMPLEX ANALYSIS AND GEOMETRY
Edited by Vincenzo Ancona and Alessandro Silva

CONTEMPORARY GEOMETRY (J.-Q. Zhong Memorial Volume)
Edited by Hung-Hsi Wu

ELLIPTIC COHOMOLOGY
Charles B. Thomas

AN INTRODUCTION TO ALGEBRAIC NUMBER THEORY
Takashi Ono

LINEAR CONTROL SYSTEMS: with solved problems and MATLAB examples
Branislav Kisačanin and Gyan C. Agarwal

MATRIX THEORY: A Second Course
James M. Ortega

PROBABILITY MEASURES ON SEMIGROUPS: Convolution Products, Random Walks, and Random Matrices
Göran Högnäs and Arunava Mukherjea

RECURRENCE IN TOPOLOGICAL DYNAMICS: Furstenberg Families and Ellis Actions
Ethan Akin

A SHORT INTRODUCTION TO INTUITIONISTIC LOGIC
Grigori Mints

TOPICS IN NUMBER THEORY
J. S. Chahal

VARIATIONS ON A THEME OF EULER: Quadratic Forms, Elliptic Curves, and Hopf Maps
Takashi Ono

A Continuation Order Plan is available for this series. A continuation order will bring delivery of each new volume immediately upon publication. Volumes are billed only upon actual shipment. For further information please contact the publisher.

Linear Control Systems

with solved problems and MATLAB examples

Branislav Kisačanin, Ph.D.
Delphi Delco Electronics Systems
Kokomo, Indiana

Gyan C. Agarwal, Ph.D.
University of Illinois at Chicago
Chicago, Illinois

Kluwer Academic / Plenum Publishers
New York, Boston, Dordrecht, London, Moscow

Library of Congress Cataloging-in-Publication Data

Kisacanin, Branislav, 1968–
Linear control systems: with solved problems and MATLAB examples/by Branislav Kisacanin, Gyan C. Agarwal.
p. cm.
Includes bibliographical references and index.
ISBN 0-306-46743-7
1. Linear control systems. 2. Linear control systems—Problems, exercises, etc. 3. MATLAB. I. Agarwal, Gyan C. II. Title.

TJ220 .K57 2002
629.8′32—dc21

2001053984

ISBN 0-306-46743-7

233 Spring Street, New York, New York 10013

http://www.wkap.nl/

10 9 8 7 6 5 4 3 2 1

A C.I.P. record for this book is available from the Library of Congress

Printed in the United States of America

To Saška & To Sadhna

Contents

Foreword

In science and engineering, a proper way to master theory is to solve relevant and meaningful problems that provide a bridge between theory and applications. Problem solving is necessary not only as a stepping stone towards the design of real systems and experiments, but also to reveal the scope, flexibility, and depth of the theoretical tools available to the designer. In this book, the authors present an excellent choice and a lucid formulation of a wide variety of problems in control engineering. In particular, their constant reliance on MATLAB in the problem-solving process is commendable, as this computational tool has become a standard and globally available control design environment.

The chapter on theoretical elements of control theory, which precedes the problem-solution part of the book, sets a proper background for problem-solving tasks. In their presentation, the authors struck the right balance in achieving a self-contained text without overwhelming the reader with detailed and exhaustive theoretical arguments.

Finally, the opening chapter on the history of automatic control is a welcome part of the book. The authors admirably describe the events and concepts that have evolved over the centuries to the present-day control theory and technology. Theoretical analysis and problem-solving processes are not only useful in understanding what the world is, but also what it can become. In attempting to predict future development in science and technology we are greatly aided by the history of scientific innovations and discoveries.

The present book is an excellent piece of work and should be on the shelf of every student and practicing engineer of automatic control.

May 2001

Dragoslav D. Šiljak
Santa Clara, California

Preface

This book is a self-contained exposition of the theory of linear control systems and the underlying mathematical apparatus. It has more than 250 solved problems, numerous illustrative examples, and over 70 figures and diagrams. In addition, MATLAB examples provide a good introduction to this powerful design and simulation tool. A historical overview of the original ideas in control theory is provided to describe the evolution of the theory from its early stages of development.

For whom is this book?

It was written for students and engineers interested in Control Systems and Signal Processing, typically the first-year graduate students of Engineering. There is more than one way this book can be used: it contains a sufficient amount of theory to allow its use as a textbook, it has many solved problems so it can be a good supplement to a more advanced text, and it has enough of both to be used as a self-study guide. A suitable choice of material can be made to fit the format of the course and preferences of the instructor and the audience.

What is in the book?

The book has three major parts:

Part I is an overview of the history and theory of the subject. It consists of Chapters 1 and 2:

- In Chapter 1 we investigate the historical development of the automatic feedback control from the Antiquity to the present day.
- In Chapter 2 we present the fundamental concepts of control theory. We start from the representation of systems using the states and continue with a discussion of the most important system properties: stability, controllability, and observability. We end that Chapter by discussing the most important design techniques: state feedback, optimal control, state observation, and state estimation (Kalman filtering).

Part II consists of Chapters 3, 4, and 5, which contain solved problems categorized by the basic type of systems (continuous and discrete) and by the topic (system representations, system properties, and design techniques):

- In Chapter 3 the reader will find solved problems about continuous-time systems. It starts with simple differential equations and some matrix theory and continues with matrix representation of simultaneous differential equations and input-output representation of control systems. Then it moves on to problems of state representation of systems, stability, controllability and observability, canonical forms, and finishes with design techniques for state feedback, optimal control, and state estimation and observation.

- Chapter 4 is very similar in format to Chapter 3 and contains solved problems on discrete-time systems. It starts with simple difference equations and input-output representation of discrete-time systems. Then, just like Chapter 3, it discusses problems of state representation, system properties, and design techniques.

- Chapter 5 contains several exercise problems.

Part III consists of three Appendixes:

- Appendix A is a quick introduction to the basic syntax and functionality of MATLAB, a powerful numerical and simulation tool for many engineering disciplines, including Control Theory.

- In Appendix B we review the mathematical tools and notation used in the book: differential and difference equations, Laplace and z transforms, and matrices and determinants. To make the exposition more interesting we also trace the origins of these branches of mathematics.

- Appendix C is a compilation of the basic notions and results in matrix theory, the most important mathematical tool used in this book. It covers similarity of matrices, important classes of matrices and their properties, and some important techniques such as singular value decomposition (SVD). Most of the results are given with proofs and some are illustrated by examples.

To make the book easier to read ...

To help the reader we included a detailed Index at the end of the book and used standard textbook conventions. The asterisk in a subsection name denotes advanced material that can be skipped during the first reading. When we reference a book or an article, we use a bracketed number, e.g., [22]. All references are listed alphabetically in the Bibliography section towards the end of the book. All definitions, theorems, problems, examples, and exercises are numerated by sections. For example, Problem 4.5.2 is the second problem in Section 5 of Chapter 4.

In addition to that, we made numerous historical remarks throughout the book. This should help the reader understand and adopt the material faster and, at the same time, provide some useful information. Also, we used the inverted pendulum on a cart to illustrate many of the concepts of modern controls. This particular system offers the best of both worlds: it is complex enough to present some challenge, yet simple enough to be visualized and intuitively understandable. Another helpful feature of the book is that most of the problems and theorems are accompanied by detailed solutions and complete proofs. Finally, the MATLAB programs from this book are available on the enclosed CD.

Suggestions? Comments?

If you have any suggestions or comments about the book, please e-mail them to: `b.kisacanin@ieee.org`

Acknowledgments

We would like to thank the publisher, Kluwer Academic / Plenum Publishers, and the Editor, Ana Bozicevic, for their interest and help in publishing this book. We would also like to acknowledge the support and encouragement from Professor Dragoslav D. Šiljak from the Santa Clara University, who wrote the Foreword and provided numerous suggestions. Since the authors of the present book met through the efforts of Professor Miodrag Radulovački from the University of Illinois at Chicago to keep the Yugoslav universities connected to the world even in the worst of times, he too deserves a lot of credit for this book. Finally, we would like to thank our wives, Saška and Sadhna, for their support and patience! Saška also helped us create the Index.

May 2001

Branislav Kisačanin
Kokomo, Indiana

Gyan C. Agarwal
Chicago, Illinois

Part I

Theory of linear control systems

Chapter 1

Historical overview of automatic control

In this Chapter we review the main results of the theory of automatic control. Our presentation follows the historical development of the control theory and assumes at least the undergraduate level of exposure to this subject. The material is organized as follows:

- Section 1.1: Automatic control before the telecommunications revolution in the 1930's
- Section 1.2: The classical period of automatic control (between the 1930's and the 1950's)
- Section 1.3: The modern control theory (after the 1950's)

The basic techniques of the modern control theory are presented in a greater detail in Chapter 2.

$$|kH(j\omega)| \gg 1 \quad \Rightarrow \quad \frac{H(j\omega)}{1 + kH(j\omega)} \approx \frac{1}{k}$$

1.1 Automatic control before the 1930's

The first significant discovery in the field of feedback control was Watt's fly-ball governor. However, the history of the automatic feedback control systems dates back much earlier. In this Section we describe the history of feedback control systems from the Antiquity until the 1930's, from ancient water-clocks to electrical power distribution.

Antiquity and Middle Ages. According to Vitruvius' *De Architectura*, among many inventions of the first of great engineers from Alexandria, Ctesibius (also spelled Ktesibios; fl. c. 270 BC), was a water-clock — clepsydra. According to Vitruvius' description, it could have looked like Figure 1.1. It used two forms of feedback control: a siphon S to periodically recycle itself and a floating valve F to ensure a constant water level in tank A, and thus a constant flow of water into tank B.

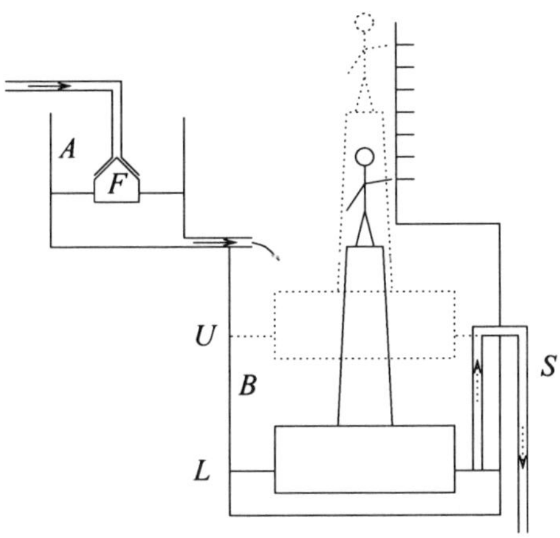

Figure 1.1: Ctesibius' clepsydra, reconstructed from a description by Vitruvius. While the floating valve F ensured a constant water flow into the tank B, the siphon S periodically reset the clock from the upper level U back to the lower level L.

Three hundred years later, the last of the great Alexandrian engineers, Hero, described a number of ingenious devices which employed feedback. His inventions also included the oldest known devices powered by steam.

The Hellenistic tradition was continued in the Arabic world through the Middle Ages. The water-clocks and other hydraulic and pneumatic devices of this period were based on the floating valve and the siphon principle.

Renaissance in engineering. Probably the first modern invention based on the automatic feedback was the 17th century Dutch windmill which was kept facing the wind by an auxiliary vane that rotated the entire upper part of the mill. Later, in England, an additional speed regulation was achieved using the shutter sails which helped compensate for the variability in the wind speed: when the wind was stronger, the shutters were automatically opened.

Precise time-keeping was another problem which required feedback mechanisms. The early clocks were driven by weights which moved under the force of gravity. Their motion was kept constant by the rotating vanes, which used the frictional resistance of air to provide the feedback. The constant motion was achieved when the resistance (proportional to the speed) was in equilibrium with the gravity.

Worth mentioning are also the 18th century Reaumur's devices for control of incubators: the temperature dependent level of mercury in a U-tube moved an arm which controlled the draft to a furnace[1]. Reaumur described the idea of the negative feedback as

> making use of these [extra] degrees of heat against themselves, so as to cause them to destroy themselves.

Watt's inventions. The first steam engine was patented in England by Savery in 1698 and improved by Newcomen in 1712 (both inventions were based on the discoveries made by Papin), but it was only after Watt had dramatically improved their efficiency that the revolution in automation really began. In 1765 Watt made his first and the most important invention, a separate steam condenser, which saved the latent heat from dissipation.

In 1788, while working on the design of the throttle valve for manual regulation of the engine, Watt learned from his partner, Boulton, about a method for changing the gap between the grindstones according to their rotation speed: the millers wanted the gap to be smaller when the stones turned faster[2]. He quickly adapted this idea to control his valve and thus invented the centrifugal governor. This invention triggered a new revolution: it was the first widely used feedback mechanism. Figure 1.2 shows its principle: as the shaft S rotates, the centrifugal force pulls the rotating masses A and B apart. This is translated into the vertical motion of the ring R whose position controls the draft to the burner via a throttle valve. This simple mechanism was used to set the steam engine's running speed.

Theoretical analysis. Watt's governor had a few flaws, and many patents were granted for attempts to correct them. But Watt's original design was very simple and for many applications it was a satisfactory solution. Among its most serious flaws were the need for careful maintenance, the lack of power to move the actuator, and "hunting," the oscillatory motion of the fly-balls. Many engineers of the 19th century noticed that their improvements on the Watt's

[1] This idea can be traced to the 17th century alchemist Drebbel.

[2] At that time Watt and Boulton were building the Albion Mills, the first steam-powered mill, hence their interest in the milling technology. Until that time steam engines were mostly used to pump water out of mines.

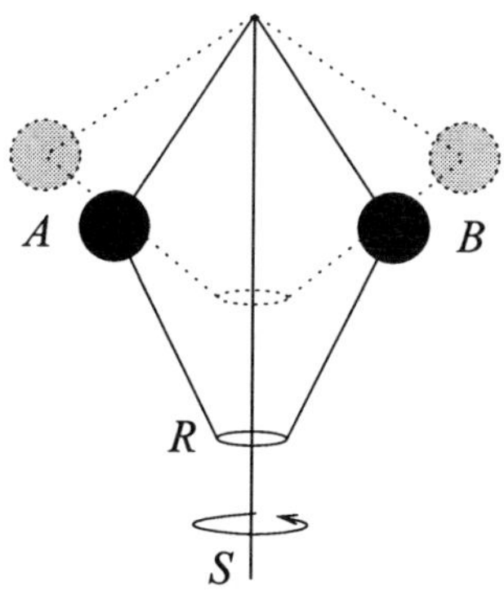

Figure 1.2: Watt's centrifugal governor: when the shaft S rotates, the masses A and B are pulled apart by the centrifugal force. This causes the vertical motion of the ring R whose position controls the draft to the burner.

design made the hunting worse, some were even saying that their governors could easily become "mad," i.e., unstable.

The first serious analysis of this phenomenon was given by Airy in 1840 in a paper which analyzed the high frequency oscillations produced by the governor which regulated the motion of a telescope. Here the problem of hunting was serious, because it adversely affected the main purpose of the instrument: one could *see* the oscillations. Airy was the first to use differential equations to describe the behavior of the governor. In a 1851 supplement to this paper he described the conditions for the stable motion of the telescope which led him to use friction to eliminate the oscillations.

Much more influential was Maxwell's 1868 paper *On Governors,* motivated by his involvement in the experiments for establishment of electrical standards. In the experiment designed by Lord Kelvin to determine the standard for the ohm, it was important to ensure uniform motion of a coil. A governor was used and thus Maxwell, who was interested in general dynamic systems at the time, became interested in its dynamics. He found the stability conditions for systems described by differential equations up to order three. Even before publishing this paper, Maxwell asked the members of the London Mathematical Society:

> if any member present could point out a method of determining in what cases all the possible [real] parts of the impossible [complex] roots of an equation are negative.

Clifford was the first to solve Maxwell's problem. He proposed that

> by forming an auxiliary equation whose roots are the sums of the roots of the original equation taken in pairs and determining the conditions of the real roots of this equation being negative we should obtain the condition required.

This was to be done by applying the Descartes' rule of signs to both the original and the auxiliary equation. Unfortunately, finding the auxiliary equation becomes increasingly difficult as the degree of the original equation increases.

Routh's criterion. Algorithmically more satisfactory criterion was established by Routh in 1874 (for equations of order 5) and in 1876 (equations of any order). In his 1877 Adams Prize winning essay *A treatise on the stability of a given state of motion,* Routh uses the Cauchy index theorem and properties of polynomials discovered by Sturm to prove what we now call the Routh's stability criterion:

Let the characteristic polynomial of a dynamic system be given by

$$a(s) = a_0 s^n + a_1 s^{n-1} + \ldots + a_{n-1} s + a_n$$

where coefficients $a_0, \ldots, a_n$ *are real and* $a_0 > 0$. *All of its roots have negative real parts if and only if*

- *All* a_i*'s are positive (this alone is a necessary but not sufficient condition for stability)*
- *The first-column coefficients in the following array are all positive:*

$$\begin{array}{ccccc} a_0 & a_2 & a_4 & a_6 & \cdots \\ a_1 & a_3 & a_5 & a_7 & \cdots \\ b_1 & b_2 & b_3 & b_4 & \cdots \\ c_1 & c_2 & c_3 & c_4 & \cdots \\ d_1 & d_2 & d_3 & d_4 & \cdots \\ \vdots & \vdots & & & \\ u_1 & u_2 & & & \\ v_1 & & & & \\ w_1 & & & & \end{array}$$

where

$$b_1 = \frac{a_1 a_2 - a_0 a_3}{a_1} \qquad b_2 = \frac{a_1 a_4 - a_0 a_5}{a_1} \qquad b_3 = \frac{a_1 a_6 - a_0 a_7}{a_1} \qquad \cdots$$

$$c_1 = \frac{b_1 a_3 - a_1 b_2}{b_1} \qquad c_2 = \frac{b_1 a_5 - a_1 b_3}{b_1} \qquad c_3 = \frac{b_1 a_7 - a_1 b_4}{b_1} \qquad \cdots$$

$$d_1 = \frac{c_1 b_2 - b_1 c_2}{c_1} \qquad d_2 = \frac{c_1 b_3 - b_1 c_3}{c_1} \qquad d_3 = \frac{c_1 b_4 - b_1 c_4}{c_1} \qquad \cdots$$

etc.

If any of the coefficients $a_1, \ldots, a_n, b_1, \ldots, w_1$ is zero, then $a(s)$ has one or more zeros with a nonnegative real part. Further analysis is possible to determine whether any of these zeros lies to the right of the imaginary axis (see [1] and [41]).

Example 1.1.1 *In order to demonstrate that* $a_i > 0$ *is only a necessary condition in Routh's analysis, let us determine a few examples of third order polynomials with positive real coefficients and roots with nonnegative real parts. In general, a third order polynomial* $a(s) = a_0 s^3 + a_1 s^2 + a_2 s + a_3$ *with* $a_0 > 0$ *has roots to the left of the imaginary axis if and only if* $a_0, a_1, a_2, a_3 > 0$ *and* $a_1 a_2 - a_0 a_3 > 0$. *Hence the condition for a third order polynomial with positive real coefficients to have a root with nonnegative real part is* $a_1 a_2 \leq a_0 a_3$. *Thus, for example,*

$$a_1(s) = s^3 + s^2 + s + 1 \qquad s_1 = -1 \qquad s_{2,3} = \pm j$$
$$a_2(s) = s^3 + s^2 + s + 2 \qquad s_1 = -1.3532 \qquad s_{2,3} = 0.1766 \pm 1.2028j \qquad \square$$

Vishnegradskii, Stodola, and Hurwitz. At first, the results obtained in Britain by Airy, Maxwell, and Routh didn't have much influence on the practical design of governors. Much more influential among the engineers was the work by Vishnegradskii. Around 1876 he gave a clear derivation of the third order governor differential equation and provided a graphical stability criterion. The fact that his work appeared in German, among other languages, was also important, because Germany and other German-speaking countries were about to become the main stage for developments in mechanical engineering.

In 1893 Stodola applied the techniques developed by Vishnegradskii to the study of regulation of water turbines to obtain the seventh order equations. He asked Hurwitz for help with the stability analysis, and as a result, Hurwitz published his stability criterion in 1895 (with a footnote saying that the criterion was applied at the Davos Spa Plant):

If the characteristic polynomial of a dynamic system is given by

$$a(s) = a_0 s^n + a_1 s^{n-1} + \ldots + a_{n-1} s + a_n$$

where coefficients $a_0, \ldots, a_n$ *are real and* $a_0 > 0$, *all of its roots have negative real parts if and only if*

$$D_k > 0 \qquad (k = 1, 2, \ldots, n)$$

where

$$D_k = \begin{vmatrix} a_1 & a_0 & 0 & 0 & 0 & \cdots & 0 \\ a_3 & a_2 & a_1 & a_0 & 0 & \cdots & 0 \\ a_5 & a_4 & a_3 & a_2 & a_1 & \cdots & 0 \\ \vdots & \vdots & \vdots & \vdots & \vdots & \ddots & \vdots \\ a_{2k-1} & a_{2k-2} & a_{2k-3} & a_{2k-4} & a_{2k-5} & \cdots & a_k \end{vmatrix}$$

and $a_j = 0$ *if* $j > n$.

Like Routh, Hurwitz derived this result using Cauchy's index theorem, but instead of using Sturm's properties of polynomials, he used Hermite's quadratic form methods. In 1911 Bompiani proved the equivalence of the two stability criteria, while in 1921 Schur gave an elementary derivation.

Lyapunov stability. A major breakthrough in theoretical mechanics and stability analysis of dynamic systems was made in 1892 by Lyapunov. His doctoral thesis *The general problem of the stability of motion* was based on the theoretical mechanics of Poincare, and contains the first general stability criteria applicable to both linear and nonlinear systems. His work remained practically unknown outside Russia until after the World War II, and we will study it in greater detail in Chapter 2.

Operational calculus. In his 1892 *Electrical Papers*, Heaviside introduced the operational calculus as a novel method for solving differential equations occurring in the theoretical analysis of telegraphy and electrical transmission. Initially dismissed by many as mathematically unfounded, his operational calculus was related to the mathematical theory of integral transforms through the work of Carson and Wiener in the 1920's. These methods are now better known as Laplace and Fourier transforms (see Appendix B.3).

Maritime applications. Improvements of the original Watt's governor allowed for the explosion in the number of applications of the automatic feedback control. To name just a few of them:

- ship steering engines (1849)
- torpedo (1866)
- stabilized passenger-ship saloon (1874)
- stabilized gun platform (1889)
- stabilized ship (1892)
- automatic ship-steering – *gyropilot* (1912)

In 1922 Minorsky published his research on ship-steering. He observed the methods of experienced helmsmen and tried to design automatic ship-steering with similar performance. He found that it was necessary to use the PID control, and also maintained that it was important to ensure not only the stability of the output variable, but also to watch the values of the internal system variables, which could become unstable or go into saturation if the output variable was forced to have a small time constant. His ideas were first implemented on the battleship *USS New Mexico*.

Methods developed for the maritime applications were very useful in the emerging airplane technology: the first airplane stabilizers appeared in 1914, while the first autopilot was made in 1926.

Electricity. The wide-spread use of electrical energy for lighting and power begun in 1870's with the electric arc lamps, which required regulation of the gap between the electrodes. Hence, the first electrical feedback systems were made. For the operation of the arc lamps it was also useful to regulate the electrical current, hence the 1880's saw the first current regulators. Soon, the arc lamps were superseded by incandescent lamps which operated best if the voltage was constant, therefore first voltage regulators were developed. Finally, in the 1920's, many smaller generating plants started to merge into national power grids, hence first frequency controllers were developed, as well as the first systems to control the stability of these complex systems.

Remarks. Until the 1930's, the development of feedback control systems was facilitated by the talent of a small number of inventors. Their methods were based mostly on practical experience from many trials, and the only available theoretical tool was the Routh-Hurwitz stability criterion. The next period of feedback control, its applications in communications, was characterized by the extensive use of Nyquist criterion, which allowed engineers to determine the stability conditions without writing differential equations.

1.2 Classical period of automatic control

The most important developments between the 1930's and the mid 1950's were closely related to communications, radar, industrial process control, and analog computing machines. Several theoretical results had a great practical impact, among them most important were the stability criterion due to Nyquist and graphical methods due to Bode, Nichols, and Evans.

Black, Nyquist, and Bode. During the 1920's H. Black worked at AT&T on the improvements of amplifiers used in long-distance telephony: the goal was to decrease the influence of nonlinearities (and thus increase the range and quality of phone calls) and to increase their bandwidth (to allow more channels to be transmitted over the same physical line). His 1927 invention of the negative feedback amplifier (see Figure 1.3) solved both problems at the same time: it used a high-gain amplifier in the negative feedback configuration and traded a part of its amplification for linearity, noise reduction, and bandwidth. Black published his invention in 1934 and explained it as follows:

> by building an amplifier whose gain is deliberately made, say 40 decibels higher than necessary, and then feeding the output back on the input in such a way as to throw away the excess gain, it has been found possible to effect extraordinary improvement in constancy of amplification and freedom from non-linearity.

If the open-loop gain $|kH(j\omega)|$ is large, then the closed-loop gain is

$$\frac{H(j\omega)}{1 + kH(j\omega)} \approx \frac{1}{k}$$

It is a constant value, even if $H(j\omega)$ varies considerably with frequency, temperature, or due to variability of components.

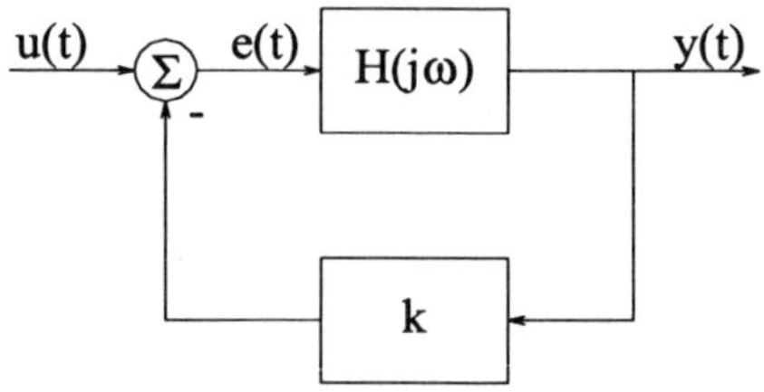

Figure 1.3: Negative feedback amplifier.

In order to completely understand the behavior of his amplifier, Black asked H. Nyquist for help, and as a result, in 1932 Nyquist published a general stability criterion. It uses measurements of the amplifier's open-loop frequency characteristic to determine the stability of the closed-loop system[3]. In our notation the Nyquist criterion is:

Plot the measured values of $-kH(j\omega)$ and their complex conjugate points in the complex plane for all frequencies from 0 to ∞. If the point $-1 + j0$ lies completely outside this curve the system is stable; if not it is unstable.

In the Nyquist's original formulation the values being plotted were the values of $kH(j\omega)$ and $(kH(j\omega))^*$, while the critical point was at $1+j0$. It was H. Bode who introduced the sign change. In 1946 W. Frey extended the formulation to cover the systems with unstable open-loop characteristics.

In 1938 A. V. Mikhailov gave a superficially similar stability criterion which required the knowledge of the characteristic polynomial of the closed-loop system:

For a closed-loop system with the characteristic polynomial

$$a(s) = a_0 s^n + a_1 s^{n-1} + \ldots + a_{n-1} s + a_n$$

plot $a(j\omega)$ in the complex plane, where ω varies from 0 to ∞. The system is stable if and only if this curve describes a positive angle of $n\pi/2$ radians around the origin, without passing through it.

In further attempts to flatten frequency responses of amplifiers and sharpen the cut-off edges in frequency responses of filters, H. Bode discovered the limits on how far one could go. In the same 1940 paper in which he introduced the famous Bode plots and the gain and phase margins, Bode described the relations between the amplitude of the system's frequency response and its phase response:

> While no unique relation between attenuation and phase can be stated for a general circuit, a unique relation does exist between any given loss characteristic and the *minimum* phase shift which must be associated with it.

Servo-mechanisms and radars. During the 1930's there had also been great advances in the theory and practice of servo-mechanisms. The design of servo-mechanisms was critical for the war effort, especially for the development of radar-controlled anti-aircraft guns. The methods based on solving or simulating differential equations were inadequate for this purpose and the frequency response methods, first used by communications engineers, were introduced to this field. Writing on the involvement of the Bell engineers in this work, W. Weaver commented:

[3]See Problem 3.4.12 for more information about this fascinating history.

> if one applies the term "signal" to the variables which describe the actual true motion of the target; and the term "noise" to the inevitable tracking errors, then the purpose of a smoothing circuit is to minimize the noise and at the same time to distort the signal as little as possible.

During the World War II important innovations came out of the MIT Radiation Lab. They designed the first completely automated (auto-track) radar system for the anti-aircraft gun control, the SCR-584, and introduced the Nichols charts (decibel-phase-angle diagrams). One of the proposed methods to predict the future target position (typically 20-30 s into the future) was the theoretical work of N. Wiener. He proposed the use of the statistical properties of the target tracking signal to estimate its future value from imperfect measurements:

Given measurements $f(t-\tau)$ for $\tau > 0$ and knowing that $f(t-\tau)$ is a sum of the actual target position $s(t-\tau)$ and the noise $n(t-\tau)$, the task is to estimate $s(t+\Delta t)$ for some $\Delta t > 0$. Wiener assumed a linear solution: $\hat{s}(t+\Delta t)$, the estimate of $s(t+\Delta t)$, was supposed to be a linear combination of the available measurements:

$$\hat{s}(t+\Delta t) = \int_a^b h(\tau) f(t-\tau)\, d\tau$$

The "coefficients" $h(t)$ in this linear combination can be viewed as the impulse response of the predictor circuit. To further simplify the solution, the best $h(t)$ is assumed to be the one which minimizes the mean-square error of estimation. The solution also assumes the knowledge of the following correlation functions: $r_{ff}(\tau)$, the auto-correlation of $f(t)$, and $r_{fs}(\tau)$, the cross-correlation of $f(t)$ and $s(t)$.

The minimization of the mean-square error

$$\lim_{T\to\infty} \frac{1}{2T} \int_{-T}^{T} \left(s(t+\Delta t) - \int_a^b h(\tau) f(t-\tau)\, d\tau \right)^2 dt$$

leads to the following integral equation of the Wiener-Hopf type:

$$\int_a^b r_{ff}(t-\tau) h(\tau)\, d\tau = r_{fs}(t+\Delta t)$$

The limits in the convolution integral are one of the following:

1. *$a = -\infty$ and $b = \infty$ (noncausal filter: interesting for image processing or off-line filtering of data)*

2. *$a = 0$ and $b = \infty$ (the case studied by Wiener: the predictor is causal and uses measurements from a semi-infinite interval)*

3. *$a = 0$ and $b < \infty$ (the case studied by N. Levinson: causal predictor with measurements only from the recent past)*

The first case can be solved using the Wiener-Khinchin theorem, which relates the correlation functions with the power spectral densities of the corresponding signals via the Fourier transform. Thus, the transfer function of the predictor is obtained as

$$H(j\omega) = \frac{S_{fs}(j\omega)}{S_{ff}(j\omega)}$$

where $S_{fs}(j\omega)$ *is the cross-power spectral density of* $f(t)$ *and* $s(t + \Delta t)$*, while* $S_{ff}(j\omega)$ *is the power spectral density of* $f(t)$*.*

The second case is the most difficult to treat theoretically: it involves factorization of functions into purely causal and stable and purely anticausal and stable parts. The transfer function of the causal Wiener predictor is given by

$$H(j\omega) = \frac{1}{S_{ff}^{+}(j\omega)} \left[\frac{S_{fs}(j\omega)}{S_{ff}^{-}(j\omega)} \right]^{+}$$

where the $+$ *and* $-$ *signs here denote the causal stable and the anticausal stable parts of the function, respectively.*

The third case was studied by Levinson in order to simplify the design of the causal Wiener filter. His assumption $b < \infty$ *was not only more realistic than* $b = \infty$*, but also led to a simpler derivation of the optimal filter coefficients. Levinson first discretized the equation for the estimate as the linear combination of the measurements:*

$$\hat{s}[k + l] = \sum_{i=0}^{M-1} h[i] f[k - i]$$

The probabilistic arguments then yielded the Yule-Walker equation

$$R_{ff} h = p$$

where R_{ff} *is the auto-correlation matrix of the sequence of measurements* $f[k]$*,* h *is the vector of predictor coefficients (in discrete-time this is the same as the impulse response of the circuit)*

$$h = [h[0] \; h[1] \; \ldots \; h[M - 1]]'$$

while p *is the cross-correlation vector of sequences* $f[k]$ *and* $s[k+l]$*. The solution is then*

$$h = R_{ff}^{-1} p$$

The Toeplitz structure of R_{ff} *can be used to simplify its inversion (Levinson's algorithm).*

Wiener's theory, when applied to actual radar-controlled guns in 1942, proved to be only marginally better than the existing techniques, most likely due to the lack of reliable data to calculate the correlation functions $r_{ff}(\tau)$ and $r_{fs}(\tau)$ and to the initial assumption that the solution was a linear function of measurements. It was not used during the World War II, but the manual originally written by Wiener in 1942 (declassified and published along with two explanatory papers by Levinson seven years later) *Extrapolation, Interpolation, and Smoothing of Stationary Time Series,* proved to be very influential among the communications engineers. Wiener's idea to use statistics to deal with noise flourished in the control theory only when in 1960 R. Kalman reformulated the problem in the language of state-space models and found a recursive solution (more will be said in Chapter 2).

After the war. Due to the importance of the automatic control for the design of anti-aircraft guns, torpedoes, guided missiles, and autopilots, this discipline greatly advanced during the war, but was shrouded in a veil of secrecy. After the war, the restrictions on publication of the war-time results and design techniques were lifted. The frequency methods emerged as a universal technique for design of a wide variety of systems: mechanical, electro-mechanical, and electronic devices and quickly found new applications in industry.

In 1948 W. Evans introduced the root-locus method, which is used both as a stability analysis tool – it graphically displays the positions of the system's poles – and a design tool – it also shows how the poles are shifted as the gain changes. The root-locus method requires the knowledge of the open-loop poles and zeros and graphically shows the positions of the closed-loop poles with the gain as parameter.

Discrete-time systems. The need to theoretically analyze the discrete-time techniques arose from the fact that most of the radar systems used pulse signals. Also, as we saw with Levinson's approach to the Wiener problem, discretization of equations can simplify the treatment of a problem, and allow the use of digital computers for numerical calculations.

Around 1942 W. Hurewicz showed how to apply the Nyquist stability analysis to sampled-data systems. This work led to the z-transform methods developed by J. Ragazzini and L. Zadeh in 1952.

It was quickly recognized that for a discrete-time system to be stable, all roots of its characteristic equation had to lie inside the unit circle in the complex plane. The general test for a polynomial with complex coefficients to have all roots inside the unit circle was given independently by I. Schur in 1917 and A. Cohn in 1922. In 1961 E. Jury gave a different criterion which becomes much simpler than the Schur-Cohn test when applied to polynomials with real coefficients:

Let the characteristic polynomial of a discrete-time system be given by

$$a(z) = a_0 z^n + a_1 z^{n-1} + \ldots + a_{n-1} z + a_n$$

where coefficients $a_0, \ldots, a_n$ are real and $a_0 > 0$. All of its roots lie inside the unit circle in the complex plane if and only if

- $|a_n| < |a_0|$
- $a(1) > 0$ *(note that $a(1) = a_0 + a_1 + \ldots + a_n$)*
- $(-1)^n a(-1) > 0$ *(note that $(-1)^n a(-1) = a_0 - a_1 + \ldots + (-1)^n a_n$)*
- *For the first-column coefficients in the following table with $2n-3$ rows:*

$$\begin{array}{cccccccc}
a_n & a_{n-1} & a_{n-2} & a_{n-3} & \cdots & a_2 & a_1 & a_0 \\
a_0 & a_1 & a_2 & a_3 & \cdots & a_{n-2} & a_{n-1} & a_n \\
b_{n-1} & b_{n-2} & b_{n-3} & b_{n-4} & \cdots & b_1 & b_0 & \\
b_0 & b_1 & b_2 & b_3 & \cdots & b_{n-2} & b_{n-1} & \\
c_{n-2} & c_{n-3} & c_{n-4} & c_{n-5} & \cdots & c_0 & & \\
c_0 & c_1 & c_2 & c_3 & \cdots & c_{n-2} & & \\
\vdots & \vdots & \vdots & \vdots & & & & \\
v_3 & v_2 & v_1 & v_0 & & & & \\
v_0 & v_1 & v_2 & v_3 & & & & \\
w_2 & w_1 & w_0 & & & & &
\end{array}$$

where

$$\begin{aligned}
b_i &= \begin{vmatrix} a_n & a_{n-1-i} \\ a_0 & a_{i+1} \end{vmatrix} & & i = 0, 1, 2, \ldots, n-1 \\
c_i &= \begin{vmatrix} b_{n-1} & b_{n-2-i} \\ b_0 & b_{i+1} \end{vmatrix} & & i = 0, 1, 2, \ldots, n-2 \\
&\vdots \\
w_i &= \begin{vmatrix} v_3 & v_{2-i} \\ v_0 & v_{i+1} \end{vmatrix} & & i = 0, 1, 2
\end{aligned}$$

the following inequalities hold:

$$|b_{n-1}| > |b_0| \qquad |c_{n-2}| > |c_0| \qquad \ldots \qquad |w_2| > |w_0|$$

Remarks. Between the 1930's and the mid 1950's the first great stimulus for the development of automatic control came from the revolution in the telecommunications. Yet, even bigger impetus was given to this discipline by the World War II. After the war the frequency methods became universally used, but soon it was realized that different techniques were necessary in order to overcome the difficulties associated with nonlinearities, model uncertainties, noise, and the fact that many systems had multiple inputs or outputs. In addition to all of this, for many systems it was critical that the control was achieved in an optimal way with respect to energy, time, or constraints on variables. Thus, in the early 1950's the stage was set for the modern control theory.

1.3 Beginnings of modern control theory

The First IFAC (International Federation of Automatic Control) Congress, held in Moscow in 1960, is usually considered to be the start of the modern era in automatic control. It brought together the researchers from many countries, both the East and the West, and allowed them to see the new research directions that had been brewing during the 1950's. Particularly influential was R. Kalman's paper [24]. In this Section we examine the development of the main ideas of the modern control theory in their historical order. It will make a lot of sense for the reader to revisit this Section while working on Chapter 2, where most of these results will be derived.

State-variable approach. H. Poincare was the first to make an extensive use of writing the higher-order differential equations as a system of first-order equations. In 1892 he introduced the phase-plane analysis of (generally nonlinear) dynamic systems:

A second-order differential equation can be rewritten as a system of two first-order equations

$$\dot{x_1} = P(x_1, x_2)$$

$$\dot{x_2} = Q(x_1, x_2)$$

Then the system trajectories can be sketched in a phase plane (the x_1-x_2 plane) from

$$\frac{dx_2}{dx_1} = \frac{\dot{x_2}}{\dot{x_1}} = \frac{Q(x_1, x_2)}{P(x_1, x_2)}$$

For years, the state variables were inherently used in automatic control, in analog computer simulations. Following the suggestion made by Lord Kelvin in 1876, the so-called Kelvin's scheme, analog computers were made using integrators, rather than differentiators (differentiators amplify the noise, while integrators tend to smooth it). The outputs of integrators completely determine the state of the system and are used as state variables.

In 1936 A. Turing was the first to use the states as the representation of a dynamic system in his automata theory. In the 1940's the state-space concept was introduced to control theory by M. A. Aizerman, A. A. Fel'dbaum, A. M. Letov, and A. I. Lur'e. Additionally, C. E. Shannon used this approach in his information theory published in 1949.

The representation of dynamic systems using state variables came to prominence in 1957, through the work of T. Bashkow in network theory and R. Bellman and R. Kalman in control theory. The state-space approach quickly found applications in aero-space and military technologies, where, for example, the trajectory of a guided missile can be controlled by several inputs.

Usually, the notation was as follows:

$$\begin{aligned} \dot{x}(t) &= Ax(t) + Bu(t) \\ y(t) &= Cx(t) \end{aligned}$$

where $u(t)$ is the $m \times 1$ input to the system, $y(t)$ is its $p \times 1$ output, while $x(t)$ is an $n \times 1$ vector whose components are the states of the system. A is an $n \times n$ matrix, while B and C are $n \times m$ and $p \times n$ matrices, respectively. Matrix A is usually called the *system* matrix, while matrices B and C are called *input* and *output* matrices.

State-variable feedback. One of the first triumphs of the state variable description of systems was the realization that while the output feedback, even if its derivatives were used, could not always stabilize the system, much less put the poles of the system to specific locations, the state-variable feedback could do it all. In 1959 J. Bertram was the first to realize that if a system realization was controllable and observable[4], than, using an appropriate state-variable feedback, any characteristic polynomial could be achieved. His reasoning was based on root-locus methods. The first direct proof of this was given by J. Rissanen in 1960. The following is the 1965 result due to R. W. Bass and I. Gura for the feedback gain vector which shifts the poles of a single-input system to the desired locations:

If a system is controllable and observable, its poles can be arbitrarily relocated using the state-variable feedback $u = -k'x$. If the characteristic polynomial of the system is

$$a(s) = s^n + a_1 s^{n-1} + \ldots + a_{n-1}s + a_n$$

while the desired characteristic polynomial of the system is

$$\alpha(s) = s^n + \alpha_1 s^{n-1} + \ldots + \alpha_{n-1}s + \alpha_n$$

then with $a' = [a_1 \ \ldots \ a_n]$ and $\alpha' = [\alpha_1 \ \ldots \ \alpha_n]$, the feedback gain that moves the poles to the desired locations is given by

$$k' = (\alpha' - a')\mathsf{a}_-^{-T}\mathcal{C}^{-1}$$

where the superscript "$-T$" indicates inverse and transpose, and

$$\mathsf{a}_- = \begin{bmatrix} 1 & 0 & \ldots & 0 & 0 \\ a_1 & 1 & \ldots & 0 & 0 \\ \vdots & \vdots & \ddots & \vdots & \vdots \\ a_{n-2} & a_{n-3} & \ldots & 1 & 0 \\ a_{n-1} & a_{n-2} & \ldots & a_1 & 1 \end{bmatrix}$$

while $\mathcal{C} = [B \ \ AB \ \ \ldots \ \ A^{n-1}B]$ is the controllability matrix of the system.

[4]We shall say more about these conditions later, in Chapter 2.

Optimal control. Another early success of the state-space approach was Kalman's procedure for design of optimal control systems [23]:

Consider a linear time-invariant system given by

$$\dot{x}(t) = Ax(t) + Bu(t)$$

with the cost function defined by

$$V(x(0), u(t)) = \int_0^\infty (x'(\tau)Qx(\tau) + u'(\tau)Ru(\tau))\, d\tau$$

The matrix Q is positive semi-definite, while R is positive definite. They determine the relative cost of state variables and of the control.

The optimal control input is given by

$$u(t) = -Kx(t)$$

with $K = R^{-1}B'P$, where P is a positive definite symmetric solution of the algebraic Riccati equation

$$PA + A'P - PBR^{-1}B'P + Q = 0$$

The closed-loop system with desirable properties is then given by

$$\dot{x}(t) = (A - BK)x(t)$$

This result was immediately applied in the aero-space programs and the military, like, for example, in planning of the optimal trajectories for space vehicles.

Other important techniques in the field of optimal control were given by R. Bellman (dynamic programming, 1952) and L. Pontryagin (the Maximum Principle, 1956). Their study is beyond the scope of this book, but let us just mention that all these techniques (Kalman's, Bellman's, and Pontryagin's) are dual of the calculus of variations, a mathematical discipline developed by Fermat, Newton, the Bernoulli's, Euler, Lagrange, Hamilton, Jacobi, Weierstrass, and Bolza.

Kalman filtering. Another great contribution due to R. Kalman was the reformulation and the solution in the framework of state-space equations of the Wiener's problem of signal estimation in noisy environment. His first paper on this subject [26] dealt with discrete-time systems, while the second paper [28], co-authored with R. Bucy, solved that problem in the continuous time. The optimal estimators developed in these papers are called the Kalman filter and the Kalman-Bucy filter, respectively. In [26], Kalman commented on the duality between the optimal control and optimal estimation:

> The new formulation of the Wiener problem brings it into contact with the growing new theory of control systems based on the "state" point of view. It turns out, ***surprisingly,*** that the Wiener problem is the *dual* of the noise-free optimal regulator problem, which has been solved by the author, using the state-transition method to great advantage. The mathematical background of the two problems is identical – this has been suspected all along, but until now the analogies have never been made explicit.

Both the optimal control and the optimal estimation reduce to the algebraic Riccati equation.

Kalman filters were quickly implemented in aero-space and military programs, because they were perfectly suited for navigation and tracking problems.

Later developments. Since the 1960's, optimal control, Kalman filtering, and theory of systems in general, found numerous applications and fruitful contacts with other sciences. Here we give a very brief account of these "later developments" in the control theory. The choice of the topics presented here is highly subjective and reflects the research interests of the present authors.

- *Robust control.* The optimal control problem which was reduced to the solution of a matrix Riccati differential equation for a fixed time problem or to a matrix Riccati algebraic equation for an infinite time problem, is known as linear-quadratic-regulation (LQR) problem. However, the LQR theory does not deal with two critical issues associated with the design of feedback-control systems in industrial control problems: sensor noise and plant uncertainty (see [13] and [34]). In 1961, Kalman and Bucy developed a state-variable version of the Wiener filter, which allowed for the optimal estimation of the system state variables from noisy measurements of the system output. The optimal estimation problem (also known as linear-quadratic-estimation (LQE) problem) was also reduced to the solution of a Riccati equation.

  Both the LQR and the LQE problems require accurate mathematical model of the system which is not routinely available and most plant engineers have no idea as to the statistical nature of the external disturbances impinging on their plant. The H_∞ optimal control is a frequency-domain optimization and synthesis theory that was developed to address the questions of plant modeling errors and unknown disturbances. The basic philosophy is to treat the worst case scenario and the optimization is based on infinite norm rather than the quadratic norm in LQR and LQE problems (see [18]).

- *Biological controls.* The importance of control systems engineering in medical and biological applications has grown because of the inherent complexity of biological systems. Although there is no formal definition of complex systems, H. A. Simon's concept of complexity is very appropriate for biological control systems (see [53]). Complex systems are composed of subsystems that in turn have their own subsystems, and so on; and the large number of parts interact in a complicated way so that it is sometimes impossible to infer the properties of the whole from the properties of the parts and their laws of interaction. Indeed, the analytical models developed, using control systems engineering tools, for the components of a biological system have had limited success in predicting the behavior of the overall system for inputs other than for which the model was developed.

The problems in medical control systems may be classified into two groups: (1) the physiological control systems in normal or pathological conditions such as control of electrolytes in the body, arterial pressure, blood sugar, body temperature, neuromuscular and motor activity, etc. and (2) the external (artificial) control systems that interface with physiological systems such as artificial kidney or hemodialyzers, heart-lung machines, cardiac pacemakers, ventilators, implantable pumps for drug delivery, etc. Regulation, control, and system stability are at the heart of the survival of all living organisms from unicellular to multicellular. W. B. Cannon (1929) differentiated the stability properties of biological systems from those of the physical systems, and introduced the term ***homeostasis*** to describe the steady states in the body that are maintained by complex, coordinated physiological reactions (see [10]). The condition of homeostasis is achieved either by regular of the inputs (control of blood sugar) or by regular of the processes (control of body temperature).

- *Man-machine systems.* Humans interact with machines in many different situations such as driving an automobile, flying an airplane, controlling a nuclear power plant, and numerous other activities. As designers of such machines, we are concerned with the way in which human functions as integral part of the man-machine system. To predict the performance of a man-machine system, some representation of the system is required that allows us to determine how independent variables affect the dependent variables. To model a man-machine system, we must depict both human and machine behavior in compatible terms. Since the tools used for such modeling are from control engineering, it is only appropriate to represent human behavior in machine-like terms, as opposed to vice versa. The basic idea is that the human acts as an error-nulling device when driving an automobile, flying an airplane, doing just about any other machine interactions. Human performance in such tracking tasks has been extensively studied and modeled using control terminology (see [52]).

Chapter 2

Modern control theory

Usually, when speaking about the "modern" automatic control, we think of that part of the control theory that relies on the state-space approach to system representation and design. This approach is particularly important for the systems with multiple inputs and outputs and for the higher-order systems in general. The "classical" control, characterized by the use of frequency domain methods, is still preferable for lower-order single-input single-output systems. Although the adjective "classical" may suggest that this approach is a matter of the past, it is certainly not. In many cases the most effective attack on a problem is made by a combined use of both frequency and state-space methods. That shouldn't be surprising, because, as T. Kailath says in [22],

> transfer functions (or high-order differential equation) descriptions and state-space (or first-order differential equation) descriptions are only two extremes of a whole spectrum of possible descriptions of finite-dimensional systems.

One should also keep in mind that what was modern back in the 1960's cannot be modern today. But the revolution caused by the introduction of the state-space methods in control theory and the influence it still has today were so big that the word "modern" has become a part of the name of the discipline ("modern controls") rather than just an adjective.

We start this Chapter by a discussion of the ways to obtain and write state-space equations (Section 2.1). We continue by examining the most important properties of linear control systems: stability, controllability, observability, and others (Section 2.2). Next, we study the relocation of system poles by the state feedback and optimal control as a special case of particular interest (Section 2.3). Finally, we study the state observers and estimators (Section 2.4).

$$\mathcal{C} = [b \quad Ab \quad A^2b \quad \ldots \quad A^{n-1}b]$$

$$\mathcal{O} = \begin{bmatrix} c' \\ c'A \\ c'A^2 \\ \vdots \\ c'A^{n-1} \end{bmatrix}$$

2.1 State-space representation

In this Section we discuss the state equations and study several convenient forms to write them in. State equations can be obtained from the input-output differential equation or by a direct analysis of the system. We also describe several important forms for state equations. Each of them has some advantage over the others and we will outline them. We end this Section with formulas for the system's transfer function and impulse response in terms of the state-space matrices and a brief discussion of discretization.

State equations. State equations provide the most complete description of a dynamic system. They not only specify the relation between the input and the output, but also tell us about the internal system properties. Most often the state equations are written in the matrix form:

continuous-time:	*discrete-time:*
$\dot{x}(t) = Ax(t) + Bu(t)$	$x[k+1] = Ax[k] + Bu[k]$
$y(t) = Cx(t)$	$y[k] = Cx[k]$

where u is the $m \times 1$ input vector to the system, y is its $p \times 1$ output vector, while x is an $n \times 1$ state vector. A is an $n \times n$ matrix, while B and C are $n \times m$ and $p \times n$ matrices, respectively. Matrix A is usually called the *system* matrix (for continuous-time systems) or the *state-transition* matrix (for discrete-time systems), while matrices B and C are called *input* and *output* matrices. When $m = 1$ and $n = 1$, we write b and c' instead of B and C, respectively.

State-space realizations of the input-output equation. If our goal is to write a state-space realization for a given input-output differential equation, for example

$$\dddot{y} + a_1\ddot{y} + a_2\dot{y} + a_3 y = b_1\ddot{u} + b_2\dot{u} + b_3 u$$

or, equivalently, for a given transfer function, in this case

$$H(s) = \frac{b_1 s^2 + b_2 s + b_3}{s^3 + a_1 s^2 + a_2 s + a_3}$$

we can use the following set of equations (the so-called *controller* form):

$$\begin{bmatrix} \dot{x}_1 \\ \dot{x}_2 \\ \dot{x}_3 \end{bmatrix} = \begin{bmatrix} -a_1 & -a_2 & -a_3 \\ 1 & 0 & 0 \\ 0 & 1 & 0 \end{bmatrix} \begin{bmatrix} x_1 \\ x_2 \\ x_3 \end{bmatrix} + \begin{bmatrix} 1 \\ 0 \\ 0 \end{bmatrix} u$$

$$y = \begin{bmatrix} b_1 & b_2 & b_3 \end{bmatrix} \begin{bmatrix} x_1 \\ x_2 \\ x_3 \end{bmatrix}$$

It is an easy exercise to see that this set of first-order equations reduces to the given differential equation. There are infinitely many other state-space representations of this differential equation, for example any other state vector given by $w(t) = Sx(t)$, where S is a nonsingular matrix, defines another realization. Such transformations are called ***similarity*** transformations (see Appendix C.3).

Another popular form is the ***observer*** realization:

$$\begin{bmatrix} \dot{x}_1 \\ \dot{x}_2 \\ \dot{x}_3 \end{bmatrix} = \begin{bmatrix} -a_1 & 1 & 0 \\ -a_2 & 0 & 1 \\ -a_3 & 0 & 0 \end{bmatrix} \begin{bmatrix} x_1 \\ x_2 \\ x_3 \end{bmatrix} + \begin{bmatrix} b_1 \\ b_2 \\ b_3 \end{bmatrix} u$$

$$y = \begin{bmatrix} 1 & 0 & 0 \end{bmatrix} \begin{bmatrix} x_1 \\ x_2 \\ x_3 \end{bmatrix}$$

The following is a summary of the most commonly used realizations and their advantages:

- controller (especially useful in the design of the state feedback to place the system poles to desired positions)
- observer (allows easy reconstruction of the system states from the inputs and the outputs)
- controllability (particularly suitable for setting the initial states)
- observability (allows for simple determination of the initial states)
- modal or parallel (useful because modes of the system are distributed to individual states)

We shall see later where the names of these realizations originate. We shall also see that the controller and the controllability realizations are always ***controllable***, while the observer and the observability realizations are always ***observable***. As always, we shall define these properties first. These realizations always exist, but unless they are both controllable and observable, there are no similarity transformations between the controllable realizations on one side and observable realizations on the other.

State equations of a system. In general, state equations are obtained from the physical laws which govern the system's dynamic behavior. Depending on the basic nature of the system, i.e., whether it is mechanical, electrical, hydraulic, thermal, or acoustic, we use conservation of energy, conservation of momentum, conservation of angular momentum, Kirchoff's laws, Bernoulli's law, etc.

The following example illustrates the process of obtaining state-space description of a moderately complex mechanical system, the inverted pendulum on a cart: we start from the physical laws which govern it and then linearize them around the desired operating point.

Example 2.1.1 *In this book we will often use the example of the inverted pendulum on a cart, shown in Figure 2.1. Besides being an interesting system in itself and having an ideal level of complexity to illustrate many important ideas presented in this book, it is also an idealized model of several important systems, for example of a standing human being or of a vertically launched rocket. The equations describing its behavior are as follows:*

$$\begin{aligned}(M+m)\ddot{z} + ml\ddot{\theta}\cos\theta - ml\dot{\theta}^2\sin\theta &= f \\ m\ddot{z}\cos\theta + ml\ddot{\theta} - mg\sin\theta &= 0\end{aligned}$$

where M and m are the masses of the cart and the bob, l is the length of the pendulum rod, z and θ are the horizontal displacement of the cart and the angle between the vertical and the pendulum rod (expressed in radians), while f is the force applied to the cart.

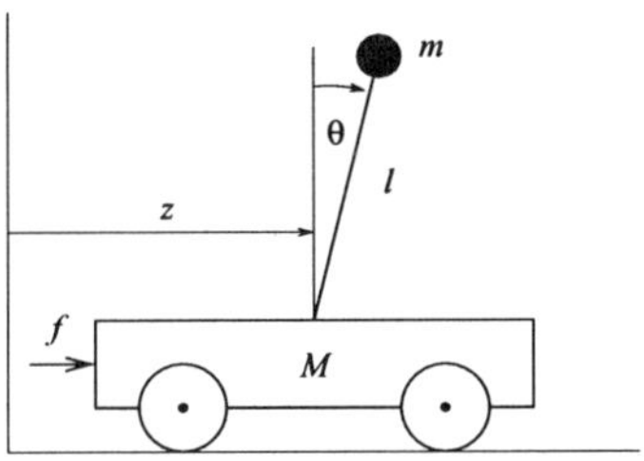

Figure 2.1: The inverted pendulum on a cart. This system will be used throughout this book to illustrate various concepts (Example 2.2.5 and Problems 3.5.11, 3.7.12, 3.9.9, 3.11.4, 3.12.2, 3.13.3, and 4.9.1).

These are two nonlinear coupled second-order differential equations. In order to write them as four linear coupled first-order equations we need to linearize them. If the goal is to stabilize the pendulum in the vertical position, we linearize in the neighborhood of $\theta = 0$, when

$$\sin\theta \approx \theta \qquad \textit{and} \qquad \cos\theta \approx 1$$

Thus we obtain

$$\dot{x} = Ax + bu$$

where

$$x = \begin{bmatrix} z \\ \theta \\ \dot{z} \\ \dot{\theta} \end{bmatrix}, \qquad A = \begin{bmatrix} 0 & 0 & 1 & 0 \\ 0 & 0 & 0 & 1 \\ 0 & -\frac{mg}{M} & 0 & 0 \\ 0 & \frac{(M+m)g}{Ml} & 0 & 0 \end{bmatrix}, \qquad \textit{and} \qquad b = \begin{bmatrix} 0 \\ 0 \\ 1/M \\ -1/Ml \end{bmatrix}$$

while $u = f$, the external force applied to control the cart and the inverted pendulum. In the rest of the book we shall assume that the measured variables are z and θ, i.e., that

$$y = Cx, \qquad \textit{where} \qquad C = \begin{bmatrix} 1 & 0 & 0 & 0 \\ 0 & 1 & 0 & 0 \end{bmatrix}$$

Solution of state-space equations. In Problems 3.5.4 and 4.5.1 we show that the solution of state-space equations, the impulse response, and the transfer function are given for continuous-time and discrete-time systems, respectively, as follows:

state-space equations:

$$\begin{aligned} \dot{x}(t) &= Ax(t) + Bu(t) \\ y(t) &= Cx(t) \end{aligned} \qquad\qquad \begin{aligned} x[k+1] &= Ax[k] + Bu[k] \\ y[k] &= Cx[k] \end{aligned}$$

solution for the state vector:

$$x(t) = e^{At}x(0) + (e^{At}B) * u(t) \qquad\qquad x[k] = A^k x_0 + (A^k B) * (u[k-1])$$

solution for the output:

$$y(t) = Ce^{At}x(0) + (Ce^{At}B) * u(t) \qquad\qquad y[k] = CA^k x_0 + (CA^k B) * (u[k-1])$$

impulse response:

$$h(t) = Ce^{At}B \qquad (t > 0) \qquad\qquad h[k] = CA^{k-1}B \qquad (k = 1, 2, \ldots)$$

transfer function:

$$H(s) = C(sI - A)^{-1}B \qquad\qquad H(z) = C(zI - A)^{-1}B$$

Markov parameters. Write the transfer function of a continuous-time system as a power series:

$$H(s) = c'(sI - A)^{-1}b = \sum_{i=1}^{\infty} h_i s^{-i}$$

Since

$$(sI - A)^{-1} = \frac{1}{s}\left(I - \frac{A}{s}\right)^{-1} = \frac{1}{s}\left(I + \frac{A}{s} + \frac{A^2}{s^2} + \ldots\right)$$

we have

$$h_i = c'A^{i-1}b \quad (i = 1, 2, \ldots)$$

These coefficients are called the *Markov parameters*. Their interpretation is as follows. Since the impulse response of the system is given by $h(t) = c'e^{At}b$ and since $H(s) = \mathcal{L}\{h(t)\}$ and $H(s) = c'(sI - A)^{-1}b$, we see that

$$h_i = \left.\frac{d^{i-1}}{dt^{i-1}}h(t)\right|_{t=0} \qquad (i = 1, 2, \ldots)$$

The Markov parameters are defined similarly for the discrete-time systems:

$$H(z) = c'(zI - A)^{-1}b = \sum_{i=1}^{\infty} h_i z^{-i}$$

In this case also $h_i = c'A^{i-1}b \ (i = 1, 2, \ldots)$, but their interpretation is much easier to find. Directly from their definition and the definition of the z-transform of the impulse response, we see that the Markov parameters of a discrete-time system are the system's impulse response:

$$h_i = h[i] \quad (i = 1, 2, \ldots)$$

The Hankel matrix of Markov parameters will often be encountered in our later discussions:

$$\mathcal{M} = \begin{bmatrix} h_1 & h_2 & \ldots & h_n \\ h_2 & h_3 & \ldots & h_{n+1} \\ \vdots & \vdots & \ddots & \vdots \\ h_n & h_{n+1} & \ldots & h_{2n-1} \end{bmatrix}$$

We shall find it interesting that this matrix can be written as a product of two important matrices:

$$\mathcal{M} = \mathcal{O}\mathcal{C} \tag{2.1}$$

where $\mathcal{O}$ and $\mathcal{C}$ are the so-called *observability* and *controllability* matrices, respectively:

$$\mathcal{O} = \begin{bmatrix} c' \\ c'A \\ \vdots \\ c'A^{n-1} \end{bmatrix} \quad \text{and} \quad \mathcal{C} = [b \;\; Ab \;\; A^2b \;\; \ldots \;\; A^{n-1}b]$$

Indeed,

$$\mathcal{O}\mathcal{C} = \begin{bmatrix} c'b & c'Ab & \ldots & c'A^{n-1}b \\ c'Ab & c'A^2b & \ldots & c'A^nb \\ \vdots & \vdots & \ddots & \vdots \\ c'A^{n-1}b & c'A^nb & \ldots & c'A^{2n-2}b \end{bmatrix} = \mathcal{M}$$

Similarity transformation. Consider a system described by $\{A_1, B_1, C_1\}$. If we introduce a change of variables in its state-space equations described by a nonsingular transformation matrix S, such that $x_2 = Sx_1$ the new realization is described by $\{A_2, B_2, C_2\}$, where

$$A_2 = SA_1S^{-1} \qquad B_2 = SB_1 \qquad C_2 = C_1S^{-1}$$

This transformation (called *similarity* transformation) does not change the characteristic polynomial of the system nor its transfer function. Indeed,

$$\begin{aligned}\det(sI - A_2) &= \det(sSS^{-1} - SA_1S^{-1}) = \det(S(sI - A_1)S^{-1}) \\ &= \det(S)\det(sI - A_1)\det(S^{-1}) = \det(sI - A_1)\end{aligned}$$

and

$$\begin{aligned}C_2(sI - A_2)^{-1}B_2 &= C_1S^{-1}(sSS^{-1} - SA_1S^{-1})^{-1}SB_1 \\ &= C_1S^{-1}S(sI - A_1)^{-1}S^{-1}SB_1 \\ &= C_1(sI - A_1)^{-1}B_1\end{aligned}$$

Since the system's impulse response is the inverse Laplace transform of the transfer function, it is also invariant under similarity transformations. Hence, so are the Markov parameters $h_i = c'A^{i-1}b$ $(i = 1, 2, \ldots)$ and the matrix $\mathcal{M}$. Another way to see that is as follows:

$$c_2'A_2^{i-1}b_2 = c_1'S^{-1}SA_1^{i-1}S^{-1}Sb_1 = c_1'A_1^{i-1}b_1$$

Eigenvalues, modes, and poles. If $\lambda_1, \ldots, \lambda_n$ are the eigenvalues of A (some of them may be repeated) then each state is a linear combination of terms

$$e^{\lambda_i t} \qquad \text{or} \qquad \lambda_i^{k-1}$$

for continuous-time and discrete-time systems, respectively, and, if λ_i is a multiple eigenvalue, of terms

$$te^{\lambda_i t}, \ \ldots, t^{m_i-1}e^{\lambda_i t} \qquad \text{or} \qquad k\lambda_i^{k-1}, \ \ldots, k^{m_i-1}\lambda_i^{k-1}$$

where m_i is the multiplicity of λ_i. These different terms are called the *modes of oscillation* or simply *modes*.

In general, an arbitrary initial condition will excite all modes of oscillation (see Problem 3.5.9 for the special initial condition which excites only one mode). Since

$$y(t) = Cx(t) \qquad \text{or} \qquad y[k] = Cx[k]$$

some of the modes may never appear in the output, regardless of the initial conditions or the input to the system. The modes that can appear in the output correspond to the *poles* of the system. Thus, the poles of the system are those eigenvalues of A that may appear in the output.

This is better illustrated in the transform domain. We do this only in the continuous-time, because the terminology and ideas are identical in the discrete-time. The transfer function is given by $H(s) = C(sI - A)^{-1}B$. If $a(s) = \det(sI - A)$ and $b(s) = C\text{adj}(sI - A)B$ then

$$H(s) = \frac{b(s)}{a(s)} = \frac{b_r(s)}{a_r(s)}$$

where $a_r(s)$ and $b_r(s)$ are coprime polynomials. The roots of $a(s)$ are the eigenvalues of A, while the roots of $a_r(s)$ are the poles of the system. If there are no cancellations between $a(s)$ and $b(s)$, then $a(s) = a_r(s)$, and the eigenvalues of A coincide with the system poles. Otherwise, all poles are the eigenvalues of A, but not all eigenvalues are poles. The systems with no cancellations between $a(s) = \det(sI - A)$ and $b(s) = C\text{adj}(sI - A)B$ are called *minimal*. We shall see in Section 2.2 that such systems have important properties. For example, they are both controllable and observable, and their internal and external stabilities are equivalent.

Finally, let us just mention that one of the reasons the modal canonical realization is so important, and certainly the reason it is called modal, is that in this realization each state has modes of oscillation corresponding to only one eigenvalue. If some of the eigenvalues are repeated, then some of the states may be linear combinations of more than one mode of oscillation, but they all correspond to the same eigenvalue.

Example 2.1.2 *Given a system in the modal form with the system matrix in Jordan form*

$$J = \begin{bmatrix} -1 & 1 & 0 & 0 & 0 \\ 0 & -1 & 0 & 0 & 0 \\ 0 & 0 & -1 & 0 & 0 \\ 0 & 0 & 0 & -2 & 0 \\ 0 & 0 & 0 & 0 & -3 \end{bmatrix}$$

the corresponding matrix exponential is found as the inverse Laplace transform of

$$(sI - J)^{-1} = \begin{bmatrix} \frac{1}{s+1} & \frac{1}{(s+1)^2} & 0 & 0 & 0 \\ 0 & \frac{1}{s+1} & 0 & 0 & 0 \\ 0 & 0 & \frac{1}{s+1} & 0 & 0 \\ 0 & 0 & 0 & \frac{1}{s+2} & 0 \\ 0 & 0 & 0 & 0 & \frac{1}{s+3} \end{bmatrix}$$

which is

$$e^{Jt} = \begin{bmatrix} e^{-t} & te^{-t} & 0 & 0 & 0 \\ 0 & e^{-t} & 0 & 0 & 0 \\ 0 & 0 & e^{-t} & 0 & 0 \\ 0 & 0 & 0 & e^{-2t} & 0 \\ 0 & 0 & 0 & 0 & e^{-3t} \end{bmatrix}$$

An arbitrary initial condition therefore excites only two modes of oscillation in $x_1(t)$ and only one mode of oscillation in each of the other states.

Discretization. In Problem 4.5.5 we show that if a continuous-time system given by

$$\begin{aligned}\dot{x}(t) &= Ax(t) + Bu(t)\\ y(t) &= Cx(t)\end{aligned}$$

is discretized using the sampling period T, then its discrete-time version is given by

$$\begin{aligned}x[k+1] &= Gx[k] + Hu[k]\\ y[k] &= Cx[k]\end{aligned}$$

where

$$G = e^{AT} \quad \text{and} \quad H = \left(\int_0^T e^{A\tau}\, d\tau\right) B$$

If A is invertible (i.e., nonsingular), then

$$H = (e^{AT} - I)A^{-1}B$$

For very small values of T these formulas can be further simplified:

$$G \approx I + AT \quad \text{and} \quad H \approx BT$$

Example 2.1.3 *If the system matrix G of a discrete-time linear system is nilpotent, i.e., if $G^m = 0$ for some $m < \infty$, then with no input to the system,*

$$x[k] = G^k x[0]$$

hence

$$x[k] \equiv 0 \quad (k \geq m)$$

Such discrete time systems are called deadbeat *systems. There is nothing similar to this in the continuous-time. If there was a continuous-time system whose discretized version is deadbeat, it would have all eigenvalues equal to $-\infty$. This is because (as we show in a Note after Problem 4.7.1) all eigenvalues of nilpotent matrices are equal to zero, and from*

$$G = e^{AT}$$

and the Cayley-Hamilton theorem we know that if λ is an eigenvalue of A then

$$\gamma = e^{\lambda T}$$

is an eigenvalue of G.

Example 2.1.4 *If a system given by*

$$A = \begin{bmatrix} -1 & 0 \\ 0 & -2 \end{bmatrix} \quad \text{and} \quad B = \begin{bmatrix} 1 \\ 1 \end{bmatrix}$$

is discretized with $T = 0.5s$*, the corresponding discrete-time matrices are*

$$G = e^{AT} = \begin{bmatrix} 0.6065 & 0 \\ 0 & 0.3679 \end{bmatrix}$$

and (since $\det(A) \neq 0$*)*

$$H = (e^{AT} - I)A^{-1}B = \begin{bmatrix} 0.3935 \\ 0.3161 \end{bmatrix}$$

The states of these systems are plotted for $x(0) = [2 \quad 3]'$ *in Figure 2.2.*

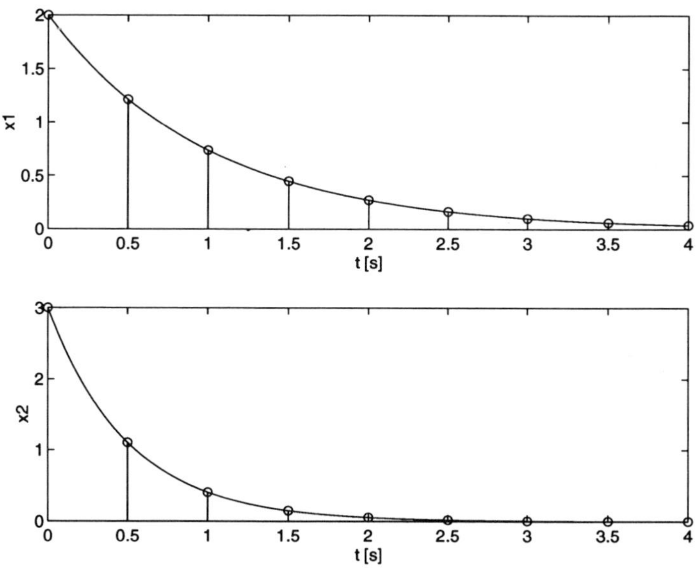

Figure 2.2: Plots of both the continuous-time and the corresponding discrete-time states.

2.2 System properties

In this Section we discuss the most important properties of linear control systems. First we study stability, especially the stability in the sense of Lyapunov, and then controllability, observability, and several related properties. We describe their relation to the canonical realizations studied in Section 2.1.

Stability

Stability is the most important property of a dynamic system. From the earliest days of control theory it has been realized that the characteristic roots of the system's differential equation must have negative real parts, otherwise the system will either oscillate, go into saturation, or blow up.

External stability. We first discuss two types of external[1] stability: the bounded-input bounded-output (BIBO) stability and the marginal stability. The definitions are general enough to apply to nonlinear as well as linear systems. The BIBO stability is defined as follows:

Definition 2.2.1 (BIBO stability) *A system is BIBO stable if its response to any bounded input is also bounded.*

In Problem 3.6.1 we show that a continuous linear time-invariant system is BIBO stable if and only if its impulse response $h(t)$ is absolutely integrable, i.e., if and only if

$$\int_0^\infty |h(t)|\, dt < \infty$$

In Problem 3.6.2 we show that for the systems with rational transfer functions this is equivalent with the requirement that all poles of the system's transfer function $H(s)$ must have negative real parts.

If we want to allow the oscillatory system behavior, the concept of marginal stability becomes useful:

Definition 2.2.2 (Marginal stability) *A system is marginally stable if its impulse response is bounded.*

In Problem 3.6.2 we show that the s-domain requirement for the marginal stability of a continuous linear time-invariant system with rational $H(s)$ is very similar to the condition for BIBO stability, with the additional "freedom" for its non-repeated poles, which now may lie on the imaginary axis.

Definition 2.2.3 (Instability) *A system is said to be unstable if it is not marginally stable.*

---

[1]Both of these types of stability are defined in terms of the system input and the system output, hence the name external.

Internal stability. Unlike the external types of stability, defined in terms of the input and the output of the system, the internal types of stability are defined in terms of the states of the system. In what follows we first define two types of internal stability. The definitions are general and apply equally to linear and nonlinear systems. Very much like the BIBO stability, the ***asymptotic stability in the sense of Lyapunov*** does not allow oscillatory modes in the system; on the other hand, the ***Lyapunov stability*** allows such modes. For most practical purposes the asymptotic stability is much more important than stability.

Consider a general nonlinear time-variable system given by the following state equation:

$$\dot{x} = f(x,t)$$

States x_e for which $f(x_e,t) = 0$ are called the *equilibrium points.* In general, a system may have anywhere from one to infinitely many equilibrium points. For example, linear systems given by $\dot{x} = Ax$ have either one or infinitely many equilibrium points: if A is nonsingular, the system has only one equilibrium point, $x_e = 0$; if A is singular, the system has infinitely many equilibrium points. Back to the general case, if for the initial condition $x(t_0)$ the solution is given by $x(t) = s(t,t_0,x(t_0))$ and if $||v||$ denotes any vector norm of vector v, we are ready to define the stability in the sense of Lyapunov:

Definition 2.2.4 (Lyapunov stability) *An equilibrium state x_e of the system described by*

$$\dot{x} = f(x,t)$$

is said to be stable in the sense of Lyapunov if for every t_0 and every $\varepsilon > 0$ there exists $\delta(\varepsilon) > 0$ such that $||x(t_0)|| < \delta$ implies that for $t > t_0$ we have $||s(t,t_0,x(t_0)) - x_e|| < \varepsilon$.

In other words, the equilibrium point x_e is stable if for each ε-neighborhood $S(\varepsilon)$ of x_e there exists a δ-neighborhood $S(\delta)$ of x_e such that if the initial condition $x(t_0)$ is in $S(\delta)$ then the system trajectory remains in $S(\varepsilon)$ at all times.

For the practical purposes the notion of the asymptotic Lyapunov stability is much more important than stability:

Definition 2.2.5 (Asymptotic Lyapunov stability) *An equilibrium x_e of a system is asymptotically stable in the sense of Lyapunov if it is stable in the sense of Lyapunov and attractive, i.e.,*

$$\lim_{t\to\infty} s(t,t_0,x(t_0)) = x_e$$

If $s(t,t_0,x(t_0))$ converges to x_e regardless of $x(t_0)$, then we say that x_e is *globally asymptotically stable.*

A simple pendulum is an example of a system having a stable but not asymptotically stable equilibrium (assuming there is no friction in the system).

Note also that the attractivity of x_e by itself (without stability) does not imply the asymptotic stability [62].

Definition 2.2.6 (Instability) *An equilibrium x_e of a system is said to be unstable if it is not stable in the sense of Lyapunov.*

The relations between asymptotic stability, stability, and instability in the sense of Lyapunov are illustrated in Figure 2.3.

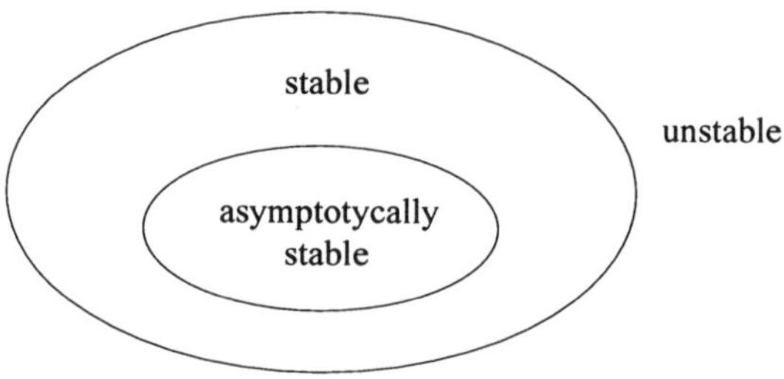

Figure 2.3: The most important type of Lyapunov stability is the *asymptotic Lyapunov stability.* Sometimes it is referred to as simply *stability* (cf. [22]). In order to emphasize the difference, some other authors call these two types of stability *asymptotic* and *weak* (cf. [44]).

It is also important to note that the trajectories of a system with an unstable equilibrium point do not have to "blow up" (although that is exactly what happens with unstable linear systems).

Example 2.2.1 *The van der Pol oscillator is described by the following nonlinear state equations:*

$$\begin{aligned}\dot{x}_1 &= x_2\\ \dot{x}_2 &= -a(x_1^2 - 1)x_2 - x_1\end{aligned}$$

Obviously, $x_e = 0$ is an equilibrium. Although the trajectories of the system do not "blow up" (as can be inferred from the simulation in Figure 2.4), the origin is an unstable equilibrium. If, for example, we use the Euclidean norm

$$||v||_2 = \sqrt{v_1^2 + v_2^2 + \ldots + v_n^2}$$

no $\delta > 0$ can be found so that if $x(0)$ is inside the circle of radius δ the trajectory remains within the circle of radius $\varepsilon = 1$, or any other circle within the limit cycle *of this system.*

However, the limit cycle is asymptotically stable (although we did not define this terminology, this statement should be clear). In general it is difficult to define the stability of nonlinear systems, because they may have a variety of equilibrium points and limit cycles at once, some stable, others not. As we shall see next, the situation for linear systems is much simpler.

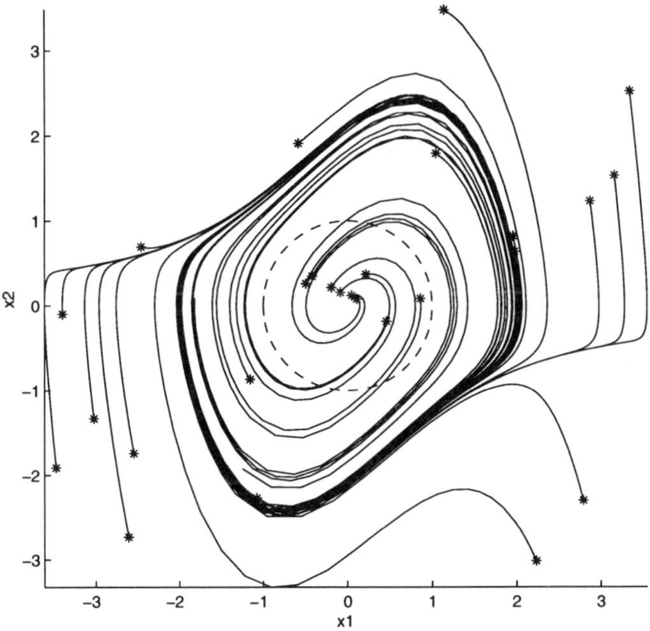

Figure 2.4: Simulation of the phase-plane for the van der Pol oscillator with $a = 0.75$. The dashed circle has radius $\varepsilon = 1$. No matter how close the initial state to the origin, the trajectory will leave this circle. Therefore the origin is an unstable equilibrium.

Next we examine these definitions in the world of continuous linear time-invariant systems, i.e., for $f(x,t) = Ax$.

Theorem 2.2.1 *A continuous linear time-invariant system given by $\dot{x} = Ax$ is (a) stable in the sense of Lyapunov if and only if all eigenvalues of A have negative real parts, except the non-repeated eigenvalues, which may lie on the imaginary axis;*
(b) asymptotically stable in the sense of Lyapunov if and only if A is Hurwitz, i.e., if and only if all eigenvalues of A have negative real parts.

Proof. Since the trajectory of the system is $x(t) = e^{At}x(0)$, each component of $x(t)$ is a sum of terms of the form $t^k e^{\lambda_i t}$, where $k = 0, 1, \ldots, m_i - 1$, and m_i is the multiplicity of λ_i, and it is easy to see that the theorem is true. □

If the origin is asymptotically stable, than according to Theorem 2.2.1, $\text{Re}\{\lambda_i\} < 0$, hence $\det(A) \neq 0$, i.e., A is nonsingular, and the origin is the only equilibrium point of this system. This justifies the following definition:

Definition 2.2.7 *A linear time-invariant system without inputs is said to be asymptotically stable if it has an asymptotically stable equilibrium point at the origin.*

The asymptotic Lyapunov stability is not always equivalent to BIBO stability. Recall that BIBO stability is determined by the poles of $H(s)$, while the asymptotic Lyapunov stability depends on the eigenvalues of the system matrix A. Since the poles of $H(s)$ are also eigenvalues of A, the asymptotic Lyapunov stability implies BIBO stability. However, unless all eigenvalues of A are poles of $H(s)$, including multiplicities, the BIBO stability does not imply the asymptotic Lyapunov stability. This happens when there are pole-zero cancellations in $c'(sI - A)^{-1}b$. We shall see later that no such cancellations occur if and only if the system is both controllable and observable. Such systems are called *minimal realizations.*

Lyapunov's second method. In his 1892 doctoral thesis, A. M. Lyapunov generalized the notion of the mechanical energy by introducing what is now known as the Lyapunov function. Lyapunov's "second" or "direct" method is still the most important technique equally applicable to linear and nonlinear, time-invariant and time-variable systems. It is based on the following theorem:

Theorem 2.2.2 (Lyapunov's theorem) *Let $V(x)$ be a continuously differentiable positive definite function of the system states $x(t)$ defined on a neighborhood D of the equilibrium point $x_e = 0$. This function may also be time-varying.*
(a) If its time derivative $\dot{V}(x)$ is negative semi-definite, than this equilibrium point is stable in the sense of Lyapunov.
(b) If $\dot{V}(x)$ is negative definite, than this equilibrium point is asymptotically stable in the sense of Lyapunov. (Such $V(x)$ is called the Lyapunov function.*)*

Proof of (a). First, we prove part (a) by showing that for any ε-neighborhood $S(\varepsilon)$ of x_e there exists a δ-neighborhood $S(\delta)$ of x_e such that if $x(t_0) \in S(\delta)$ then the system trajectory remains in $S(\varepsilon)$ at all times.

Recall that "$V(x)$ is positive definite" means $V(0) = 0$ and $V(x) > 0$ for $x \neq 0$. Similarly, the phrase "$\dot{V}(x)$ is negative definite" means $\dot{V}(0) = 0$ and $\dot{V}(x) < 0$ for $x \neq 0$, while "$\dot{V}(x)$ is negative semi-definite" means that $\dot{V}(0) = 0$ and $\dot{V}(x) \leq 0$ for all x.

Consider any $\varepsilon > 0$ such that $S(\varepsilon) \subset D$. Denote by a the minimum value of $V(x)$ on the boundary of $S(\varepsilon)$, i.e.,

$$a = \min_{\|x\|=\varepsilon} V(x)$$

Since $V(x)$ is positive definite and $\varepsilon > 0$ we know that $a > 0$.

Now define a new set, D_a, as a connected set of states x such that $V(x) < a$. If there is more than one such set, select the one containing the origin. In the following we will show by contradiction that D_a is a subset of $S(\varepsilon)$. First of all, we know that they have at least one point in common, the origin. Now assume D_a is not completely in the interior of $S(\varepsilon)$. Then there exists point $P \in D_a$ on the boundary of $S(\varepsilon)$. For that point we would have $V(x_P) \geq a$, because a was defined as the minimum value of $V(x)$ on the boundary of $S(\varepsilon)$. But this contradicts the definition of D_a, which requires that $V(x_P) < a$. Hence, D_a is in the interior of $S(\varepsilon)$. All these sets are illustrated in Figure 2.5.

Since $\dot{V}(x) \leq 0$, any system trajectory originating in D_a always remains in D_a and hence in $S(\varepsilon)$. This property of D_a is very important for the rest of the proof. Since $V(x)$ is continuous, there exists $\delta > 0$ such that if $\|x\| < \delta$ (i.e., $x \in S(\delta)$) then $V(x) < a$ (i.e., $x \in D_a$). In other words, there exists δ such that $S(\delta) \subseteq D_a$. Obviously, any trajectory originating in $S(\delta)$ will never leave D_a, and thus it will never leave $S(\varepsilon)$. We have just demonstrated that if the conditions of part (a) of the theorem are satisfied, then for any $\varepsilon > 0$ there exists $\delta > 0$ such that if $\|x(t_0)\| < \delta$ then $\|x(t)\| < \varepsilon$, i.e., the origin is stable in the sense of Lyapunov.

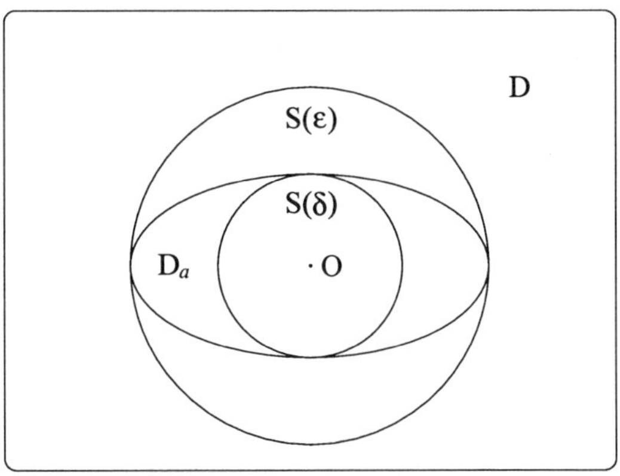

Figure 2.5: Illustration of the sets used in the proof of Theorem 2.2.2. All these sets are open, hence their boundaries, which represent the sets in this figure, touch each other and can even completely overlap.

Proof of (b). To prove part (b) we continue exactly where we left off with part (a). We already know that if $\dot{V}(x)$ is negative definite then the origin is stable (this is a special case of part (a), because negative definiteness is a special case of negative semi-definiteness). To finish part (b) we need to demonstrate that if $\dot{V}(x)$ is negative definite then the origin is an attractive equilibrium point, i.e.,

$$\lim_{t \to \infty} x(t) = 0$$

We start by showing that under conditions of part (b) we can write

$$\lim_{t\to\infty} V(x) = 0$$

Since $V(x)$ is monotone decreasing and bounded from below by zero, it is convergent, i.e.,

$$\lim_{t\to\infty} V(x) = c \geq 0$$

We use the method of contradiction to show that $c = 0$. Assume that $c > 0$. Since $V(x)$ is monotone decreasing, this means that $V(x)$ cannot attain values below c. But if we denote the slowest rate of decrease of $V(x)$ between $V(x(t_0))$ and c by $-\gamma$ (since c is assumed to be non-zero, we know that $-\gamma < 0$ because $\dot{V}(x)$ is negative definite), then

$$V(x(t)) = V(x(t_0)) + \int_{t_0}^{t} \dot{V}(x(\tau))\, d\tau \leq V(x(t_0)) - \gamma t$$

We can see that in a finite amount of time $V(x)$ will break the barrier and go below c. This happens no later than at time t_c given by

$$t_c = \frac{V(x(t_0)) - c}{\gamma} < \infty$$

This is a clear contradiction to the assumption that $V(x) \geq c$. Hence, under the conditions of part (b) of the theorem, $V(x)$ converges to zero.

To show that this implies the origin is attractive, we use contradiction again. Assume that $x(t)$ converges to some point other than the origin or that it does not converge at all. Then, since $V(x)$ is continuous and $V(x) = 0$ only for $x = 0$, this function would not converge to zero either.

Thus, the origin is not only stable, but also attractive, and hence it is an asymptotically stable equilibrium point. □

As we showed in part (a) of Theorem 2.2.2, if $\dot{V}(x)$ is negative semi-definite, the origin is stable. But, if $\dot{V}(x)$ does not vanish along any system trajectory $x(t)$, the origin is asymptotically stable in the sense of Lyapunov. If $\dot{V}(x)$ is positive (semi-)definite, we can similarly conclude that the origin is unstable. If $\dot{V}(x)$ is indefinite, so is our knowledge of the system stability: we need to try some other candidate for the Lyapunov function.

Lyapunov's stability for linear systems. Let us apply Theorem 2.2.2 to linear time-invariant systems. We shall first consider the continuous-time case and then describe how the same ideas apply to the discrete-time case. In the Lyapunov stability analysis it is assumed that there is no input to the system, hence in general

$$\dot{x} = Ax$$

For linear time-invariant systems it is sufficient to reduce the choice of candidate Lyapunov functions $V(x)$ to positive definite quadratic forms. Any positive definite quadratic form can be written as

$$V(x) = x'Px$$

where P is a positive definite symmetric matrix. Then the time derivative of $V(x)$ is given by

$$\dot{V}(x) = x'(A'P + PA)x$$

If $Q = -(A'P + PA)$ is positive definite, then $\dot{V}(x)$ is negative definite, and the system is asymptotically stable in the sense of Lyapunov. If Q is positive semi-definite, the system is guaranteed to be stable. Since the Lyapunov's theorem provides only sufficient stability conditions, the system still may be asymptotically stable. To prove this we need to find a better choice for P. An alternative approach in cases with Q a positive semi-definite matrix is to try to show that for any system trajectory we have $\dot{V}(x) \not\equiv 0$. That would also prove that the system is asymptotically stable.

Example 2.2.2 *A simple pendulum with friction is described by the following two equations (assuming small amplitude oscillations, so that* $\sin x_1 \approx x_1$*):*

$$\begin{aligned} \dot{x}_1 &= x_2 \\ \dot{x}_2 &= -\frac{g}{l}x_1 - \frac{k}{m}x_2 \end{aligned}$$

A natural candidate for the Lyapunov function $V(x_1, x_2)$ *is the total energy of the system:*

$$V(x_1, x_2) = \frac{1}{2}mglx_1^2 + \frac{1}{2}ml^2x_2^2$$

In that case

$$\dot{V}(x_1, x_2) = mglx_1\dot{x}_1 + ml^2x_2\dot{x}_2 = -kl^2x_2^2$$

Obviously, $\dot{V}(x_1, x_2)$ *is negative semi-definite, which guarantees only stability. But a pendulum with friction is asymptotically stable. Hence we either need a better candidate for the Lyapunov function or we need to show that on the system trajectories* $\dot{V}(x_1, x_2) \equiv 0$ *only when both* $x_1 = 0$ *and* $x_2 = 0$*. Since* $\dot{V}(x_1, x_2) = -kl^2x_2^2$*, in order for this time derivative to be zero and stay zero,* x_2 *must be zero and it has to stay there, so* $\dot{x}_2 = 0$*. But then the system equations imply that* x_1 *must also be zero. This proves the asymptotic stability of a simple pendulum with friction.*

Showing that $\dot{V}(x) \not\equiv 0$ on the system trajectories or finding an appropriate matrix P may be difficult. Fortunately, there is also a third approach, based on the following result due to Lyapunov:

Theorem 2.2.3 *Matrix A is Hurwitz if and only if for any given positive definite symmetric matrix Q there exists a positive definite symmetric matrix P such that $A'P + PA = -Q$. If A is Hurwitz, P is unique.*

Proof. If A is Hurwitz, i.e., if all of its eigenvalues have negative real parts, then the following integral exists:

$$P = \int_0^\infty e^{A't} Q e^{At}\, dt$$

It is easy to see that this P satisfies the Lyapunov equation $A'P + PA = -Q$:

$$\begin{aligned} A'P + PA &= \int_0^\infty A' e^{A't} Q e^{At}\, dt + \int_0^\infty e^{A't} Q e^{At} A\, dt \\ &= \int_0^\infty \frac{d}{dt}\left(e^{A't} Q e^{At}\right) dt \\ &= \left(e^{A't} Q e^{At}\right)\Big|_0^\infty = -Q \end{aligned}$$

As long as Q is positive definite and symmetric, so is this integral. In addition, this is the only solution of the Lyapunov equation. This can be proved in many ways, for example by contradiction. If $P_1 \neq P_2$ are two solutions of $A'P + PA = -Q$, then $A'P_1 + P_1 A = -Q = A'P_2 + P_2 A$, hence

$$A'(P_1 - P_2) + (P_1 - P_2)A = 0$$

and

$$\frac{d}{dt}\left(e^{A't}(P_1 - P_2)e^{At}\right) = 0$$

Therefore $e^{A't}(P_1 - P_2)e^{At} = const$. In special cases $t = 0$ and $t \to \infty$, since A is Hurwitz, we have

$$P_1 - P_2 = \lim_{t\to\infty} e^{A't}(P_1 - P_2)e^{At} = 0$$

This is in contradiction with the initial assumption $P_1 \neq P_2$, so the solution must be unique.

The fact that the existence of positive definite solution P of $A'P + PA = -Q$, where Q is any positive definite symmetric matrix, implies that A is Hurwitz, follows directly from Theorems 2.2.1 and 2.2.2 with $V(x) = x'Px$ and $\dot{V}(x) = -x'Qx$. □

In this approach we start from any positive definite matrix Q. The matrix equation we need to solve is called the Lyapunov matrix equation. Since P can be assumed to be symmetric before we solve the equation, this matrix equation represents a system of $\frac{n(n+1)}{2}$ linear equations in coefficients of P, where n is the system order.

Example 2.2.3 *Let us apply this approach to the pendulum with friction described in Example 2.2.2. If the time derivative of* $V(x_1, x_2)$ *is given by, for example,*

$$\dot{V}(x_1, x_2) = -gklx_1^2 - kl^2x_2^2$$

the corresponding matrix Q *is*

$$Q = \begin{bmatrix} gkl & 0 \\ 0 & kl^2 \end{bmatrix}$$

With

$$A = \begin{bmatrix} 0 & 1 \\ -g/l & -k/m \end{bmatrix}$$

the Lyapunov equation $A'P + PA = -Q$ *has a unique symmetric solution*

$$P = \begin{bmatrix} mgl + \frac{k^2l^2}{2m} & \frac{1}{2}kl^2 \\ \\ \frac{1}{2}kl^2 & ml^2 \end{bmatrix}$$

It is easy to verify that P *is positive definite:*

$$mgl + \frac{k^2l^2}{2m} > 0 \qquad \text{and} \qquad \det(P) = m^2gl^3 + \frac{1}{4}k^2l^4 > 0$$

Note that it is sufficient to solve the Lyapunov equation for only one positive definite matrix Q*. We shall often do that for* $Q = I$*, the identity matrix.*

Stability and linearization. Historically, one of the first applications of the Lyapunov stability theory was in the theoretical investigations of linearization and stability. Since many linear models are actually linearized models of nonlinear systems, this is a very important topic.

If a nonlinear system is given by

$$\dot{x}(t) = f(x(t))$$

its linearized model around the equilibrium point x_e is found by writing $x(t) = x_e + z(t)$, where $z(t)$ denotes perturbations around the equilibrium. The linearized model is obtained from the Taylor series expansion of $f(x(t))$ around $x = x_e$:

$$f(x(t)) = f(x_e + z(t)) = f(x_e) + Az(t) + r(z(t))$$

where A is the Jacobian matrix of $f(x(t))$

$$A_{ij} = \left.\frac{\partial f_i(x)}{\partial x_j}\right|_{x=x_e}$$

while $r(z(t))$ is the linearization error such that $r(z) = O(z^2)$, i.e.,

$$\lim_{\|z\| \to 0} \frac{\|r(z)\|}{\|z\|} = 0$$

Since $\dot{z}(t) = \dot{x}(t)$ and $f(x_e) = 0$, neglecting any terms of second or higher orders yields the linearized model

$$\dot{z}(t) = Az(t)$$

Lyapunov and Poincare proved the following theorem:

Theorem 2.2.4 *If the linearized system is asymptotically stable, then the original nonlinear system is also asymptotically stable under sufficiently small perturbations.*

Proof. Suppose A is a stability matrix and consider the following candidate for the Lyapunov function of the original nonlinear system

$$V(z(t)) = z'(t)Pz(t)$$

where P is the unique symmetric positive definite solution of the Lyapunov equation $AP + PA' = -I$. Then

$$\begin{aligned} \dot{V}(z(t)) &= -z'(t)z(t) + 2r'(z(t))Pz(t) \\ &= -z'(t)z(t)\left(1 - \frac{2r'(z(t))Pz(t)}{z'(t)z(t)}\right) \end{aligned}$$

Since $r(z) = O(z^2)$, for sufficiently small perturbations $z(t)$, this time derivative is negative, hence the nonlinear system is asymptotically stable. □

Stability of discrete-time systems. The fundamental ideas of the stability theory are the same for the discrete-time systems as for their continuous-time cousins. The only two differences are:

- *The locations of the system eigenvalues allowed for stability.* For example, a discrete-time system is asymptotically stable if and only if all roots of its characteristic equation lie inside the unit circle, i.e., if and only if $|\lambda_i| < 1$ $(i = 1, 2, \ldots, n)$.
- *The form of the Lyapunov equation.* To derive the discrete-time Lyapunov equation, consider a discrete-time linear time-invariant system given by $x[k+1] = Ax[k]$. If $V(x[k])$ is a positive definite quadratic form represented using a symmetric positive definite matrix P, i.e., if $V(x[k]) = x'[k]Px[k]$, then

$$\Delta V(x[k]) = V(x[k+1]) - V(x[k]) = x'(A'PA - P)x$$

With $Q = -(A'PA - P)$, the requirement for asymptotic stability is that Q must be positive definite or at least positive semi-definite with the condition that $\Delta V(x[k]) \not\equiv 0$ along any possible system trajectory. Therefore, the discrete-time Lyapunov equation is

$$A'PA - P = -Q$$

Controllability, observability, and minimallity

The most important properties of dynamical systems are stability, controllability and observability. Unlike stability, which was among the first topics to be studied in control theory, the investigation of controllability and observability begun only with the emergence of state-space approach, in the late 1950's.

We start by defining controllability and observability and continue by investigation of several equivalent conditions that guarantee them. Finally we discuss the minimallity of systems. Through these discussions we also introduce several commonly used system realizations: controllability, controller, observability, observer, and modal forms and explore the duality between controllability and observability. We define these properties for both continuous- and discrete-time systems. We will emphasize any differences, but if no special reference is made to either type of systems, the reader may safely assume we refer to both.

State controllability. The question of whether or not we can drive a system from any given state to any desired state, and do that in a finite amount of time, arises in many diverse control problems: setting up the initial conditions in simulations; determining if the system itself allows any control law to be effective; system stabilization; optimal control; minimallity of systems.

Definition 2.2.8 (State controllability) *A system is state controllable if a proper input can drive it from any given state to any desired state in a finite amount of time.*

In Problems 3.7.1 and 4.7.3 we prove the following theorem

Theorem 2.2.5 *A system described by $\{A, B, C\}$ is state controllable if and only if $\rho(\mathcal{C}) = n$, where $\mathcal{C}$ is the controllability matrix given by*

$$\mathcal{C} = [B \quad AB \quad A^2B \quad \dots \quad A^{n-1}B]$$

and n is the order of the system.

For single-input systems this is equivalent to $\det(\mathcal{C}) \neq 0$.

Controllability-from-the-origin and controllability-to-the-origin. A system is *controllable-from-the-origin* or *controllable p.s.f.o.* (pointwise state from the origin) if an appropriate input can drive the system from the origin to any desired state in a finite amount of time. This property is equivalent to state controllability and is often called *reachability*.

A system is said to be *controllable-to-the-origin* or *controllable p.s.t.o.* (pointwise state to the origin) if an appropriate input can drive all of its states to the origin in a finite amount of time. For continuous time systems this property is equivalent to the state controllability. For discrete-time systems (as we show in Problems 4.7.1 and 4.7.2) state controllability is sufficient but not necessary for this property. If $\det(A) \neq 0$, these two properties are equivalent in the discrete-time too. For historical reasons this property is often called *controllability*, but we do not use this name in order to avoid any confusion.

Controller canonical realization. The controller realization is one of several most popular realizations, commonly called canonical forms. In order to make our discussion definite, we shall concentrate here on the continuous-time systems, but everything applies equally to the discrete-time systems as well. The controller realization of

$$H(s) = \frac{b(s)}{a(s)} = \frac{b_1 s^{n-1} + b_2 s^{n-2} + \ldots + b_{n-1}s + b_n}{s^n + a_1 s^{n-1} + a_2 s^{n-2} + \ldots + a_{n-1}s + a_n}$$

is defined by

$$A_c = \begin{bmatrix} -a_1 & \ldots & -a_{n-1} & -a_n \\ 1 & \ldots & 0 & 0 \\ \vdots & \ddots & \vdots & \vdots \\ 0 & \ldots & 1 & 0 \end{bmatrix} \quad \text{and} \quad b_c = \begin{bmatrix} 1 \\ 0 \\ \vdots \\ 0 \end{bmatrix}$$

Matrix A_c is a *companion* matrix. General properties of companion matrices are discussed in Appendix C.3.

The components of $c_c' = [c_{c1} \; c_{c2} \; \ldots \; c_{cn}]$ are uniquely determined because

$$\begin{aligned} Y(s) &= c_{c1}X_1(s) + c_{c2}X_2(s) + \ldots + c_{cn}X_n(s) \\ &= \frac{\left(c_{c1}/s + c_{c2}/s^2 + \ldots + c_{cn}/s^n\right) U(s)}{1 + a_1/s + a_2/s^2 + \ldots + a_n/s^n} \\ &= \frac{c_{c1}s^{n-1} + c_{c2}s^{n-2} + \ldots + c_{cn}}{s^n + a_1 s^{n-1} + a_2 s^{n-2} + \ldots + a_n} U(s) \end{aligned}$$

therefore

$$c_c' = \left[\begin{array}{cccc} b_1 & b_2 & \ldots & b_n \end{array} \right]$$

Example 2.2.4 *In Figure 2.6 we show the signal flow diagram of the controller realization of* $H(s) = \frac{s^2+2}{s^3+7s^2+14s+8}$.

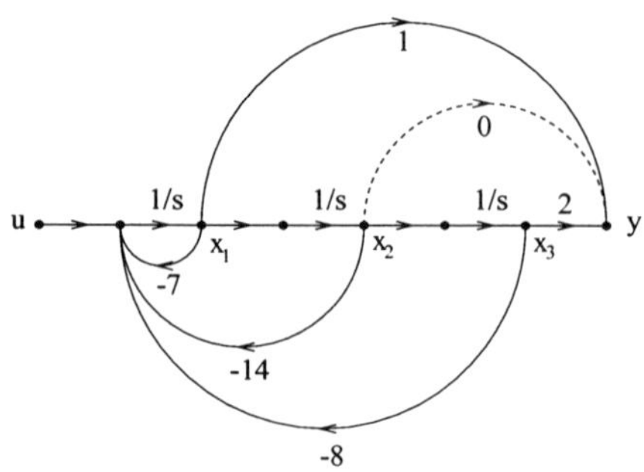

Figure 2.6: Controller realization of $H(s) = \frac{s^2+2}{s^3+7s^2+14s+8}$.

Controllability canonical realization. Another canonical form is the controllability form, defined by

$$A_{co} = \begin{bmatrix} 0 & \dots & 0 & -a_n \\ 1 & \dots & 0 & -a_{n-1} \\ \vdots & \ddots & \vdots & \vdots \\ 0 & \dots & 1 & -a_1 \end{bmatrix} \quad \text{and} \quad b_{co} = \begin{bmatrix} 1 \\ 0 \\ \vdots \\ 0 \end{bmatrix}$$

In order to determine c_{co} we first show that the controller and the controllability forms are similar through a transformation given by[2] $S = \mathsf{a}_-^T$, where

$$\mathsf{a}_- = \begin{bmatrix} 1 & 0 & \dots & 0 & 0 \\ a_1 & 1 & \dots & 0 & 0 \\ \vdots & \vdots & \ddots & \vdots & \vdots \\ a_{n-2} & a_{n-3} & \dots & 1 & 0 \\ a_{n-1} & a_{n-2} & \dots & a_1 & 1 \end{bmatrix}$$

Indeed, from $b_{co} = Sb_c$ we find that the first column of S is $[1 \; 0 \; \dots \; 0]'$. Using this, from $A_{co}S = SA_c$ we can easily find a recursive relation for the jth column of S in terms of the $(j-1)$th column:

$$\begin{bmatrix} \sigma_{1,j} \\ \sigma_{2,j} \\ \sigma_{3,j} \\ \vdots \\ \sigma_{n,j} \end{bmatrix} = \begin{bmatrix} a_{j-1} \\ \sigma_{1,j-1} \\ \sigma_{2,j-1} \\ \vdots \\ \sigma_{n-1,j-1} \end{bmatrix} \qquad (j = 2, 3, \dots, n)$$

Then $c'_{co} = c'_c S^{-1} = [b_1 \; \dots \; b_n]\mathsf{a}_-^{-T}$. Since $[h_1 \; \dots \; h_n]\mathsf{a}_-^T = [b_1 \; \dots \; b_n]$ is just another way of writing the definition of Markov parameters and $\det(\mathsf{a}_-) \neq 0$

$$\sum_{i=1}^{\infty} h_i s^{-i} = \frac{b(s)}{a(s)}$$

we find that

$$c'_{co} = [h_1 \;\; h_2 \;\; \dots \;\; h_n] \tag{2.2}$$

Controllability and similarity. Let us now show that if a realization given by $\{A_1, B_1, C_1\}$ undergoes a nonsingular similarity transformation into $\{A_2, B_2, C_2\}$, the (un)controllability of the system is not affected. This is because

$$\begin{aligned} \rho(\mathcal{C}_2) &= \rho\left([\, B_2 \quad A_2B_2 \quad \dots \quad A_2^{n-1}B_2 \,]\right) \\ &= \rho\left([\, SB_1 \quad SA_1S^{-1}SB_1 \quad \dots \quad SA_1^{n-1}S^{-1}SB_1 \,]\right) \\ &= \rho(S\mathcal{C}_1) \;=\; \rho(\mathcal{C}_1) \qquad \text{(since } S \text{ is nonsingular)} \end{aligned}$$

[2] In order to simplify the notation, we occasionally use the superscript T to denote the matrix transpose (cf. Appendix B.4).

This analysis gives us a hint on how to determine the similarity transformation between the two given order-n controllable realizations of $H(s)$. We have just seen that if we assume that this transformation exists, then

$$\mathcal{C}_2 = S\mathcal{C}_1 \qquad \text{i.e.,} \qquad S = \mathcal{C}_2\mathcal{C}_1^{-1}$$

We can show its existence for any two controllable, same order realizations of $H(s)$ directly, by showing that with $S = \mathcal{C}_2\mathcal{C}_1^{-1}$ we have $A_2 = SA_1S^{-1}$, $B_2 = SB_1$, and $C_2 = C_1S^{-1}$, but the following proof is simpler. In Problem 3.8.5 we show that $S_{co1} = \mathcal{C}_1^{-1}$ transforms a controllable, order-n realization $\{A_1, B_1, C_1\}$ of

$$H(s) = \frac{b(s)}{a(s)} = \frac{b_1s^{n-1} + b_2s^{n-2} + \ldots + b_{n-1}s + b_n}{s^n + a_1s^{n-1} + a_2s^{n-2} + \ldots + a_{n-1}s + a_n}$$

into the controllability form. Obviously, there is only one such order-n form for any transfer function. Therefore, any other controllable, order-n realization $\{A_2, B_2, C_2\}$ of that transfer function will be transformed into the same controllability realization with $S_{co2} = \mathcal{C}_2^{-1}$. Finally, to go directly from $\{A_1, B_1, C_1\}$ to $\{A_2, B_2, C_2\}$ we can use

$$S = S_{co2}^{-1}S_{co1} = \mathcal{C}_2\mathcal{C}_1^{-1} \tag{2.3}$$

Controllability matrices $\mathcal{C}_c$ and $\mathcal{C}_{co}$. The formula for the similarity transformation between any two controllable realizations can be used to determine the controllability matrices for the controller and the controllability forms.

- Controllability matrix of the controllability form: It is easy to see that $\mathcal{C}_{co} = I$, either by looking at the first column of the powers of A_{co} or even simpler, from the fact that the transformation into the controllability form is given by $\mathcal{C}_1^{-1}$ and, by the general formula we just proved, it is given by $\mathcal{C}_{co}\mathcal{C}_1^{-1}$. Therefore

$$\mathcal{C}_{co} = I \tag{2.4}$$

- Controllability matrix of the controller form: Likewise, knowing that the similarity transformation from the controller to the controllability form is given by a_-^T and that by the general formula it should be $\mathcal{C}_{co}\mathcal{C}_c^{-1}$, where $\mathcal{C}_{co} = I$, we find that

$$\mathcal{C}_c = \mathsf{a}_-^{-T} \tag{2.5}$$

The similarity transformation from any controllable realization into the corresponding controller form is then given by

$$S = \mathcal{C}_c\mathcal{C}_1^{-1} \qquad (\mathcal{C}_c = \mathsf{a}_-^{-T})$$

We illustrate these relations in Example 2.2.5:

Example 2.2.5 *In Problem 3.7.12 we show that the inverted pendulum on a cart (first described in Example 2.1.1) with* $m = 0.102\,kg$, $g = 9.81\,m/s^2$, $M = 1\,kg$, *and* $l = 0.5\,m$, *i.e., with*

$$A = \begin{bmatrix} 0 & 0 & 1 & 0 \\ 0 & 0 & 0 & 1 \\ 0 & -1 & 0 & 0 \\ 0 & 21.6 & 0 & 0 \end{bmatrix} \quad b = \begin{bmatrix} 0 \\ 0 \\ 1 \\ -2 \end{bmatrix} \quad \text{and} \quad C = \begin{bmatrix} 1 & 0 & 0 & 0 \\ 0 & 1 & 0 & 0 \end{bmatrix}$$

is controllable because its controllability matrix

$$\mathcal{C} = [b \;\; Ab \;\; A^2b \;\; A^3b] = \begin{bmatrix} 0 & 1 & 0 & 2 \\ 0 & -2 & 0 & -43.2 \\ 1 & 0 & 2 & 0 \\ -2 & 0 & -43.2 & 0 \end{bmatrix}$$

has a full rank.

The characteristic polynomial of this system is

$$a(s) = \det(sI - A) = s^4 - 21.6s^2$$

Therefore

$$\mathsf{a}_- = \begin{bmatrix} 1 & 0 & 0 & 0 \\ a_1 & 1 & 0 & 0 \\ a_2 & a_1 & 1 & 0 \\ a_3 & a_2 & a_1 & 1 \end{bmatrix} = \begin{bmatrix} 1 & 0 & 0 & 0 \\ 0 & 1 & 0 & 0 \\ -21.6 & 0 & 1 & 0 \\ 0 & -21.6 & 0 & 1 \end{bmatrix}$$

The transformation into the controller form is given by

$$S = \mathsf{a}_-^{-T}\mathcal{C}^{-1} = \begin{bmatrix} 0 & 0 & 0 & -0.5000 \\ 0 & -0.5000 & 0 & 0 \\ 0 & 0 & -0.0510 & -0.0255 \\ -0.0510 & -0.0255 & 0 & 0 \end{bmatrix}$$

Indeed,

$$SAS^{-1} = \begin{bmatrix} 0 & 21.6 & 0 & 0 \\ 1 & 0 & 0 & 0 \\ 0 & 1 & 0 & 0 \\ 0 & 0 & 1 & 0 \end{bmatrix}, \quad Sb = \begin{bmatrix} 1 \\ 0 \\ 0 \\ 0 \end{bmatrix}, \quad \text{and} \quad CS^{-1} = \begin{bmatrix} 0 & 1 & 0 & -19.6 \\ 0 & -2 & 0 & 0 \end{bmatrix}$$

Note that

$$H(s) = C(sI - A)^{-1}b = \frac{\begin{bmatrix} s^2 - 19.6 \\ -2s^2 \end{bmatrix}}{s^4 - 21.6s^2}$$

Finally, we can calculate $\mathcal{C}_c$ *to verify it is equal to* a_-^{-T}:

$$\mathcal{C}_c = [b_c \;\; A_cb_c \;\; A_c^2b_c \;\; A_c^3b_c] = \begin{bmatrix} 1 & 0 & 21.6 & 0 \\ 0 & 1 & 0 & 21.6 \\ 0 & 0 & 1 & 0 \\ 0 & 0 & 0 & 1 \end{bmatrix}$$

which is indeed a_-^{-T}.

State observability. Very much like the basic problem of controllability (determining whether or not an input can be designed to appropriately control the system), the question of whether or not we can determine the states from the output measurements arises in many practical and theoretical problems: determination of initial conditions of the individual states from the initial conditions for the system output; state observation; state estimation; minimallity.

Definition 2.2.9 (Observability) *A system is said to be observable if its state at some time t_0 can be determined from the values of the system's output over a finite time interval $[t_0, t_f]$.*

In Problem 3.7.5 we prove the following theorem

Theorem 2.2.6 *The necessary and sufficient condition for observability of the system given by $\{A, B, C\}$ is $\rho(\mathcal{O}) = n$, where $\mathcal{O}$ is the observability matrix given by*

$$\mathcal{O} = \begin{bmatrix} C \\ CA \\ \vdots \\ CA^{n-1} \end{bmatrix}$$

and n is the order of the system.

For single-output systems this is the same as $\det(\mathcal{O}) \neq 0$.

Duality. It is very interesting and extremely useful to note the duality between controllability and observability.

Theorem 2.2.7 *A system given by $\{A, B, C\}$ is observable if and only if its dual system, described by $\{A', C', B'\}$, is controllable.*

Proof. Compare the observability and controllability conditions for systems $\{A, B, C\}$ and $\{A', C', B'\}$, respectively. □

This theorem will allow us to prove many theorems for observability by simply noting that they are dual to the theorems already proven for controllability.

Constructibility. What is the dual property to the controllability-to-the-origin? Recall that in the continuous-time the controllability-to-the-origin was equivalent to $\rho(\mathcal{C}) = n$, while in the discrete-time it was equivalent to $\rho(\mathcal{C}) = n$ if and only if $\det(A) \neq 0$. Otherwise, $\rho(\mathcal{C}) = n$ was only a sufficient condition, it was not necessary.

It turns out that the dual to this situation in the "observability world" is the property called constructibility. A system is said to be *constructible* if its state at some time t_f can be determined from the values of the system's output over a finite time interval $[t_0, t_f]$. Note that for observability we must be able to determine the states from the ***future*** values of the output, while for constructibility this is done from the ***past*** values of the output. Relations between these properties are illustrated in Figure 2.7.

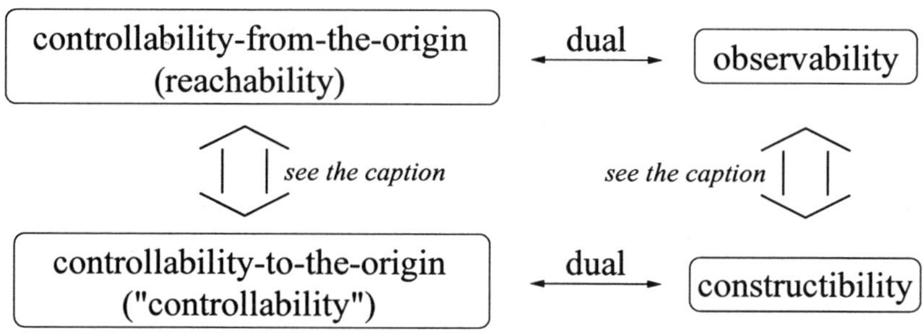

Figure 2.7: Illustration of the equivalence and duality relations between various system properties. For the discrete-time systems the equivalences hold if and only if the system matrix is nonsingular, otherwise they are only downward oriented implications. For continuous-time systems they are always equivalences.

Observable realizations. Just like in the case of controllability, any two order-n observable realizations $\{A_1, B_1, C_1\}$ and $\{A_2, B_2, C_2\}$ of $H(s)$ are similar, and the corresponding transformation is found from $\mathcal{O}_2 = \mathcal{O}_1 S^{-1}$, therefore

$$S = \mathcal{O}_2^{-1}\mathcal{O}_1 \tag{2.6}$$

The dual realizations to the controllability and controller realizations are the ***observability*** and ***observer*** realizations, respectively. According to our duality relations the observability realization is given by

$$A_{ob} = A'_{co} \qquad B_{ob} = C'_{co} \quad \text{and} \quad C_{ob} = B'_{co}$$

while the observer realization is given by

$$A_o = A'_c \qquad B_o = C'_c \quad \text{and} \quad C_o = B'_c \ .$$

For the observability realization we have

$$\mathcal{O}_{ob} = I \tag{2.7}$$

while for the observer realization

$$\mathcal{O}_o = a_-^{-1} \tag{2.8}$$

where

$$a_- = \begin{bmatrix} 1 & 0 & \dots & 0 & 0 \\ a_1 & 1 & \dots & 0 & 0 \\ \vdots & \vdots & \ddots & \vdots & \vdots \\ a_{n-2} & a_{n-3} & \dots & 1 & 0 \\ a_{n-1} & a_{n-2} & \dots & a_1 & 1 \end{bmatrix}$$

Minimallity and joint controllability and observability. Up to this point we have considered the controllability of realizations of $H(s)$ separately from their observability. The realizations that are both controllable and observable are the most desirable and we will now explore some of their properties. By now we have learned that for any $H(s)$ there are infinitely many controllable realizations and infinitely many observable realizations.

The next question is if a controllable realization of $H(s)$ can be observable while another is not? In general, the answer is yes, as we shall see in the next example. However, if we limit the discussion to the realizations of the same order, the answer is no: the set of controllable realizations of $H(s)$ is either identical to or completely disjoint from the set of its observable realizations.

Example 2.2.6 *Consider the transfer function given by*

$$H(s) = \frac{s+5}{s^3 + 8s^2 + 17s + 10}$$

An order 3 *controller realization of* $H(s)$ *is given by*

$$\begin{bmatrix} \dot{x}_1 \\ \dot{x}_2 \\ \dot{x}_3 \end{bmatrix} = \begin{bmatrix} -8 & -17 & -10 \\ 1 & 0 & 0 \\ 0 & 1 & 0 \end{bmatrix} \begin{bmatrix} x_1 \\ x_2 \\ x_3 \end{bmatrix} + \begin{bmatrix} 1 \\ 0 \\ 0 \end{bmatrix} u$$

$$y = \begin{bmatrix} 0 & 1 & 5 \end{bmatrix} \begin{bmatrix} x_1 \\ x_2 \\ x_3 \end{bmatrix}$$

The controllability matrix in this case is

$$\mathcal{C} = \begin{bmatrix} 1 & -8 & 47 \\ 0 & 1 & -8 \\ 0 & 0 & 1 \end{bmatrix}$$

and since $\det(\mathcal{C}) = 1 \neq 0$, *this realization is controllable. Its observability matrix is*

$$\mathcal{O} = \begin{bmatrix} 0 & 1 & 5 \\ 1 & 5 & 0 \\ -3 & -17 & -10 \end{bmatrix}$$

and since $\det(\mathcal{O}) = 0$, *this realization is not observable. If we note that the numerator and the denominator of* $H(s)$ *have a common factor, we can write*

$$H(s) = \frac{s+5}{(s+1)(s+2)(s+5)} = \frac{1}{(s+1)(s+2)} = \frac{1}{s^2+3s+2}$$

This is the irreducible form of $H(s)$ *and the corresponding realizations are said to be minimal realizations of* $H(s)$. *The minimum order controller realization of* $H(s)$ *is*

$$\begin{bmatrix} \dot{x}_1 \\ \dot{x}_2 \end{bmatrix} = \begin{bmatrix} -3 & -2 \\ 1 & 0 \end{bmatrix} \begin{bmatrix} x_1 \\ x_2 \end{bmatrix} + \begin{bmatrix} 1 \\ 0 \end{bmatrix} u$$

$$y = \begin{bmatrix} 0 & 1 \end{bmatrix} \begin{bmatrix} x_1 \\ x_2 \end{bmatrix}$$

The controllability and observability matrices in this case are

$$\mathcal{C} = \begin{bmatrix} 1 & -3 \\ 0 & 1 \end{bmatrix} \quad \textit{and} \quad \mathcal{O} = \begin{bmatrix} 0 & 1 \\ 1 & 0 \end{bmatrix}$$

respectively. Since $\det(\mathcal{C}) = 1 \neq 0$ *and* $\det(\mathcal{O}) = -1 \neq 0$, *this realization is both controllable and observable.*

We shall soon see that there is a very tight connection between minimallity and joint controllability and observability. In fact we shall see that they are equivalent, but let us finish what we have already begun. In order to show that among the realizations of $H(s)$ having equal orders either all or none of the controllable realizations are observable, recall the following two facts:

- All controllable realizations of $H(s)$ with equal orders are related through similarity transformations.
- Similarity transformations preserve observability.

Therefore, if any controllable order-n realization of $H(s)$ is observable, so are all other controllable order-n realizations. Also, if any controllable order-n realization of $H(s)$ is not observable, none of them are. By duality, if any observable order-n realization of $H(s)$ is controllable so are all other such realizations and if any observable order-n realization of $H(s)$ is not controllable, neither are the others.

Furthermore, if there exists an order-n realization of $H(s)$ that is both controllable and observable, then any other order-n realization of $H(s)$ is both controllable and observable. Indeed, recall that for a given $H(s)$ and given order of realizations n, there is a unique Hankel matrix of Markov parameters

$$\mathcal{M} = \begin{bmatrix} h_1 & h_2 & \cdots & h_n \\ h_2 & h_3 & \cdots & h_{n+1} \\ \vdots & \vdots & \ddots & \vdots \\ h_n & h_{n+1} & \cdots & h_{2n-1} \end{bmatrix}$$

Recall also that for any order-n realization of $H(s)$ we have $\mathcal{O}\mathcal{C} = \mathcal{M}$. Hence, if $\{A_1, B_1, C_1\}$ is jointly controllable and observable order-n realization of $H(s)$ and $\{A_2, B_2, C_2\}$ is any other order-n realization of $H(s)$, then from

$$\mathcal{O}_1\mathcal{C}_1 = \mathcal{M} = \mathcal{O}_2\mathcal{C}_2$$

we see that the nonsingularity of $\mathcal{C}_1$ and $\mathcal{O}_1$ implies the nonsingularity of $\mathcal{C}_2$ and $\mathcal{O}_2$. Therefore, any other order-n realization of $H(s)$ is jointly controllable and observable if one of them is.

Are there any realizations of $H(s)$ which are neither controllable nor observable? The following example answers that question affirmatively.

Example 2.2.7 *Consider the following transfer function:*

$$H(s) = \frac{2s^2 + 12s + 16}{s^3 + 8s^2 + 19s + 12}$$

The following realization is neither controllable nor observable:

$$\begin{bmatrix} \dot{x}_1 \\ \dot{x}_2 \\ \dot{x}_3 \end{bmatrix} = \begin{bmatrix} -1 & 0 & 0 \\ 0 & -3 & 0 \\ 0 & 0 & -4 \end{bmatrix} \begin{bmatrix} x_1 \\ x_2 \\ x_3 \end{bmatrix} + \begin{bmatrix} 1 \\ 1 \\ 0 \end{bmatrix} u$$

$$y = \begin{bmatrix} 1 & 1 & 0 \end{bmatrix} \begin{bmatrix} x_1 \\ x_2 \\ x_3 \end{bmatrix}$$

Note that the irreducible form of $H(s)$ is $\frac{s+2}{(s+1)(s+3)}$.

Let us formally define the minimallity and use the previous discussion to prove several important results about minimallity, controllability, and observability.

Definition 2.2.10 (Minimallity) *A realization is said to be minimal if no other realization of the same transfer function has lower order.*

One of the most important properties of minimal realizations is the equivalence of internal types of stability to their external counterparts. This follows from the fact that the realization $\{A, b, c'\}$ is minimal if and only if there are no pole-zero cancellations between $b(s) = c'\text{adj}(sI - A)b$ and $a(s) = \det(sI - A)$, and therefore the poles and eigenvalues of the system coincide, along with their multiplicities.

We already have one practical criterion for minimallity: *Realization $\{A, b, c'\}$ is minimal if and only if there are no pole-zero cancellations between $b(s) = c'\,adj(sI - A)b$ and $a(s) = \det(sI - A)$.* Another is to use the following theorem, which establishes the second most important property of minimal realizations, the equivalence between minimallity and joint controllability and observability.

Theorem 2.2.8 *The realization $\{A, b, c'\}$ of $H(s)$ is minimal if and only if it is both controllable and observable.*

Proof. We shall prove here that the order-n controller realization $\{A_c, b_c, c_c'\}$ is observable if and only if there are no pole-zero cancellations between $b(s) = c_c'\text{adj}(sI - A_c)b_c$ and $a(s) = \det(sI - A_c)$. From what we discussed earlier, this will mean that the set of controllable order-n realizations is equal to the set of order-n observable realizations, and furthermore, it covers the set of all order-n realizations of $H(s)$.

The proof[3] is based on the "shifting" property of the companion matrices:

$$e_i' A_c = e_{i-1}' \quad (2 \leq i \leq n)$$

where e_i' is the ith row of the identity matrix. By the way, $e_1' A_c = [-a_1 \ldots - a_n]$.

[3]We shall see a different proof of this theorem in Example 2.2.9.

This property is used to show that $e_i' b(A_c) = c_c' A_c^{n-i}$ $(i = 1, \ldots, n)$. First we prove the special case $i = n$:

$$\begin{aligned} e_n' b(A_c) &= b_1 e_n' A_c^{n-1} + b_2 e_n' A_c^{n-2} + \ldots + b_{n-1} e_n' A_c + b_n e_n' \\ &= b_1 e_1' + b_2 e_2' + \ldots + b_{n-1} e_{n-1}' + b_n e_n' \\ &= [b_1 \ b_2 \ \ldots \ b_{n-1} \ b_n] \\ &= c_c' \end{aligned}$$

Using this result and the fact that A_c commutes with $b(A_c)$ we find that for $k = 0, \ldots, n-1$

$$e_{n-k}' b(A_c) = e_n' A_c^k b(A_c) = e_n' b(A_c) A_c^k = c_c' A_c^k$$

Therefore

$$\mathcal{O}_c = \tilde{I} b(A_c) \tag{2.9}$$

where $\tilde{I} = [e_n \ \ldots \ e_1]$ is the flipped identity matrix.

Now, $\det(\mathcal{O}_c) \neq 0$ if and only if $\det(b(A_c)) \neq 0$. Since the determinant of a matrix is equal to the product of its eigenvalues and the eigenvalues of $b(A_c)$ are $b(\lambda_i)$ $(i = 1, \ldots, n)$, where λ_i are the eigenvalues of A_c, we can write

$$\det(b(A_c)) = b(\lambda_1) \ldots b(\lambda_n)$$

Hence $\det(\mathcal{O}_c) \neq 0$ if and only if $b(\lambda_i) \neq 0$ $(i = 1, \ldots, n)$. By definition of eigenvalues $a(\lambda_i) = 0$ $(i = 1, \ldots, n)$, therefore the controller form is observable if and only if $b(s) = c_c' \mathrm{adj}(sI - A_c) b_c$ and $a(s) = \det(sI - A_c)$ have no common factors. □

Example 2.2.8 *Let us verify that the controller form is observable if and only if there are no pole-zero cancellations between $b(s)$ and $a(s)$ for the simple case when $n = 2$. In this case*

$$A_c = \begin{bmatrix} -a_1 & -a_2 \\ 1 & 0 \end{bmatrix} \quad b_c = \begin{bmatrix} 1 \\ 0 \end{bmatrix} \quad c_c' = [\, b_1 \ \ b_2 \,]$$

The observability matrix is

$$\mathcal{O}_c = \begin{bmatrix} b_1 & b_2 \\ -a_1 b_1 + b_2 & -a_2 b_1 \end{bmatrix}$$

and its determinant is

$$\det(\mathcal{O}_c) = -a_2 b_1^2 + a_1 b_1 b_2 - b_2^2$$

On the other hand $a(s) = s^2 + a_1 s + a_2$ and $b(s) = b_1 s + b_2$ have a common factor if and only if the zero of $b(s)$

$$s_b = -\frac{b_2}{b_1}$$

is also a zero of $a(s)$, i.e., if and only if

$$b_2^2 - a_1 b_1 b_2 + a_2 b_1^2 = 0$$

Obviously, then and only then $\det(\mathcal{O}_c) = 0$.

*** *Uncontrollable and unobservable realizations*[4].** Whenever we need to analyze a general system that may be uncontrollable or unobservable, it is useful to know that there are similarity transformations into the standard forms which separate controllable from uncontrollable and observable from unobservable subsystems. We first discuss the decomposition of a general system into controllable and uncontrollable subsystems, then the dual decomposition into observable and unobservable subsystems. There is also a general decomposition theorem (due to Gilbert and Kalman) which allows us to identify the following four subsystems: (1) controllable and observable; (2) controllable but unobservable; (3) observable but uncontrollable; and (4) uncontrollable and unobservable. We will not consider this most general case.

Let $\{A, b, c'\}$ be such that $\rho(\mathcal{C}) = r < n$. Then it can be shown (see for example [22], p. 131) that there exists a nonsingular transformation matrix which leads to a new realization $\{A_{\bar{c}}, b_{\bar{c}}, c'_{\bar{c}}\}$ such that

$$A_{\bar{c}} = \begin{bmatrix} A_{cc} & | & A_{c\bar{c}} \\ -- & + & -- \\ 0 & | & A_{\bar{c}\bar{c}} \end{bmatrix} \qquad b_{\bar{c}} = \begin{bmatrix} b_{cc} \\ -- \\ 0 \end{bmatrix} \qquad c'_{\bar{c}} = [\ c'_{cc} \ \ | \ \ c'_{\bar{c}\bar{c}}\ ]$$

where A_{cc} is $r \times r$, b_{cc} is $r \times 1$, c'_{cc} is $1 \times r$, and the subsystem $\{A_{cc}, b_{cc}, c'_{cc}\}$ is controllable. It is easy to verify that the transfer function of this controllable subsystem is equal to the transfer function of the original system:

$$c'_{\bar{c}}(sI - A_{\bar{c}})^{-1} b_{\bar{c}} = [\ c'_{cc} \ \ c'_{\bar{c}\bar{c}}\ ] \begin{bmatrix} (sI - A_{cc})^{-1} b_{cc} \\ 0 \end{bmatrix} = c'_{cc}(sI - A_{cc})^{-1} b_{cc}$$

If we partition the state vector

$$x_{\bar{c}} = \begin{bmatrix} x_{cc} \\ -- \\ x_{\bar{c}\bar{c}} \end{bmatrix}$$

then the r states in x_{cc} are said to be controllable, while the $n - r$ states in $x_{\bar{c}\bar{c}}$ are said to be uncontrollable.

The dual standard form for unobservable systems is given by

$$A_{\bar{o}} = \begin{bmatrix} A_{oo} & | & 0 \\ -- & + & -- \\ A_{\bar{o}o} & | & A_{\bar{o}\bar{o}} \end{bmatrix} \qquad b_{\bar{o}} = \begin{bmatrix} b_{oo} \\ -- \\ b_{\bar{o}\bar{o}} \end{bmatrix} \qquad c'_{\bar{o}} = [\ c'_{oo} \ \ | \ \ 0\ ]$$

where A_{oo} is $r \times r$, b_{oo} is $r \times 1$, c'_{oo} is $1 \times r$, and the subsystem $\{A_{oo}, b_{oo}, c'_{oo}\}$ is observable. With

$$x_{\bar{o}} = \begin{bmatrix} x_{oo} \\ -- \\ x_{\bar{o}\bar{o}} \end{bmatrix}$$

the r states in x_{oo} are said to be observable, while the $n - r$ states in $x_{\bar{o}\bar{o}}$ are said to be unobservable.

[4] Titles of this and several other Subsections in this Section are marked by asterisks (*) to let the reader know that the marked material may be skipped in the first reading.

***PBH tests.** We shall use the standard forms for uncontrollable and unobservable systems in the proof of the PBH controllability and observability criteria. They were first discovered by Gilbert in 1963 for the case of diagonalizable systems, and were later generalized by Popov (1966), Belevitch (1968), and Hautus (1969), thus the name, PBH tests. They are very powerful theoretical and numerical tool and, as T. Kailath says in [22],

> In fact, when faced with problems of checking for controllability and/or observability, it is a good heuristic rule to first try to apply the PBH tests.

Theorem 2.2.9 (PBH eigenvector tests)

Controllability: *A pair $\{A, b\}$ is controllable if and only if none of the left eigenvectors of A are orthogonal to b.*

Observability: *A pair $\{c', A\}$ is observable if and only if none of the right eigenvectors of A are orthogonal to c'.*

Proof. We shall give a detailed proof of the controllability test and call on the duality to prove the observability test.

"$\Rightarrow$" If there exists $q' \neq 0$ such that

$$q'A = q'\lambda \quad \text{and} \quad q'b = 0$$

then $q'Ab = \lambda q'b = 0$, $q'A^2b = \lambda^2 q'b = 0$, etc., hence

$$q'\mathcal{C} = 0$$

Since $q' \neq 0$ this means that $\mathcal{C}$ has less than a full rank.

"$\Leftarrow$" Assume the pair $\{A, b\}$ is uncontrollable and, without loss of generality, assume it is in the standard form for uncontrollable systems (otherwise it can be transformed into one by a nonsingular transformation)

$$A = \begin{bmatrix} A_{cc} & | & A_{c\bar{c}} \\ -- & + & -- \\ 0 & | & A_{\bar{c}\bar{c}} \end{bmatrix} \qquad b = \begin{bmatrix} b_{cc} \\ -- \\ 0 \end{bmatrix}$$

where A_{cc} is $r \times r$, b_{cc} is $r \times 1$, and r is defined by $\rho(\mathcal{C}) = r < n$. One possible choice for a left eigenvector of A which is orthogonal to b is

$$q' = [\, 0 \;\; | \;\; q'_{\bar{c}\bar{c}} \,]$$

where $q'_{\bar{c}\bar{c}}$ is any left eigenvector of $A_{\bar{c}\bar{c}}$. □

Theorem 2.2.10 (PBH rank tests)

Controllability: *A pair $\{A, b\}$ is controllable if and only if*

$$\rho([\, sI - A \quad b \,]) = n \qquad \textit{for all } s$$

where n is the order of A.

Observability: *A pair $\{c', A\}$ is observable if and only if*

$$\rho\left(\begin{bmatrix} sI - A \\ c' \end{bmatrix}\right) = n \qquad \textit{for all } s$$

where n is the order of A.

Proof. Again, we shall give a detailed proof of the controllability test and use duality to prove the observability test. Let $\rho([sI - A \quad b]) = n$ for all s. This is equivalent to saying that there does not exist a row vector q' such that $q'[sI - A \quad b] = 0$ for any s, i.e., $q'A = q's$ and $q'b = 0$, and this, according to Theorem 2.2.9, is equivalent to controllability of the pair $\{A, b\}$. □

Note that in applying the PBH rank tests it is easy to show the full rank for all s that are not eigenvalues of A, so the main task is to show that the rank remains n even when s takes the values of the eigenvalues of A.

****PBH tests for MIMO systems.*** The PBH tests presented in Theorems 2.2.9 and 2.2.10 were formulated for single-input and single-output systems. The following are more general formulations, given here without a proof:

Theorem 2.2.11 (PBH eigenvector tests for MIMO systems)

Controllability: *A pair $\{A, B\}$ is controllable if and only if none of the left eigenvectors of A are orthogonal to all columns of B.*

Observability: *A pair $\{C, A\}$ is observable if and only if none of the right eigenvectors of A are orthogonal to all rows of C.*

Theorem 2.2.12 (PBH rank tests for MIMO systems)

Controllability: *A pair $\{A, B\}$ is controllable if and only if*

$$\rho([\, sI - A \quad B \,]) = n \qquad \textit{for all } s$$

where n is the order of A.

Observability: *A pair $\{C, A\}$ is observable if and only if*

$$\rho\left(\begin{bmatrix} sI - A \\ C \end{bmatrix}\right) = n \qquad \textit{for all } s$$

where n is the order of A.

***Applications of PBH tests.** Here we show several examples of the power of PBH tests. First we give a much shorter proof of the fact that the controller realization is observable if and only if there are no cancellations between $b(s)$ and $a(s)$ (cf. the proof of Theorem 2.2.8).

Example 2.2.9 (Observability of controller form) *Let us use the PBH eigenvector test for observability to show that the order-n controller realization of $H(s)$ is observable if and only if there are no cancellations between $b(s) = c_c' adj(sI - A_c)b_c$ and $a(s) = \det(sI - A_c)$.*

From Problem C.3.3 we know that the companion matrices have only one independent eigenvector associated with each eigenvalue λ (regardless of their multiplicity):

$$p = \begin{bmatrix} \lambda^{n-1} & \lambda^{n-2} & \dots & \lambda^2 & \lambda & 1 \end{bmatrix}'$$

Thus $c'p = 0$ implies $b(\lambda) = 0$, and since by definition $a(\lambda) = 0$, we see that the controller form is observable if and only if there are no common factors between $b(s)$ and $a(s)$.

Let us determine the conditions for state controllability of a system in the modal (parallel) form. The modal form is another important realization, characterized by the fact that its system matrix is, in general, in the Jordan form. In most practical cases the systems have distinct eigenvalues, so diagonal system matrices are the most important special case. We shall consider both.

Example 2.2.10 (Controllability of modal form) *First assume the system has n distinct eigenvalues, and has only one input to it, i.e., its system matrix is diagonal*

$$A_d = \text{diag}(\lambda_1, \dots, \lambda_n)$$

with $\lambda_i \neq \lambda_j$ for $i \neq j$ and the input matrix is $n \times 1$ (hence we write b_d instead of B_d): $b_d = [b_{d1} \ \dots \ b_{dn}]'$. Then the controllability matrix is

$$\mathcal{C} = [b_d \quad A_d b_d \quad A_d^2 b_d \quad \dots \quad A_d^{n-1} b_d] = \begin{bmatrix} b_{d1} & b_{d1}\lambda_1 & \dots & b_{d1}\lambda_1^{n-1} \\ b_{d2} & b_{d2}\lambda_2 & \dots & b_{d2}\lambda_2^{n-1} \\ \vdots & \vdots & \ddots & \vdots \\ b_{dn} & b_{dn}\lambda_n & \dots & b_{dn}\lambda_n^{n-1} \end{bmatrix}$$

Therefore

$$\det(\mathcal{C}) = b_{d1} b_{d2} \dots b_{dn} V(\lambda_1, \dots, \lambda_n)$$

where $V(\lambda_1, \dots, \lambda_n) = \prod_{j>i}(\lambda_j - \lambda_i)$ is the Vandermonde determinant. Since the eigenvalues are distinct, $V(\lambda_1, \dots, \lambda_n) \neq 0$, and the system is controllable if and only if none of the components of b_d are zero.

Hence, the modal realization of a single-input system with distinct eigenvalues is controllable if and only if all components of its input vector are non-zero.

This result is consistent with the following intuitive argument: One cannot expect to control the system unless all of its states depend on the input. However, as we shall see in the following, this is not true in the case of multiple eigenvalues.

The previous result can be generalized in two ways: to cover the systems with repeated eigenvalues and for multi-input systems. To attempt to prove the more general criteria directly, like we did with the distinct eigenvalue case, would be very complicated. As we shall see, the PBH rank test allows a rather simple proof.

The controllability criterion for the general single-input modal form is: The modal realization of a single-input system is controllable if and only if: (1) no two Jordan blocks in the system matrix correspond to the same eigenvalue and (2) the elements of the input vector corresponding to the last rows of Jordan blocks in the system matrix are not zero.

The proof follows directly from the Corollary C.3.4 and from the PBH rank test for controllability. The Corollary C.3.4 states that all companion matrices are similar to Jordan matrices made of Jordan blocks having distinct eigenvalues. To prove part (2) we use the PBH rank test. It requires that $\rho([\ sI - A_J \quad b_J\ ]) = n$ *for all* s*, where* n *is the order of* A_J*. This is trivially satisfied when* s *is not an eigenvalue of* A_J*. When* $s = \lambda_i$*, where* λ_i *is any eigenvalue of* A_J*, the Jordan blocks corresponding to other eigenvalues are contributing the full rank. The critical block is the one corresponding to* λ_i *so consider that Jordan block and the corresponding components of* b_J *(in order to simplify the notation we assume this block is* 3×3*). For* $s = \lambda_i$ *the rank of*

$$\begin{bmatrix} 0 & -1 & 0 & b_{J_i1} \\ 0 & 0 & -1 & b_{J_i2} \\ 0 & 0 & 0 & b_{J_i3} \end{bmatrix}$$

must be 3 *and this will be true if and only if* $b_{J_i3} \neq 0$*. This completes the proof.*

Finally, the most general formulation: The modal realization of a system is controllable if and only if: (1) no two Jordan blocks in the system matrix are associated with the same eigenvalue and (2) the elements of rows of the input matrix corresponding to the last rows of Jordan blocks in the system matrix are not all zero.

The proof also follows directly from the Corollary C.3.4 and the multi-input version of the PBH rank test for controllability.

* ***Stabilizability and detectability.*** In the modal form each state corresponds to one or more modes associated with the same eigenvalue. Since we can classify the states in the modal form as being either controllable or uncontrollable and at the same time as either observable or unobservable, we can do the same for the eigenvalues and the corresponding modes.

The PBH eigenvector tests allow us to extend this classification to any system. Referring to the PBH eigenvector test for controllability we say: *If some left eigenvector corresponding to the eigenvalue* λ *is orthogonal to the input vector* b*, this eigenvalue and the corresponding modes are uncontrollable. Otherwise they are controllable.* Similarly we can classify the eigenvalues as observable or unobservable: *If some right eigenvector corresponding to the eigenvalue* λ *is orthogonal to the output vector* c'*, this eigenvalue and the corresponding modes are said to be unobservable. Otherwise they are observable.*

All this will be very important in Sections 2.3 and 2.4, where we discuss system stabilization using the state feedback. The main conditions for this technique to work will be controllability (so that the control can be effective) and observability (so that we can reconstruct the states required for state feedback) of at least the unstable modes. If an unstable mode is controllable it is said to be *stabilizable.* If an unstable mode is observable it is said to be *detectable.*

Transform domain criteria for controllability and observability. Earlier we showed that the joint controllability and observability of the realization represented by $\{A, b, c\}$ is equivalent to the coprimeness of $b(s) = c'\text{adj}(sI-A)b$ and $a(s) = \det(sI-A)$. We can say that this is the transform domain test for the joint controllability and observability. Here we shall derive separate transform domain criteria for controllability and observability.

Theorem 2.2.13 *Realization $\{A, b, c'\}$ is controllable if and only if the elements of the vector* $\text{adj}(sI-A)b$ *have no common factor with* $\det(sI-A)$. *Similarly, it is observable if and only if the elements of the vector* $c'\text{adj}(sI-A)$ *have no common factor with* $\det(sI-A)$.

Proof. Here again we give a detailed proof for the controllability test only and call on the duality to prove the observability condition:

"$\Rightarrow$" If the system is controllable it can be transformed into the controller form $\{A_c, b_c, c_c'\}$ by a nonsingular similarity transformation. It is fairly easy to see that the first row of $\text{adj}(sI-A_c)$ is equal to $[s^{n-1} \; s^{n-2} \; \dots \; s \; 1]$. Since $b_c = [1 \; 0 \; \dots \; 0 \; 0]'$ we have

$$\text{adj}(sI-A_c)b_c = \begin{bmatrix} s^{n-1} \\ s^{n-2} \\ \vdots \\ s \\ 1 \end{bmatrix}$$

The components of this vector have no common factor among themselves, and thus no common factor with $a(s)$. If S was the transformation from $\{A, b, c'\}$ to $\{A_c, b_c, c_c'\}$ then

$$\text{adj}(sI-A) = (sI-A)^{-1}a(s) = S(sI-A_c)^{-1}a(s)S^{-1} = S\text{adj}(sI-A_c)S^{-1}$$

Therefore $\text{adj}(sI-A)b = S\text{adj}(sI-A_c)b_c$ which guarantees that the components of $\text{adj}(sI-A)b$ have no common factors with $a(s)$. Indeed, if the components of $\text{adj}(sI-A)b$ had a common factor with $a(s)$, so would any of their linear combinations, and in particular the components of $\text{adj}(sI-A_c)b_c = S^{-1}\text{adj}(sI-A)b$ would have a common factor with $a(s)$, which is not the case.

"$\Leftarrow$" If the system is uncontrollable, it can be transformed into the standard form for uncontrollable systems. There we can write

$$\text{adj}(sI-A_{\bar{c}})b_{\bar{c}} = \begin{bmatrix} \text{adj}(sI-A_{cc})b_{cc}a_{\bar{c}\bar{c}}(s) \\ ----- \\ 0 \end{bmatrix} \quad \text{and} \quad a(s) = a_{cc}(s)a_{\bar{c}\bar{c}}(s)$$

where $a_{cc}(s) = \det(sI-A_{cc})$ and $a_{\bar{c}\bar{c}}(s) = \det(sI-A_{\bar{c}\bar{c}})$. In this case there is a common factor $a_{\bar{c}\bar{c}}(s)$ between the components of $\text{adj}(sI-A)b$ and $a(s)$. $\square$

2.3 State feedback and optimal control

In this Section we describe a linear controller based on the state feedback. We show that its two main parts, the feedback gain and the state observer, can be designed independently. The design of the observer parameters is described in Section 2.4, while the feedback gain design is explained in this Section. Two cases are of particular interest: moving the eigenvalues to desired locations and choosing the optimal feedback gain to minimize the cost of the control.

System control using the state feedback

As we show in Problem 3.4.15, one of the main problems with using the output feedback is that the poles cannot be relocated to a specific set of desired locations, or even worse, sometimes it may be impossible to stabilize the system. Since the states offer the most complete description of the system, we cannot hope to achieve more than by feeding some function of the states back to the input. It turns out that for a controllable system it is sufficient to feed back an appropriate linear combination of the states in order to relocate the eigenvalues to any desired set of locations. This, of course, also means that we can stabilize the system. If the states are not directly available, we have to reconstruct them from the system input and the output. As shown in Section 2.2, this is possible if and only if the system is observable.

The general idea. The block diagram in Figure 2.8 depicts a typical linear controller for a system described by its state vector $x(t)$. Based on the knowledge of the system parameters A, B, C, a state observer is designed to calculate $r(t)$,

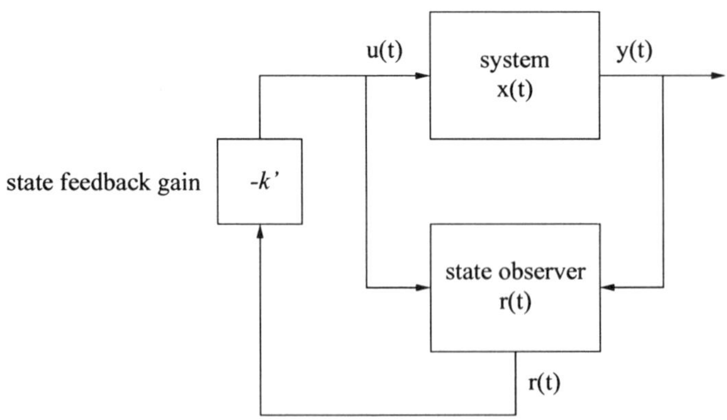

Figure 2.8: A typical configuration of a linear controller consisting of the state observer and the feedback gain.

a reconstruction of the original states $x(t)$, using the system input $u(t)$ and the system output $y(t)$. This signal is then fed back to the input using the feedback gain k. We shall see later that if the system is controllable we can determine k so that the eigenvalues of the system are moved to any desired locations. Also, we shall see that k can be chosen to minimize the cost of control, typically a quadratic function of the system states and the control input. In Section 2.4 we shall see that the observer can be designed so that the reconstruction $r(t)$ of the original states $x(t)$ converges to $x(t)$ very rapidly, so that there are no problems due to the fact that we are feeding back $r(t)$ instead of $x(t)$. Actually, we shall see that if the signals are noisy, we can use the estimate of the states, and still achieve very good control. But first, let us analyze the linear controller in some more detail.

Independence of feedback gain and observer gain design. A greater detail of the linear controller is shown in Figure 2.9.

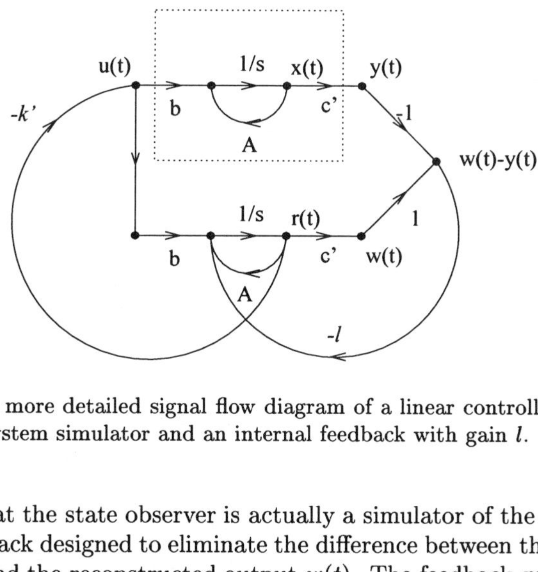

Figure 2.9: A more detailed signal flow diagram of a linear controller. The observer consists of a system simulator and an internal feedback with gain l.

We see that the state observer is actually a simulator of the system with an internal feedback designed to eliminate the difference between the actual system output $y(t)$ and the reconstructed output $w(t)$. The feedback gain l determines the rate of convergence between $w(t)$ and $y(t)$ and, since the system is assumed to be observable, between $r(t)$ and $x(t)$.

In Problem 3.11.2 we show that the characteristic equation of the augmented system is

$$\det(A - bk')\det(A - lc') = 0$$

This shows a complete autonomy between the controller (characterized by the system matrix $A - bk'$) and the observer (characterized by $A - lc'$). Thus we can separate the design of the feedback gain k from the design of the observer. As a consequence we can assume that the states are directly available to be fed back and calculate k to relocate the eigenvalues or minimize the control cost. Then we can design the observer, i.e., the observer feedback gain l, to achieve the desired convergence rate between $r(t)$ and $x(t)$. This separability of eigenvalues also guarantees that if the two subsystems are stable the whole system will also be stable. This is very convenient, because, in general, stability of parts of a system does not guarantee the stability of the whole system.

Properties of the state feedback

Next we briefly investigate the effects of the state feedback on the properties of the system: controllability, observability, eigenvalues, and zeros.

Eigenvalues under state feedback. A system is said to possess *modal controllability* if its eigenvalues can be moved to arbitrary new locations by the use of the appropriate state feedback. We shall see soon that this property is equivalent to the state controllability, i.e., that the eigenvalues of the system can be arbitrarily relocated if and only if the system is controllable, i.e., if and only if the controllability matrix

$$\mathcal{C} = [b \;\; Ab \;\; A^2b \;\; \dots \;\; A^{n-1}b]$$

has a full rank. Of course, there is an implicit assumption here about the availability of the states. In other words, the system also has to be observable, hence its observability matrix

$$\mathcal{O} = \begin{bmatrix} c' \\ c'A \\ \vdots \\ c'A^{n-1} \end{bmatrix}$$

must also have a full rank.

This is the most useful feature of the state feedback and it is in the base of the modern control theory.

Zeros under state feedback. Unfortunately, the state feedback does not offer a complete design freedom. Here we show that the state feedback cannot be used to move the zeros of the system. Not only that we cannot move them exactly where we want them to be, but we cannot move them at all. The only thing that can happen is the pole-zero cancellation, should some of the new eigenvalue locations coincide with some of the zeros of the system.

To show that, consider a system in the controller form[5] $\{A_c, b_c, c_c'\}$, where A_c is the companion matrix with $-[a_1 \, \dots \, a_n]$ at the top row while $c_c' = [b_1 \, \dots \, b_n]'$. As always, the a_i's and the b_i's are the coefficients of the denominator and the

[5] We consider the controller form without a loss of generality, because in the context of state feedback we usually consider controllable systems, and they can be nonsingularly transformed into the controller form and back.

numerator of the transfer function, respectively. After the feedback, the system is still in the controller form, now given by $\{A_c - b_c k_c', b_c, c_c'\}$. Obviously, the coefficients of the numerator of the transfer function have not changed, hence the zeros are still exactly where they were before the application of the state feedback.

Controllability under state feedback. In Problem 3.9.1 we show that if a system given by $\{A, b, c'\}$ is (un)controllable than the closed-loop system is also (un)controllable for any feedback gain vector k. The closed-loop system matrix is given by

$$A_f = A - bk'$$

while the controllability matrix of the closed-loop system can be determined from the following identity:

$$\mathcal{C} = \mathcal{C}_f D$$

where

$$\mathcal{C} = [b \; Ab \; A^2b \; \ldots \; A^{n-1}b] \qquad \mathcal{C}_f = [b \; A_f b \; A_f^2 b \; \ldots \; A_f^{n-1}b]$$

and

$$D = \begin{bmatrix} 1 & k'b & k'Ab & \ldots & k'A^{n-2}b \\ 0 & 1 & k'b & \ldots & k'A^{n-3}b \\ 0 & 0 & 1 & \ldots & k'A^{n-4}b \\ \vdots & \vdots & \vdots & \ddots & \vdots \\ 0 & 0 & 0 & \ldots & 1 \end{bmatrix}$$

Since $\det(D) \neq 0$, the controllability of the system is invariant under the state feedback.

Observability under state feedback. Unlike controllability, observability can be lost due to the state feedback. One example of this undesired property of the state feedback is given in Problem 3.9.2.

If a system described by $\{A, b, c'\}$ is minimal, i.e., both controllable and observable, than after the addition of the state feedback $u(t) = -k'x(t)$ we obtain the system $\{A - bk', b, c'\}$, which is controllable and has the same zeros as the original system. Therefore, it is observable if and only if the new eigenvalues do not coincide with any of the zeros. (Otherwise there will be pole-zero cancellations, and since the controllability is preserved, it is the observability that is lost due to cancellations.)

Stabilizability. If the purpose of the state feedback is only to stabilize the system, then the controllability is a too strong requirement. Indeed, a system whose unstable modes are controllable, while the stable modes may or may not be controllable, is stabilizable, even though it may not be completely controllable.

Feedback gain design

In the rest of this Section we describe the design of the feedback gain k. The first case assumes that we know where we want to move the eigenvalues of the system. The second method allows us to determine the optimal feedback gain which minimizes the quadratic cost function.

Relocation of eigenvalues. If our goal is to move the eigenvalues of the system from their original values $\lambda_1, \ldots, \lambda_n$ to an arbitrary desired set of locations $\mu_1, \ldots, \mu_n$ we use the result of Problem 3.9.7, where we show that this is possible if and only if the original system is controllable (again, it is tacitly assumed that the states are either directly available or that the system is observable):

If the characteristic polynomial of the original system is

$$a(s) = s^n + a_1 s^{n-1} + \ldots + a_{n-1}s + a_n$$

while the characteristic polynomial of the desired closed-loop system is

$$\alpha(s) = s^n + \alpha_1 s^{n-1} + \ldots + \alpha_{n-1}s + \alpha_n$$

then with

$$a' = [a_1 \;\ldots\; a_n] \quad \text{and} \quad \alpha' = [\alpha_1 \;\ldots\; \alpha_n]$$

we can use the Bass-Gura formula

$$k' = (\alpha' - a')\mathcal{C}_c\mathcal{C}^{-1}$$

or the Ackermann formula

$$k' = [0 \;\ldots\; 0\; 1]\mathcal{C}^{-1}\alpha(A)$$

or the Mayne-Murdoch formula (valid for systems with distinct eigenvalues only)

$$k_i b_i = \frac{\prod_j(\lambda_i - \mu_j)}{\prod_{i \neq j}(\lambda_i - \lambda_j)} \quad (i = 1, \ldots, n)$$

Optimal control. The selection of the closed-loop eigenvalues is a trade-off between the price of control and the settling time. Indeed, as we move the closed-loop eigenvalues $\mu_1, \ldots, \mu_n$ towards $-\infty$, the settling times get shorter. On the other hand, from the Mayne-Murdoch formula we see that as the differences between the open- and closed-loop eigenvalues increase, the corresponding components of the feedback gain vector increase, hence the cost of control rises.

To resolve this trade-off Kalman [23] introduced a quadratic cost function to associate weights with each of the states and also with the control input:

continuous-time:

$$J = \int_0^{\infty} (x'(t)Qx(t) + u'(t)Ru(t))\, dt$$

discrete-time:

$$J = \sum_{k=0}^{\infty} (x'[k]Qx[k] + u'[k]Ru[k])$$

The solution of the resulting *linear-quadratic-regulation* (LQR) problem is derived in Problems 3.10.1 and 4.10.2, where we show that the optimal feedback is given by $u^*(t) = -Kx(t)$, i.e., $u^*[k] = -Fx[k]$, respectively, where K and F are given by:

continuous-time:

$$K = R^{-1}B'P$$

discrete-time:

$$F = (B'PB + R)^{-1}B'PA$$

where P's are the positive definite solutions of the algebraic Riccati equations:

continuous-time:

$$PA + A'P - PBR^{-1}B'P + Q = 0$$

discrete-time:

$$P = A'P(I - B(B'PB + R)^{-1}B'P)A + Q$$

The minimum cost is then given by:

continuous-time:

$$J^* = x'(0)Px(0)$$

discrete-time:

$$J^* = x'[0]Px[0]$$

For examples of optimal control design and derivations of the above results, we refer the reader to Sections 3.10 and 4.10.

2.4 State observers and estimators

Until now, we have always assumed the states to be readily available for feedback purposes. In cases when they are not, and if the system is observable, we can design a state observer to reconstruct the states from the system input and output. We start this Section by considering design methodologies for state observers.

If the system is affected by noise with known statistical properties, we can design an optimal state estimator. This problem is dual to that of the optimal feedback of Section 2.3. For discrete-time systems such estimators are called Kalman filters, while for continuous-time systems they are called Kalman-Bucy filters. We study these techniques in the second part of this Section.

State observers

State observers are necessary in cases when the states of the system are not directly available. For example, the sensors for some of the states are too expensive, or the system is at a remote location and only its output can be measured. In such cases, if the system is observable, i.e., if its states can be determined from its input and output, we rely on a model of the system, most likely implemented as an analog computer or a program on a digital computer. If none of the states are available we have to use a full order observer, i.e., the order of the observer will be equal to the order of the system. However, if some of the states are available, we can determine the remaining states using the so-called reduced-order observer.

Full-order observer. In Figure 2.10 we show a signal flow diagram for a typical full-order observer. In Problem 3.11.2 we prove that the convergence to zero of the error between the reconstructed and actual states is determined by the roots of the observer characteristic equation

$$\det(A - lc') = 0$$

By the duality between controllability and observability we can find the dual of the Bass-Gura formula, and use it to calculate the vector l which puts the eigenvalues of the observer in specified locations:

$$l = \mathcal{O}^{-1}\mathcal{O}_o(\alpha - a)$$

where α is a vector of coefficients of the desired characteristic polynomial of the observer, while a is a vector of coefficients of the characteristic polynomial of the original system.

In the design of the feedback vector in Section 2.3 we were concerned with the price of the states and of control signals. Here we don't have such worries, simply because the simulation is implemented on a computer and the fuel for electrons is really cheap! However, putting the eigenvalues of the observer too far

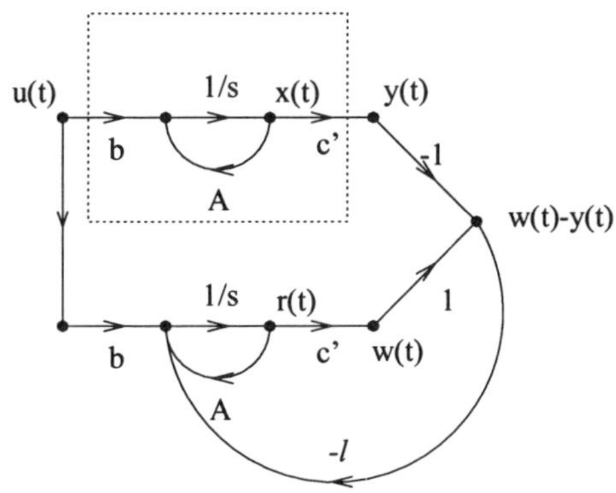

Figure 2.10: A typical system observer. Vector l is designed so that the observer error converges to zero as fast as possible without increasing the noise sensitivity too much.

to the left[6] increases its sensitivity to noise. Thus, we again have two opposing requirements: we need the observer error to converge to zero fast, but we don't want to introduce too much noise by being too fast. To resolve this situation we can use the quadratic cost function approach of Section 2.3. We shall not pursue this issue any further in the present book.

Reduced-order observer. If some of the states are available, we can reduce the order of the observer and reconstruct only the unavailable states. In Sections 3.13 and 4.13 we describe the design and implementation of the reduced-order observers for continuous-time and discrete-time systems, respectively.

State estimators

Historically, the discrete-time case of the state estimation problem was solved first. There is an almost romantic story about how Kalman got the idea for this revolutionary discovery. After solving the problem of optimal feedback in 1958, firmly establishing the importance of the state-space approach in the control theory, on one snowy night in late November 1958 Kalman traveled by train from Princeton (where he presented his work to NASA scientists), to Baltimore (where he worked at the time, in the Research Institute for Advanced Studies). Suddenly, the train got halted for about an hour and a question popped up in Kalman's mind: "Why not try the state-space approach on the Wiener's celebrated problem of estimating the system dynamics from noisy measurements?" As it often happens, asking the right question was practically a solution itself, because, as Kalman was soon to discover, this formulation of the Wiener's problem was the dual of the LQR problem he previously had solved himself!

[6] Or too close to zero for discrete-time systems.

Thus, the Kalman filter was born as a solution to the *linear-quadratic-estimation* (LQE) problem. After that, Kalman worked with Bucy on the continuous-time case, and the resulting estimator has become known as the Kalman-Bucy filter. These two techniques were published in [26] and [28], respectively.

In the following we summarize the main formulas for the steady-state versions of Kalman-Bucy and Kalman filters. All derivations and several examples are given in Sections 3.12 and 4.12.

Kalman-Bucy filter. Consider a system described by

$$\begin{aligned}\dot{x}(t) &= Ax(t) + Bu(t) + w(t)\\ y(t) &= Cx(t) + e(t)\end{aligned}$$

where the system noise $w(t)$ and the measurement noise $e(t)$ are zero-mean, white, and Gaussian. Furthermore, $w(t)$ is uncorrelated with $e(t)$. If the noise covariances are

$$\begin{aligned}E\left[w(t)w'(\tau)\right] &= Q\,\delta(t-\tau)\\ E\left[e(t)e'(\tau)\right] &= R\,\delta(t-\tau)\end{aligned}$$

the steady-state Kalman-Bucy estimator is given by

$$\dot{r}(t) = Ar(t) + Bu(t) + L(y(t) - Cr(t))$$

where $L = PC'R^{-1}$, and P is a solution of the algebraic Riccati equation

$$AP + PA' - PC'R^{-1}CP + Q = 0$$

Kalman filter. Consider a system described by

$$\begin{aligned}x[k+1] &= Ax[k] + Bu[k] + w[k]\\ y[k] &= Cx[k] + e[k]\end{aligned}$$

where the system noise $w[k]$ and the measurement noise $e[k]$ are zero-mean, white, and Gaussian. Furthermore, $w[k]$ is uncorrelated with $e[k]$. If the noise covariances are

$$\begin{aligned}E\{w[k]w'[l]\} = Q\delta[k-l]\\ E\{e[k]e'[l]\} = R\delta[k-l]\end{aligned}$$

and for the initial value $x[0]$ we have $E\{x[0]\} = x_0$ and $E\{(x[0]-x_0)(x[0]-x_0)'\} = P_0$, then the steady-state Kalman estimator is given by

$$\hat{x}[k] = z[k] + L_k(y[k] - Cz[k])$$

where

$$z[k] = A\hat{x}[k-1] + Bu[k-1] \qquad (z[0] = x_0)$$

and the gain L is given by $L = NC'(R + CNC')^{-1}$, where N is a solution of the algebraic Riccati equation

$$N = Q + ANA' - ANC'(R + CNC')^{-1}CNA$$

Part II

Solved problems

Chapter 3

Continuous linear systems

This Chapter contains solved problems about continuous-time linear control systems. It begins with the background material on linear differential equations and matrices (Sections 3.1, 3.2, and 3.3). It continues with a discussion of the advantages of the state-space representation of linear systems over their input-output representation (Sections 3.4 and 3.5). In Sections 3.6 and 3.7 we investigate three fundamental properties of systems: stability, state controllability, and state observability. In Section 3.8 we examine the canonical forms of linear systems and their properties. Section 3.9 shows that by using the state feedback we can arbitrarily place the poles of the system. The condition for this so-called modal controllability is, quite amazingly, the state controllability and observability. Next, in Section 3.10, we describe how the feedback gain should be picked so that the quadratic optimality is achieved. In Section 3.11 we explain the design of the state observers. In Section 3.12 we investigate how to pick the observer gain so that the effects of noise are minimized in a mean-square sense. The result is the Kalman-Bucy filter. Finally, in Section 3.13, we describe the reduced-order observers.

$$
\begin{aligned}
x(t) &= Ax(t) + Bu(t) \\
y(t) &= Cx(t) + Du(t)
\end{aligned}
$$

3.1 Simple differential equations

This Section should refresh the reader's memory about the two most common paths to solution of the linear differential equations with constant coefficients: the time-domain convolution and the Laplace transform. It also describes the usefulness of the Dirac's delta impulse.

Problem 3.1.1 Show that the solution of the inhomogeneous differential equation

$$\dot{x}(t) = ax(t) + f(t), \qquad \text{with } x(0) = x_0$$

is given by

$$x(t) = x_0 e^{at} + e^{at} * f(t)$$

where $*$ denotes convolution:

$$e^{at} * f(t) = \int_0^t e^{a(t-\tau)} f(\tau)\, d\tau$$

Solution: If $f(t) \equiv 0$ the equation is homogeneous, and the solution is clearly $x(t) = x_0 e^{at}$. Actually, it is obvious that if $x(t) = x_0 e^{at}$ then $\dot{x}(t) = ax(t)$, with $x(0) = x_0$, but is the other direction as easy to see? Consider the following argument: From $\dot{x}(t) = ax(t)$ we find:

$$\begin{aligned} \dot{x}(t) &= ax(t) \\ \ddot{x}(t) &= a^2 x(t) \\ &\vdots \end{aligned}$$

hence, according to the Maclaurin series expansion of $x(t)$,

$$\begin{aligned} x(t) &= x(0) + \frac{\dot{x}(0)}{1!}t + \frac{\ddot{x}(0)}{2!}t^2 + \dots \\ &= x_0 \left(1 + \frac{at}{1!} + \frac{a^2t^2}{2!} + \dots\right) \\ &= x_0 e^{at} \end{aligned}$$

In general, when $f(t) \not\equiv 0$, introduce a change of variable:

$$x(t) = e^{at} z(t), \qquad (x(0) = z(0))$$

when

$$\dot{x}(t) = ae^{at} z(t) + e^{at} \dot{z}(t)$$

Now we can write $e^{at}\dot{z}(t) = f(t)$, i.e.,

$$\dot{z}(t) = e^{-at} f(t)$$

which finally implies

$$x(t) = e^{at}z(t) = e^{at}\left(z(0) + \int_0^t \dot{z}(\tau)\,d\tau\right) = e^{at}\left(z(0) + \int_0^t e^{-a\tau}f(\tau)\,d\tau\right)$$

i.e.,

$$x(t) = \underbrace{x_0 e^{at}}_{\text{homogeneous part}} + \underbrace{\int_0^t e^{a(t-\tau)}f(\tau)\,d\tau}_{\text{non-homogeneous part}}$$

Note: *We can differentiate this expression to convince ourselves that it is indeed a solution:*

$$\dot{x}(t) = ax_0e^{at} + a\int_0^t e^{a(t-\tau)}f(\tau)\,d\tau + f(t) = ax(t) + f(t)$$

Problem 3.1.2 Use the Laplace transform to solve the equation from the previous problem:

$$\dot{x}(t) = ax(t) + f(t), \qquad \text{with } x(0) = x_0$$

Solution: Taking the Laplace transform of both sides yields

$$sX(s) - x_0 = aX(s) + F(s)$$

hence

$$X(s) = \frac{x_0}{s-a} + \frac{F(s)}{s-a}$$

The homogeneous part of the solution is therefore

$$x_h(t) = x_0e^{at}$$

while the non-homogeneous part is:

$$x_{nh}(t) = e^{at} * f(t)$$

Note: *Usually we do not calculate the convolution, but rather use the tables of Laplace transform pairs to invert* $\frac{F(s)}{s-a}$.

Problem 3.1.3 For each multiplicity-m root a of the characteristic equation of the higher order differential equation, the homogeneous part of the solution contains the following term(s)

$$\alpha_0e^{at} + \alpha_1te^{at} + \alpha_2t^2e^{at} + \ldots + \alpha_{m-1}t^{m-1}e^{at}$$

where $\alpha_0, \ldots, \alpha_{m-1}$ are constants which depend on the initial conditions.

First apply and check the above procedure and then derive it for the following homogeneous equations:

a) $\ddot{x}(t) = 5\dot{x}(t) - 6x(t), \quad \dot{x}(0) = 5,\ x(0) = 0$

b) $\ddot{x}(t) = 4\dot{x}(t) - 4x(t), \quad \dot{x}(0) = 2,\ x(0) = -1$

c) Repeat part b) using the Laplace transform.

Solution: a) The characteristic equation for this equation is

$$r^2 - 5r + 6 = 0$$

and since its roots are

$$r_1 = 2 \quad \text{and} \quad r_2 = 3$$

the solution is of the form

$$x(t) = \alpha e^{2t} + \beta e^{3t}$$

where α and β can be determined from the initial conditions:

$$\left.\begin{array}{rcl} \alpha + \beta & = & 0 \\ 2\alpha + 3\beta & = & 5 \end{array}\right\} \quad \Rightarrow \quad \alpha = -5 \quad \text{and} \quad \beta = 5$$

It is easy to verify that $x(t) = -5e^{2t} + 5e^{3t}$ satisfies both the equation and the initial conditions.

In order to derive the "usual suspects" (αe^{2t} and βe^{3t}) we shall rewrite the equation so that it reduces to the trivial form $\dot{y}(t) = ay(t)$. With the characteristic equation in mind

$$(r-2)(r-3) = 0$$

which can be rewritten as

$$r^2 - 3r = 2(r-3)$$

we write

$$\ddot{x}(t) = 5\dot{x}(t) - 6x(t) \quad \Leftrightarrow \quad \underbrace{\ddot{x}(t) - 3\dot{x}(t)}_{\dot{y}(t)} = 2(\underbrace{\dot{x}(t) - 3x(t)}_{y(t)})$$

With a new variable: $y(t) = \dot{x}(t) - 3x(t)$ the equation becomes

$$\dot{y}(t) = 2y(t) \quad \text{with} \quad y(0) = \dot{x}(0) - 3x(0) = 5$$

Hence

$$y(t) = 5e^{2t}$$

This now yields a non-homogeneous differential equation in $x(t)$:

$$\dot{x}(t) = 3x(t) + 5e^{2t}$$

whose solution is (see Problem 3.1.1)

$$
\begin{aligned}
x(t) &= x(0)e^{3t} + \int_0^t 5e^{3(t-\tau)}e^{2\tau}\, d\tau \\
&= -5e^{2t} + 5e^{3t}
\end{aligned}
$$

b) This time the characteristic equation is

$$r^2 - 4r + 4 = 0 \quad \text{hence} \quad r_{1,2} = 2$$

and

$$x(t) = \alpha e^{2t} + \beta t e^{2t}$$

We can derive this as follows:

$$\ddot{x}(t) = 4\dot{x}(t) - 4x(t) \quad \Leftrightarrow \quad \ddot{x}(t) - 2\dot{x}(t) = 2(\dot{x}(t) - 2x(t))$$

therefore

$$\dot{x}(t) - 2x(t) = \underbrace{(\dot{x}(0) - 2x(0))}_{4}e^{2t}$$

i.e.,

$$
\begin{aligned}
x(t) &= x(0)e^{2t} + \int_0^t 4e^{2(t-\tau)}e^{2\tau}\, d\tau \\
&= -e^{2t} + 4te^{2t}
\end{aligned}
$$

c) Applying the Laplace transform to the equation from part b)

$$\ddot{x}(t) = 4\dot{x}(t) - 4x(t), \quad \dot{x}(0) = 2, x(0) = -1$$

we obtain $s^2X(s) - sx(0) - \dot{x}(0) = 4sX(s) - 4x(0) - 4X(s)$, hence

$$X(s) = \frac{-s+6}{(s-2)^2} = \frac{-1}{(s-2)} + \frac{4}{(s-2)^2}$$

and finally

$$x(t) = -e^{2t} + 4te^{2t}$$

Problem 3.1.4 Consider a linear motion of a ball, whose mass is m. If the velocity before it was hit by a hammer at $t = 0$ was $v(t) = v_1$ $(t < 0)$, while the velocity afterwards was $v(t) = v_2$ $(t > 0)$, describe the forces acting on the ball as a function of time.

Solution: If we were interested in the details of the velocity changes around $t = 0$, and if we were able to measure the velocity with such a fine time resolution, we would probably find the time dependence of $p(t) = mv(t)$ and $F(t) = \frac{dp(t)}{dt}$ as in Figure 3.1-a.

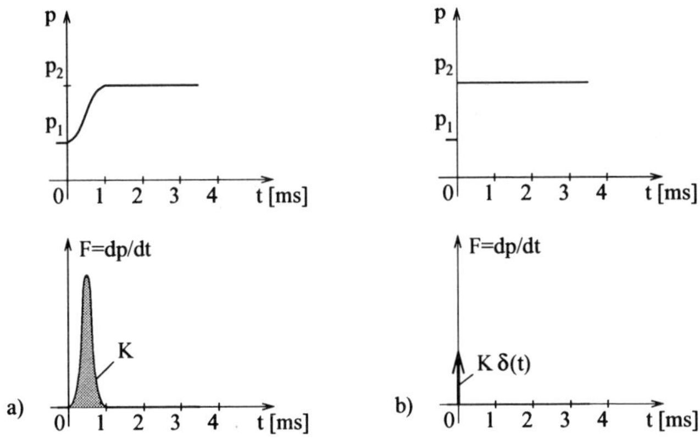

Figure 3.1: a) Linear momentum of the ball and the force causing these changes. b) Due to our ignorance about the details of the event, or just for mathematical simplicity, we often use the Dirac's $\delta(t)$ for the idealized representation of very short events whose effects are measurable. Note that $K = \Delta p = p_2 - p_1$.

Note that since $F(t) = \frac{dp(t)}{dt}$, the change of the linear momentum of the ball can be written as

$$\Delta p = p_2 - p_1 = \int_{(1)}^{(2)} F(t)\,dt$$

Therefore, the area under the curve $F(t)$ is equal to the change of the linear momentum $\Delta p = m(v_2 - v_1)$. The importance of this remark will be clearer at the end of this problem.

However, very often we are not interested in, or we are not able to achieve, such a fine time resolution in measuring the velocity. Thus, in order to represent this brief event whose consequences are measurable, we use the Dirac's delta distribution, defined[1] by

$$\delta(t) = 0 \quad (t \neq 0) \qquad \text{and} \qquad \int_{-\infty}^{\infty} \delta(t)\,dt = \int_{0-}^{0+} \delta(t)\,dt = 1 \tag{3.1}$$

Since the hit of a hammer is indeed a very fast event with lasting consequences, it is a good candidate for idealized description using the Dirac's $\delta(t)$ (see Figures 3.1-a and 3.1-b). From what we said earlier, it is obvious that we can write

$$F(t) = K\delta(t), \qquad \text{where } K = \Delta p = m(v_2 - v_1)$$

Let us check if everything agrees (recall the note about the area under $F(t)$):

$$\int_{(1)}^{(2)} F(t)\,dt = \int_{(1)}^{(2)} K\delta(t)\,dt = K \underbrace{\int_{(1)}^{(2)} \delta(t)\,dt}_{=1} = K$$

[1]See Problem 3.1.5 for more about this very useful mathematical object.

Problem 3.1.5 Describe the most important properties of the Dirac's delta distribution.

Solution: The Dirac's delta is not a function in the standard sense, because it cannot be defined at $t = 0$. The only thing we know about it at $t = 0$ is that it makes the integral in equation (3.1) equal to 1, a rather remarkable property.

Its integral is an important function

$$\int_{-\infty}^{t} \delta(\tau)\, d\tau = u_H(t)$$

the so-called Heaviside's step function defined by

$$u_H(t) = \begin{cases} 0, & t \leq 0 \\ 1, & t > 0 \end{cases}$$

This makes it plausible to write

$$\delta(t) = \frac{d}{dt} u_H(t)$$

although this is not mathematically correct, at least not with the standard definition of the derivative in mind. These notions and relations are completely redefined in the field of mathematics called the theory of distributions, so that the standard functions and derivatives become special cases of a more general theory.

Another important property of $\delta(t)$ is that it is a neutral element (unity) for the convolution operation:

$$f(t) * \delta(t) \;=\; \int_{-\infty}^{\infty} f(t-\tau)\delta(\tau)\, d\tau \;=\; f(t)$$

just like the Kronecker's delta[2] is the unity for discrete convolution

$$f[k] * \delta[k] \;=\; \sum_{i=-\infty}^{\infty} f[k-i]\,\delta[i] \;=\; f[k]$$

From the following property

$$\int_{-\infty}^{\infty} f(\tau)\delta(\tau)\, d\tau = f(0)$$

it follows that the Laplace transform of the Dirac's $\delta(t)$ is

$$\mathcal{L}_{-}\{\delta(t)\} = \int_{0-}^{\infty} e^{-st}\delta(t)\, dt = 1$$

Note that here we used the $\mathcal{L}_{-}$ Laplace transform, which is usually used when there are impulsive functions or its derivatives at the origin. See Problem 3.1.6 for comparisons of the two unilateral Laplace transforms: $\mathcal{L}_{+}$ and $\mathcal{L}_{-}$.

[2]Kronecker's delta, $\delta[k]$, is a sequence of zeros, with the only exception at $k = 0$, where it is equal to 1:

$$\delta[k] = \begin{cases} 1, & k = 0 \\ 0, & k \neq 0 \end{cases}$$

Problem 3.1.6 Describe the differences between the two unilateral Laplace transforms $\mathcal{L}_+$ and $\mathcal{L}_-$.

Solution: Recall that the definitions are as follows:

$$\mathcal{L}_+\{f(t)\} = \int_{0^+}^{\infty} f(t)e^{-st}\,dt \qquad \text{and} \qquad \mathcal{L}_-\{f(t)\} = \int_{0^-}^{\infty} f(t)e^{-st}\,dt$$

This difference in definitions can be seen only for functions whose integrals between 0^- and 0^+ are finite, i.e., for impulsive functions. Thus all differences between the two transforms reduce to the following

$$\mathcal{L}_+\{\delta(t)\} = \mathcal{L}_+\left\{\frac{d}{dt}u_H(t)\right\} = s\cdot s^{-1} - u_H(0^+) = 0$$

$$\mathcal{L}_-\{\delta(t)\} = \mathcal{L}_-\left\{\frac{d}{dt}u_H(t)\right\} = s\cdot s^{-1} - u_H(0^-) = 1$$

where $u_H(t)$ is the Heaviside's step function earlier defined by

$$u_H(t) = \begin{cases} 0, & t \le 0 \\ 1, & t > 0 \end{cases}$$

In applications of the Laplace transform this difference means that if the differential equation has impulsive functions or its derivatives at $t = 0$, and the initial conditions are given for $t = 0^-$, then there is no need to calculate the conditions for $t = 0^+$. Rather, we can directly proceed with the application of the $\mathcal{L}_-$ transform, and avoid often very tedious calculations.

Problem 3.1.7 Solve the equation

$$\ddot{x} + 3\dot{x} + 2x = \delta(t) + 2\dot{\delta}(t), \quad x(0^-) = 1, \quad \dot{x}(0^-) = 2$$

Solution: In this case we have to apply the $\mathcal{L}_-$ transform because of the Dirac's delta and its derivative at $t = 0$:

$$s^2X(s) - sx(0^-) - \dot{x}(0^-) + 3(sX(s) - x(0^-)) + 2X(s) = 1 + 2s$$

Hence $(s+1)(s+2)X(s) = 3s + 6$ and finally

$$X(s) = \frac{3}{s+1} \quad \Rightarrow \quad x(t) = 3e^{-t} \ \ (t > 0)$$

Matlab note: *The partial fraction decomposition can be performed by equating the coefficients of the corresponding powers of s in* $\frac{s+5}{(s+1)(s+2)} = \frac{A}{s+1} + \frac{B}{s+2}$, *or by the residue formulas, or simply by using the* MATLAB *command* `residue`. *Let us mention two other useful* MATLAB *commands for manipulating polynomials:* `poly` *and* `roots`.

Numerical integration of ordinary differential equations (ODE) can be done through the use of MATLAB *commands* `ode23` *and* `ode45`.

Problem 3.1.8 Describe the differences in applications of Fourier and Laplace transforms.

Solution: Typically, Fourier analysis is used when only steady state solution is of interest. For example, the frequency characteristics of a filter tell us how the filter changes the amplitude and the phase of a sinusoidal signal. The implicit assumption is that the input to the filter has been present since $t = -\infty$, so that all the transients have had enough time to die out. Here we find another implicit assumption, that the system is stable, otherwise the transients would never die out.

However, if we wish to analyze the behavior of the filter right after the sinusoidal signal has been applied (for convenience we denote that moment $t = 0$), we need to use the Laplace transform. The Laplace transform is also convenient for the stability analysis of linear systems.

Problem 3.1.9 Write the following second-order differential equation as a system of two first-order differential equations and write them in a matrix form:

$$\ddot{x}(t) + 4\dot{x}(t) + 3x(t) = f(t)$$

Solution: If we use $v(t) = \dot{x}(t)$, the above equation can be written as

$$\begin{aligned} \dot{x}(t) &= v(t) \\ \dot{v}(t) &= -3x(t) - 4v(t) + f(t) \end{aligned}$$

i.e.,

$$\begin{bmatrix} \dot{x}(t) \\ \dot{v}(t) \end{bmatrix} = \begin{bmatrix} 0 & 1 \\ -3 & -4 \end{bmatrix} \begin{bmatrix} x(t) \\ v(t) \end{bmatrix} + \begin{bmatrix} 0 \\ f(t) \end{bmatrix}$$

Note 1: *In this notation the initial conditions are*

$$\begin{bmatrix} x(0) \\ v(0) \end{bmatrix} = \begin{bmatrix} x(0) \\ \dot{x}(0) \end{bmatrix}$$

Note 2: *The eigenvalues of this matrix are the same as the eigenvalues of the original second-order equation. This remains true if we use any other appropriate change of variable, for example with $w(t) = \dot{x}(t) + x(t)$ the system becomes*

$$\begin{bmatrix} \dot{x}(t) \\ \dot{w}(t) \end{bmatrix} = \begin{bmatrix} 1 & 1 \\ -8 & -5 \end{bmatrix} \begin{bmatrix} x(t) \\ w(t) \end{bmatrix} + \begin{bmatrix} 0 \\ f(t) \end{bmatrix}$$

Much more about this will be investigated in the following sections.

3.2 Matrix theory

In this Section we investigate several properties of matrices which will be used later in this Chapter. More about matrices can be found in Appendix C.

Problem 3.2.1 Let all eigenvalues of $A_{n\times n}$ be distinct and let q_i be a right eigenvector $(n \times 1)$ of A associated with the i-th eigenvalue, i.e.,

$$Aq_i = \lambda_i q_i$$

Define matrix Q as the matrix whose i-th column is the i-th eigenvector q_i:

$$Q = [q_1\ q_2\ \dots\ q_n]$$

and also define P as the inverse[3] of Q, i.e.,

$$P = Q^{-1}$$

If the i-th row of P is p_i', i.e.,

$$P = \begin{bmatrix} p_1' \\ p_2' \\ \vdots \\ p_n' \end{bmatrix}$$

show that p_i' is a left eigenvector $(1 \times n)$ of A corresponding to the i-th eigenvalue, i.e.,

$$p_i'A = \lambda_i p_i'$$

Solution: Since $\det(Q) \neq 0$, in order to prove that $p_i'A = \lambda_i p_i',\ (i = 1, 2, \dots, n)$, i.e.,

$$PA = \Lambda P$$

where $\Lambda = \text{diag}(\lambda_1, \lambda_2, \dots, \lambda_n)$, it is sufficient to prove that

$$PAQ = \Lambda$$

Indeed,

$$PAQ = Q^{-1}[Aq_1 \dots Aq_n] = Q^{-1}[\lambda_1 q_1 \dots \lambda_n q_n] = Q^{-1}Q\text{diag}(\lambda_1, \dots, \lambda_n) = \Lambda$$

Note: *Matrices with repeated eigenvalues may or may not be diagonalizable. In general, we can write $PAQ = J$, where J is a matrix in the Jordan canonical form, P and Q are matrices of left and right principal vectors (eigenvectors and, if necessary, generalized eigenvectors), and $P = Q^{-1}$. Diagonalization is a special case of similarity transformations (cf. Appendix* C.3*).*

---

[3]Since the eigenvalues of $A_{n\times n}$ are assumed to be distinct, A has n linearly independent eigenvectors, hence Q is invertible (cf. Appendix C.1).

Problem 3.2.2 Determine the eigenvalues and right and left eigenvectors of a 2×2 matrix

$$A = \begin{bmatrix} 1 & 2 \\ 3 & 0 \end{bmatrix}$$

Solution: From $\det(\lambda I - A) = 0$ we get $\lambda^2 - \lambda - 6 = 0$, which implies $\lambda_1 = -2$ and $\lambda_2 = 3$.

From

$$\begin{bmatrix} 1 & 2 \\ 3 & 0 \end{bmatrix} \begin{bmatrix} x \\ y \end{bmatrix} = -2 \begin{bmatrix} x \\ y \end{bmatrix}$$

we see that for the right eigenvector corresponding to $\lambda_1 = -2$ we can take any non-zero vector with components such that $3x = -2y$. Thus any $x \neq 0$ is acceptable. In order to avoid fractions, let us pick $x = 2$. Then $y = -3$.

$$q_1 = \begin{bmatrix} 2 \\ -3 \end{bmatrix}$$

Similarly, for the right eigenvector corresponding to $\lambda_2 = 3$ we can pick

$$q_2 = \begin{bmatrix} 1 \\ 1 \end{bmatrix}$$

Now[4]

$$Q^{-1} = \begin{bmatrix} 2 & 1 \\ -3 & 1 \end{bmatrix}^{-1} = \frac{1}{5} \begin{bmatrix} 1 & -1 \\ 3 & 2 \end{bmatrix}$$

Let us check that the rows of $P = Q^{-1}$ are indeed the left eigenvectors of A:

$$\begin{bmatrix} 1/5 & -1/5 \end{bmatrix} \begin{bmatrix} 1 & 2 \\ 3 & 0 \end{bmatrix} = -2 \begin{bmatrix} 1/5 & -1/5 \end{bmatrix}$$

$$\begin{bmatrix} 3/5 & 2/5 \end{bmatrix} \begin{bmatrix} 1 & 2 \\ 3 & 0 \end{bmatrix} = 3 \begin{bmatrix} 3/5 & 2/5 \end{bmatrix}$$

Matlab note: *The main feature of* MATLAB *is the easiness of matrix calculations and manipulations. Even its name derives from this property:* <u>MAT</u>RIX <u>LAB</u>ORATORY. *Some of the most useful matrix commands in* MATLAB *are* `inv` *and* `eig`.

---

[4]The formula for the inverse of a 2×2 matrix is easy to remember:

$$\begin{bmatrix} a & b \\ c & d \end{bmatrix}^{-1} = \frac{\begin{bmatrix} d & -b \\ -c & a \end{bmatrix}}{ad - bc}$$

If we remember that $ad - bc$ is the determinant of the matrix being inverted, this formula can often help us remember how to apply the general formula for matrix inversion:

$$X^{-1} = \frac{\mathrm{adj}(X)}{\det(X)}$$

Problem 3.2.3 Show that the state transition matrix (in mathematics called the fundamental solution)

$$\phi(t) = I + At + \frac{(At)^2}{2!} + \frac{(At)^3}{3!} + \dots$$

satisfies the following matrix differential equation

$$\dot{X}(t) = AX(t), \qquad \text{with } X(0) = I$$

where both $X(t)$ and A are $n \times n$.

Solution: If $n = 1$, obviously $1 + At + \frac{(At)^2}{2!} + \frac{(At)^3}{3!} + \ldots = e^{At}$ and

$$\frac{d}{dt}e^{At} = Ae^{At} \quad \text{and} \quad e^{A \cdot 0} = 1$$

In general, for $n \geq 1$

$$\begin{aligned} \frac{d}{dt}\phi(t) &= \frac{d}{dt}\left(I + At + \frac{(At)^2}{2!} + \dots\right) \\ &= A + A^2t + \frac{A^3t^2}{2!} + \dots \\ &= A\left(I + At + \frac{(At)^2}{2!} + \dots\right) \\ &= A\phi(t) \end{aligned}$$

while for $t = 0$ this series reduces to I. Hence we proved that $\phi(t)$ is the solution of the given matrix differential equation. This is a good motivation to formally write

$$e^{At} = I + At + \frac{(At)^2}{2!} + \frac{(At)^3}{3!} + \dots$$

i.e.,

$$\phi(t) = e^{At}$$

Note: *This result and the new notation are important because if we wish to solve a homogeneous vector differential equation (i.e., a homogeneous system of coupled scalar differential equations) with arbitrary initial conditions*

$$\dot{x}(t) = Ax(t), \qquad \text{with } x(0) = \begin{bmatrix} x_{01} \\ \vdots \\ x_{0n} \end{bmatrix}$$

we can use the linearity and write

$$x(t) = \phi(t)x(0)$$

This is so because $\phi_i(t)$, the i-th column of $\phi(t)$, is a solution of

$$\dot{z}(t) = Az(t), \qquad \text{with } z(0) = e_i$$

where $e_i = [0 \ \dots \ 0 \ 1 \ 0 \ \dots \ 0]'$, with the only 1 at the i-th position. Then

$$x(t) = x_1\phi_1(t) + \ldots + x_n\phi_n(t) = \phi(t)x(0) = e^{At}x(0)$$

Problem 3.2.4 Determine e^{At} for

$$A = \begin{bmatrix} 1 & 1 \\ 0 & 1 \end{bmatrix}$$

Solution: It is easy to prove (by mathematical induction) that

$$A^k = \begin{bmatrix} 1 & k \\ 0 & 1 \end{bmatrix}$$

Hence

$$\begin{aligned} e^{At} &= \sum_{k=0}^{\infty} \frac{A^k t^k}{k!} \\ &= \begin{bmatrix} \sum_0^\infty t^k/k! & \sum_0^\infty k t^k/k! \\ 0 & \sum_0^\infty t^k/k! \end{bmatrix} \\ &= \begin{bmatrix} e^t & te^t \\ 0 & e^t \end{bmatrix} \end{aligned}$$

Note: *The purpose of this problem was to show that*

$$e^{\left[\begin{smallmatrix} 1 & 1 \\ 0 & 1 \end{smallmatrix}\right]t} \neq \begin{bmatrix} e^t & e^t \\ 1 & e^t \end{bmatrix}$$

as one might have naively suspected.

Also note that in this problem it was easy to determine e^{At} *due to the simplicity of* A^k. *Later we shall see techniques which allow us to determine* e^{At} *in other cases too.*

Problem 3.2.5 Let A be $n \times n$ matrix with distinct eigenvalues. If Q is a matrix of its right eigenvectors, and $\Lambda = \text{diag}(\lambda_1, \lambda_2, \ldots, \lambda_n)$ is a diagonal matrix of its eigenvalues, then

$$e^{At} = Qe^{\Lambda t}Q^{-1}$$

Also show that

$$e^{\Lambda t} = \begin{bmatrix} e^{\lambda_1 t} & 0 & \ldots & 0 \\ 0 & e^{\lambda_2 t} & \ldots & 0 \\ \vdots & \vdots & \ddots & \vdots \\ 0 & 0 & \ldots & e^{\lambda_n t} \end{bmatrix}$$

Solution: With $\phi(t) = e^{At} = \sum \frac{(At)^k}{k!}$, and $A = Q\Lambda Q^{-1}$, we find

$$\begin{aligned} \phi(t) &= I + At + \frac{(At)^2}{2!} + \ldots \\ &= I + Q\Lambda Q^{-1}t + \frac{(Q\Lambda Q^{-1}t)^2}{2!} + \ldots \\ &= QQ^{-1} + Q\Lambda Q^{-1}t + \frac{Q\Lambda Q^{-1}Q\Lambda Q^{-1}t^2}{2!} + \ldots \\ &= Q\left(I + \Lambda t + \frac{(\Lambda t)^2}{2!} + \ldots\right)Q^{-1} \\ &= Qe^{\Lambda t}Q^{-1} \end{aligned}$$

Now for the arbitrary initial conditions we can write

$$x(t) = \phi(t)x(0) = Qe^{\Lambda t}Q^{-1}x(0)$$

It is obvious that $e^{\Lambda t} = \text{diag}(e^{\lambda_1 t}, e^{\lambda_2 t}, \ldots, e^{\lambda_n t})$, because

$$\Lambda^k = \text{diag}(\lambda_1^k, \lambda_2^k, \ldots, \lambda_n^k)$$

Note: *In general, if A is not diagonalizable and $Q^{-1}AQ = J$ is its Jordan form, then $\phi(t) = Qe^{Jt}Q^{-1}$, where e^{Jt} is in general an upper-triangular matrix.*

Problem 3.2.6 Given a matrix A

$$A = \begin{bmatrix} -6 & 2 \\ -6 & 1 \end{bmatrix}$$

calculate the eigenvalues λ_1 and λ_2, and the corresponding eigenvectors q_1 and q_2. Form a matrix $Q = [q_1\ q_2]$, find Q^{-1}, calculate $Q^{-1}AQ$, and e^{At}.

Solution: To find the eigenvalues write

$$\det(A - \lambda I) = 0 \quad \Rightarrow \quad \lambda^2 + 5\lambda + 6 = 0 \quad \Rightarrow \quad \lambda_1 = -3,\ \lambda_2 = -2$$

The corresponding eigenvectors are found from

$$(A - \lambda I)q = 0$$

For $\lambda = \lambda_1 = -3$ we have

$$\begin{bmatrix} -6-(-3) & 2 \\ -6 & 1-(-3) \end{bmatrix} \begin{bmatrix} q_{11} \\ q_{21} \end{bmatrix} = \begin{bmatrix} 0 \\ 0 \end{bmatrix}$$

i.e., $-3q_{11} + 2q_{21} = 0$, so we can choose

$$q_1 = \begin{bmatrix} q_{11} \\ q_{21} \end{bmatrix} = \begin{bmatrix} 2 \\ 3 \end{bmatrix}$$

Similarly, for $\lambda = \lambda_2 = -2$ we have $-4q_{12} + 2q_{22} = 0$, and we can pick

$$q_2 = \begin{bmatrix} q_{12} \\ q_{22} \end{bmatrix} = \begin{bmatrix} 1 \\ 2 \end{bmatrix}$$

Now

$$Q = \begin{bmatrix} 2 & 1 \\ 3 & 2 \end{bmatrix}$$

while

$$Q^{-1} = \begin{bmatrix} 2 & -1 \\ -3 & 2 \end{bmatrix} / (4-3) = \begin{bmatrix} 2 & -1 \\ -3 & 2 \end{bmatrix}$$

Now we can calculate $Q^{-1}AQ$:

$$\begin{bmatrix} 2 & -1 \\ -3 & 2 \end{bmatrix}\begin{bmatrix} -6 & 2 \\ -6 & 1 \end{bmatrix}\begin{bmatrix} 2 & 1 \\ 3 & 2 \end{bmatrix} = \begin{bmatrix} 2 & -1 \\ -3 & 2 \end{bmatrix}\begin{bmatrix} -6 & -2 \\ -9 & -4 \end{bmatrix} = \begin{bmatrix} -3 & 0 \\ 0 & -2 \end{bmatrix}$$

As expected

$$Q^{-1}AQ = \text{diag}(\lambda_1, \lambda_2)$$

Finally,

$$e^{At} = Qe^{\Lambda t}Q^{-1} = \begin{bmatrix} 2 & 1 \\ 3 & 2 \end{bmatrix}\begin{bmatrix} e^{-3t} & 0 \\ 0 & e^{-2t} \end{bmatrix}\begin{bmatrix} 2 & -1 \\ -3 & 2 \end{bmatrix}$$

hence

$$e^{At} = \begin{bmatrix} (-3e^{-2t} + 4e^{-3t}) & (2e^{-2t} - 2e^{-3t}) \\ (-6e^{-2t} + 6e^{-3t}) & (4e^{-2t} - 3e^{-3t}) \end{bmatrix}$$

Problem 3.2.7 Describe how the Cayley-Hamilton (C-H) theorem can be used in determination of e^{At}, the state transition matrix of A. This method applies to other matrix functions as well and is often called the Sylvester interpolation.

Solution: If $a(\lambda) = 0$ is the characteristic equation of an $n \times n$ matrix A, then, according to the Cayley-Hamilton theorem,

$$a(A) = A^n + a_1 A^{n-1} + \ldots + a_n I = 0$$

Note: *$a(\lambda)$ is a polynomial of order n. If A has repeated eigenvalues then there may exists a polynomial $b(\lambda)$ such that $b(A) = 0$ and $\deg(b) < n$. Such polynomial exists if at least one of the repeated eigenvalues appears in more than one Jordan block in the Jordan form of A.*

Using $A^n = -(a_1 A^{n-1} + \ldots + a_n I)$ the expression for e^{At}

$$e^{At} = I + At + \frac{(At)^2}{2!} + \frac{(At)^3}{3!} + \ldots$$

can be written using only n terms:

$$e^{At} = \alpha_0(t)I + \alpha_1(t)A + \ldots + \alpha_{n-1}(t)A^{n-1}$$

How are the coefficients $\alpha_0(t), \ldots, \alpha_{n-1}(t)$ determined? Since $a(\lambda_i) = 0$ for all $i = 0, \ldots, n-1$, the expressions

$$e^{\lambda_i t} = 1 + \lambda_i + \frac{(\lambda_i t)^2}{2} + \ldots$$

can be simplified to the same form as e^{At}:

$$e^{\lambda_i t} = \alpha_0(t) + \alpha_1(t)\lambda_i + \ldots + \alpha_{n-1}(t)\lambda_i^{n-1} \qquad (i = 0, \ldots, n-1) \tag{3.2}$$

If A has n distinctive eigenvalues λ_i, the above represents n equations in n unknowns $\alpha_0(t), \ldots, \alpha_{n-1}(t)$. It has a unique solution because the system determinant is the Vandermonde determinant

$$V(\lambda_1, \ldots, \lambda_n) = \prod_{j>i} (\lambda_j - \lambda_i) \neq 0$$

If A has multiple eigenvalues the corresponding equation in (3.2) can be differentiated over λ_i to obtain new independent equations and finally a solution (see Problem 3.2.10). Is the system determinant in such cases always non-zero? Here is a hint:

$$V_{k_1,\ldots,k_r}(\lambda_1, \ldots, \lambda_r) = \left(\prod_i (0!\,1!\,2! \ldots (k_i - 1)!) \right) \left(\prod_{j>i} (\lambda_j - \lambda_i)^{k_i k_j} \right)$$

Problem 3.2.8 Using the C-H theorem find the state transition matrix for

$$A = \begin{bmatrix} -6 & 2 \\ -6 & 1 \end{bmatrix}$$

Solution: From the Cayley-Hamilton theorem we know that instead of writing $\phi(t) = e^{At}$ as an infinite series

$$e^{At} = I + At + \frac{(At)^2}{2!} + \frac{(At)^3}{3!} + \ldots$$

we can use the characteristic equation of A

$$\lambda^2 + 5\lambda + 6 = 0$$

to express A^2 in terms of A and I:

$$A^2 = -5A - 6I$$

and therefore to eliminate second, third, and all other powers of A from the expression for $\phi(t)$. Thus

$$\phi(t) = e^{At} = \alpha(t)I + \beta(t)A$$

The parameters $\alpha(t)$ and $\beta(t)$ can be found by writing the same equations for the eigenvalues of A:

$$\begin{aligned} \alpha(t) - 3\beta(t) &= e^{-3t} \\ \alpha(t) - 2\beta(t) &= e^{-2t} \end{aligned}$$

hence

$$\begin{aligned} \alpha(t) &= 3e^{-2t} - 2e^{-3t} \\ \beta(t) &= e^{-2t} - e^{-3t} \end{aligned}$$

so that

$$\phi(t) = (3e^{-2t} - 2e^{-3t})I + (e^{-2t} - e^{-3t})A = \begin{bmatrix} (-3e^{-2t} + 4e^{-3t}) & (2e^{-2t} - 2e^{-3t}) \\ (-6e^{-2t} + 6e^{-3t}) & (4e^{-2t} - 3e^{-3t}) \end{bmatrix}$$

Of course, the application of formula $\phi(t) = Q^{-1}e^{\Lambda t}Q$ yields the same result.

Problem 3.2.9 Use the Laplace transform to determine e^{At} for

$$A = \begin{bmatrix} -6 & 2 \\ -6 & 1 \end{bmatrix}$$

Solution: We shall show in Problem 3.3.3 that e^{At} and $(sI - A)^{-1}$, the so-called resolvent matrix, are a Laplace transform pair:

$$\mathcal{L}\left\{e^{At}\right\} = (sI - A)^{-1} \quad \text{i.e.,} \quad \mathcal{L}^{-1}\left\{(sI - A)^{-1}\right\} = e^{At}$$

We obtain the same solution as in Problem 3.2.8 from the following:

$$(sI - A)^{-1} = \begin{bmatrix} s+6 & -2 \\ 6 & s-1 \end{bmatrix}^{-1} = \frac{\begin{bmatrix} s-1 & 2 \\ -6 & s+6 \end{bmatrix}}{(s+2)(s+3)} = \begin{bmatrix} \left(\frac{-3}{s+2} + \frac{4}{s+3}\right) & \left(\frac{2}{s+2} + \frac{-2}{s+3}\right) \\ \left(\frac{-6}{s+2} + \frac{6}{s+3}\right) & \left(\frac{4}{s+2} + \frac{-3}{s+3}\right) \end{bmatrix}$$

Problem 3.2.10 Given

$$A = \begin{bmatrix} -1 & 0 & 0 \\ 0 & -1 & 0 \\ 1 & 0 & -1 \end{bmatrix}$$

find its eigenvalues, eigenvectors, and the state transition matrix $\phi(t) = e^{At}$.

Solution: It is easy to see that A has a triple eigenvalue $\lambda_{1,2,3} = -1$. Since

$$\nu(A - \lambda I) = n - \rho(A - \lambda I) = 3 - 1 = 2$$

there are only two independent eigenvectors corresponding to the eigenvalue:

$$\begin{bmatrix} 0 & 0 & 0 \\ 0 & 0 & 0 \\ 1 & 0 & 0 \end{bmatrix} \begin{bmatrix} a \\ b \\ c \end{bmatrix} = \begin{bmatrix} 0 \\ 0 \\ 0 \end{bmatrix} \quad \Rightarrow \quad a = 0$$

hence we are free to pick b and c independently, so we can find two independent eigenvectors, for example:

$$q_1 = \begin{bmatrix} 0 \\ 1 \\ 0 \end{bmatrix} \quad \text{and} \quad q_2 = \begin{bmatrix} 0 \\ 0 \\ 1 \end{bmatrix}$$

while the third principal vector is a generalized eigenvector. To find it we write

$$(A - \lambda I)^2 q_3 = 0 \qquad \text{and} \qquad (A - \lambda I) q_3 = q_2$$

This reduces to

$$q_3 = \begin{bmatrix} 1 \\ 0 \\ 0 \end{bmatrix}$$

Note that now

$$Q = \begin{bmatrix} 0 & 0 & 1 \\ 1 & 0 & 0 \\ 0 & 1 & 0 \end{bmatrix}$$

so

$$Q^{-1} = \begin{bmatrix} 0 & 1 & 0 \\ 0 & 0 & 1 \\ 1 & 0 & 0 \end{bmatrix}$$

and

$$J = Q^{-1}AQ = \begin{bmatrix} -1 & 0 & 0 \\ 0 & -1 & 1 \\ 0 & 0 & -1 \end{bmatrix}$$

a Jordan matrix similar to A.

Since A is 3×3 and $\lambda = -1$ is a triple eigenvalue, in order to determine the coefficients in

$$e^{At} = \alpha(t)I + \beta(t)A + \gamma(t)A^2$$

we form the three equations by writing

$$e^{\lambda t} = \alpha(t) + \beta(t)\lambda + \gamma(t)\lambda^2$$

and the first and the second derivatives over λ:

$$te^{\lambda t} = \beta(t) + 2\gamma(t)\lambda$$

and

$$t^2 e^{\lambda t} = 2\gamma(t)$$

When we solve this system with $\lambda = -1$, we finally get

$$e^{At} = \begin{bmatrix} e^{-t} & 0 & 0 \\ 0 & e^{-t} & 0 \\ te^{-t} & 0 & e^{-t} \end{bmatrix}$$

Note: *Since*

$$J = \begin{bmatrix} -1 & 0 & 0 \\ 0 & -1 & 1 \\ 0 & 0 & -1 \end{bmatrix}$$

we have

$$J^k = \begin{bmatrix} (-1)^k & 0 & 0 \\ 0 & (-1)^k & (-1)^{k-1}k \\ 0 & 0 & (-1)^k \end{bmatrix} \quad \textit{and} \quad e^{Jt} = \begin{bmatrix} e^{-t} & 0 & 0 \\ 0 & e^{-t} & te^{-t} \\ 0 & 0 & e^{-t} \end{bmatrix}$$

hence (cf. Problem 3.2.5*)*

$$e^{At} = Qe^{Jt}Q^{-1} = \begin{bmatrix} e^{-t} & 0 & 0 \\ 0 & e^{-t} & 0 \\ te^{-t} & 0 & e^{-t} \end{bmatrix}$$

Problem 3.2.11 Show that $\phi(t)$ is invertible for all real t.

Solution: Since in general $\phi(t) = Qe^{Jt}Q^{-1}$, we can write

$$\det(\phi(t)) = \det(Q)\det(e^{Jt})\det(Q^{-1}) = e^{\lambda_1 t}\dots e^{\lambda_n t} > 0$$

Therefore $\phi(t)$ is invertible for all real t.

Problem 3.2.12 Show that $\phi(t_1)\phi(t_2) = \phi(t_1+t_2)$. In particular, show that $\phi^{-1}(t) = \phi(-t)$.

Solution: To show that $\phi(t_1)\phi(t_2) = \phi(t_1 + t_2)$ we can multiply two infinite series

$$\left(I + At_1 + \frac{(At_1)^2}{2!} + \dots\right)\left(I + At_2 + \frac{(At_2)^2}{2!} + \dots\right)$$

to get the following:

$$\phi(t_1)\phi(t_2) = Qe^{Jt_1}Q^{-1}Qe^{Jt_2}Q^{-1} = Qe^{J(t_1+t_2)}Q^{-1} = \phi(t_1 + t_2)$$

The special case of this formula is $\phi(t)\phi(-t) = \phi(0) = I$, which implies $\phi^{-1}(t) = \phi(-t)$.

Problem 3.2.13 For the matrix differential equation

$$\frac{d}{dt}M(t) = AM(t) + M(t)B, \quad M(0) = C$$

where A, B, and C are $n \times n$ constant matrices and M is an $n \times n$ matrix, show that the solution is given in the form

$$M(t) = e^{At}Ce^{Bt}$$

Solution: $M(t) = e^{At}Ce^{Bt}$ is a solution of the given equation because it satisfies the initial condition:

$$M(0) = e^{A\cdot 0}Ce^{B\cdot 0} = I\cdot C\cdot I = C$$

and the equation itself:

$$\frac{d}{dt}M(t) = \frac{d}{dt}(e^{At}Ce^{Bt}) = Ae^{At}Ce^{Bt} + e^{At}CBe^{Bt} = Ae^{At}Ce^{Bt} + e^{At}Ce^{Bt}B$$

We are allowed to do the last step because B and e^{Bt} commute for all B.

Problem 3.2.14 Prove, assuming all inverses exist, the following identities for resolvent matrices:

$$(sI - A)^{-1} - (sI - B)^{-1} = (sI - A)^{-1}(A - B)(sI - B)^{-1}$$

and

$$(sI - A)^{-1} - (vI - A)^{-1} = (sI - A)^{-1}(v - s)(vI - A)^{-1}$$

Solution: To get the idea for the proof of the first identity, multiply it by $(sI - B)$ from the right and by $(sI - A)$ from the left to obtain the identity $A - B = A - B$. Hence, we can derive the first identity as follows:

$$A - B = A - B$$

$$(sI - B) - (sI - A) = A - B$$

$$(sI - A)^{-1}(sI - B) - I = (sI - A)^{-1}(A - B)$$

$$(sI - A)^{-1} - (sI - B)^{-1} = (sI - A)^{-1}(A - B)(sI - B)^{-1}$$

We can prove the second identity as follows:

$$v - s = v - s$$

$$(vI - A) - (sI - A) = (v - s)I$$

$$(sI - A)^{-1}(vI - A) - I = (sI - A)^{-1}(v - s)$$

$$(sI - A)^{-1} - (vI - A)^{-1} = (sI - A)^{-1}(v - s)(vI - A)^{-1}$$

Problem 3.2.15 With the notation as in Problem 3.2.1, including the assumption that A has distinct eigenvalues, prove that

$$(sI - A)^{-1} = \sum_{k=1}^{n} \frac{R_k}{s - \lambda_k}, \qquad \text{where } R_k = q_k p'_k$$

Note that R_k is $n \times n$.

Solution I: To prove the above formula we shall need the following properties of eigenvectors and matrices formed from them:

If $R_k = q_k p'_k$, then:

- From $QP = I$, i.e., $\sum_k q_{ik} p_{kj} = \delta_{ij}$, and from $(R_k)_{ij} = q_{ik} p_{kj}$, we find

$$\sum_{k=1}^{n} R_k = I$$

- From

$$\sum_{k=1}^{n} \lambda_k R_k = \sum_{k=1}^{n} \lambda_k q_k p'_k = \sum_{k=1}^{n} A q_k p'_k = A \sum_{k=1}^{n} q_k p'_k = A \sum_{k=1}^{n} R_k = A$$

  we see that

$$\sum_{k=1}^{n} \lambda_k R_k = A$$

- Since $PQ = I$, we have $p'_k q_m = \delta_{km}$. Hence $R_k R_m = q_k p'_k q_m p'_m = q_k p'_m \delta_{km} = R_k \delta_{km}$, so we can write

$$R_k R_m = R_k \delta_{km}$$

Now we can proceed with the proof:

$$(sI - A)^{-1} = \left(s \sum_{k=1}^{n} R_k - \sum_{k=1}^{n} \lambda_k R_k \right)^{-1} = \left(\sum_{k=1}^{n} (s - \lambda_k) R_k \right)^{-1} = \sum_{k=1}^{n} \frac{R_k}{s - \lambda_k}$$

The last step is justified by the following:

$$\left(\sum_{k=1}^{n} (s - \lambda_k) R_k \right) \left(\sum_{m=1}^{n} \frac{R_m}{s - \lambda_m} \right) = \sum_{k=1}^{n} \sum_{m=1}^{n} \frac{s - \lambda_k}{s - \lambda_m} \underbrace{R_k R_m}_{R_k \delta_{km}} = \sum_{k=1}^{n} R_k = I$$

Solution II: Multiply both sides by $(sI - A)$,

$$\begin{aligned} I &= \sum_{k=1}^{n} \frac{(sI - A) q_k p'_k}{s - \lambda_k} \\ &= \sum_{k=1}^{n} \frac{(s q_k - \lambda_k q_k) p'_k}{s - \lambda_k} \\ &= \sum_{k=1}^{n} q_k p'_k \end{aligned}$$

As we saw at the beginning of Solution I, this last sum is equal to I.

Note: *When A has repeated eigenvalues, it may or may not be diagonalizable. If it is not diagonalizable, then $\sum \lambda_k R_k \neq A$, and besides simple terms with linear denominators $(s - \lambda_k)$, we also get higher-order terms, with denominators $(s - \lambda_k)^2$, $(s - \lambda_k)^3$, etc.*

3.3 Systems of linear differential equations

This Section introduces the matrix notation for the systems of linear differential equations. The results of Section 3.1 are generalized.

Problem 3.3.1 Write the following system of equations in a matrix form:

$$\begin{aligned}\dot{u}(t) &= 3u(t) - 3v(t) - 2w(t) + \sin t \\ \dot{v}(t) &= 2u(t) - 4v(t) + 8w(t) + \cos t \\ \dot{w}(t) &= 2u(t) + 3v(t) + 3w(t) + 1\end{aligned}$$

Solution: If we write

$$x(t) = \begin{bmatrix} u(t) \\ v(t) \\ w(t) \end{bmatrix} \quad \text{and} \quad f(t) = \begin{bmatrix} \sin t \\ \cos t \\ 1 \end{bmatrix}$$

the system can be written as

$$\dot{x}(t) = Ax(t) + f(t)$$

where

$$A = \begin{bmatrix} 3 & -3 & -2 \\ 2 & -4 & 8 \\ 2 & 3 & 3 \end{bmatrix}$$

Problem 3.3.2 Show that the solution of the inhomogeneous vector differential equation

$$\dot{x}(t) = Ax(t) + f(t), \qquad \text{with } x(0) = \begin{bmatrix} x_{01} \\ \vdots \\ x_{0n} \end{bmatrix}$$

where A is $n \times n$, while $x(t)$ and $f(t)$ are $n \times 1$, is given by

$$x(t) = \phi(t)x(0) + \phi(t) * f(t)$$

where (as in Section 3.2) $\phi(t) = I + At + \frac{(At)^2}{2!} + \frac{(At)^3}{3!} + \ldots$ i.e., $\phi(t) = e^{At}$, and $*$ denotes convolution:

$$\phi(t) * f(t) = \int_0^t \phi(t-\tau) f(\tau)\, d\tau$$

Solution: Introduce a change of variables analogous to the change usually made in the scalar case:

$$x(t) = \phi(t)z(t), \qquad (x(0) = z(0))$$

when we can formally write

$$\dot{x}(t) = \dot{\phi}(t)z(t) + \phi(t)\dot{z}(t)$$

This is justified by differentiation of

$$x_i(t) = \sum_j \phi_{ij}(t) z_j(t)$$

Since $\dot{\phi}(t) = A\phi(t)$, now we can write $\phi(t)\dot{z}(t) = f(t)$, i.e.,

$$\dot{z}(t) = \phi^{-1}(t) f(t) = \phi(-t) f(t)$$

This finally implies

$$x(t) = \phi(t) z(t) = \phi(t) \left(z(0) + \int_0^t \dot{z}(\tau)\, d\tau \right) = \phi(t) \left(z(0) + \int_0^t \phi(-\tau) f(\tau)\, d\tau \right)$$

i.e.,

$$x(t) = \underbrace{\phi(t)x(0)}_{\text{homogeneous part}} + \underbrace{\int_0^t \phi(t-\tau) f(\tau)\, d\tau}_{\text{non-homogeneous part}}$$

Problem 3.3.3 Apply the Laplace transform to the vector differential equation

$$\dot{x}(t) = Ax(t) + f(t), \qquad \text{with } x(0) = \begin{bmatrix} x_{01} \\ \vdots \\ x_{0n} \end{bmatrix}$$

where A is $n \times n$, while $x(t)$ and $f(t)$ are $n \times 1$.

Solution: Keeping in mind that this vector differential equation is actually a system of scalar differential equations, we can write

$$sX(s) - x(0) = AX(s) + F(s)$$

where $X(s)$ and $F(s)$ are $n \times 1$ vectors whose components are the Laplace transforms of the corresponding components of vectors $x(t)$ and $f(t)$.

Now we can see that

$$(sI - A)X(s) = x(0) + F(s)$$

i.e.,

$$X(s) = \Phi(s)x(0) + \Phi(s)F(s)$$

where

$$\Phi(s) = (sI - A)^{-1}$$

is the so-called resolvent matrix. Since the inverse Laplace transform of a product is a time-domain convolution, taking the inverse Laplace transform we obtain

$$x(t) = \phi(t)x(0) + \int_0^t \phi(t-\tau) f(\tau)\, d\tau$$

Note: *Compare these expressions to the final expressions in Problems* 3.1.2 *and* 3.3.2. *This is another justification to write*

$$\phi(t) = e^{At}$$

Problem 3.3.4 Write the equation from Problem 3.1.7

$$\ddot{x} + 3\dot{x} + 2x = \delta(t) + 2\dot{\delta}(t), \quad x(0^-) = 1, \quad \dot{x}(0^-) = 2$$

as a system of two first-order equations, and use the formalism developed above to solve it.

Solution: Let $w_1(t) = \dot{x}(t)$ and $w_2(t) = x(t)$. Then $w_1(0^-) = \dot{x}(0^-) = 2$ and $w_2(0^-) = x(0^-) = 1$, and

$$\dot{w}_1(t) + 3w_1(t) + 2w_2(t) = \delta(t) + 2\dot{\delta}(t)$$

$$\dot{w}_2(t) = w_1(t)$$

Hence, we can write

$$\begin{bmatrix} \dot{w}_1(t) \\ \dot{w}_2(t) \end{bmatrix} = \begin{bmatrix} -3 & -2 \\ 1 & 0 \end{bmatrix} \begin{bmatrix} w_1(t) \\ w_2(t) \end{bmatrix} + \begin{bmatrix} \delta(t) + 2\dot{\delta}(t) \\ 0 \end{bmatrix}$$

i.e.,

$$\dot{w} = \underbrace{\begin{bmatrix} -3 & -2 \\ 1 & 0 \end{bmatrix}}_{A} w + \underbrace{\begin{bmatrix} \delta(t) + 2\dot{\delta}(t) \\ 0 \end{bmatrix}}_{f(t)}$$

Therefore

$$\begin{aligned} W(s) &= (sI - A)^{-1}(w(0^-) + F(s)) \\ &= \begin{bmatrix} s+3 & 2 \\ -1 & s \end{bmatrix}^{-1} \left(\begin{bmatrix} 2 \\ 1 \end{bmatrix} + \begin{bmatrix} 1 + 2s \\ 0 \end{bmatrix} \right) \\ &= \frac{1}{(s+1)(s+2)} \begin{bmatrix} s & -2 \\ 1 & s+3 \end{bmatrix} \begin{bmatrix} 2s+3 \\ 1 \end{bmatrix} \\ &= \begin{bmatrix} \frac{2s-1}{s+1} \\ \frac{3}{s+1} \end{bmatrix} \end{aligned}$$

The solution we are looking for is

$$x(t) = w_2(t) = \mathcal{L}_-^{-1} \left\{ \frac{3}{s+1} \right\} = 3e^{-t} \quad (t > 0)$$

Matlab note: MATLAB *commands* `ode23` *and* `ode45` *can be used to numerically integrate systems of ordinary differential equations.*

3.4 Input-output representation

This Section is here to refresh the reader's memory about some of the many transfer function techniques available in analysis and design of linear systems. More importantly, it should illustrate some of the problems that are a lot easier to solve using the state-space methods, or at least require some insights from that approach.

Problem 3.4.1 Determine the output of a system described by

$$\dot{y}(t) + 2y(t) = u(t) \qquad (t > 0)$$

$$y(0) = 5$$

when

a) $u(t) = \cos 3t$
b) $u(t) = e^{-t}$
c) $u(t) = e^{-(2-\varepsilon)t}$ where ε is a small positive number
d) $u(t) = e^{-2t}$

Solution: The homogeneous part of the solution is the same for all four cases. Since the root of the characteristic equation (also called the *pole* of the system) is $a = -2$

$$y_h(t) = y(0)e^{at} = 5e^{-2t}$$

a) The non-homogeneous part (also called the *particular* solution) is as in Problem 3.1.1

$$y_{nh}(t) = \int_0^t e^{a(t-\tau)}u(\tau)\,d\tau = e^{-2t}\int_0^t e^{2\tau}\cos 3\tau\,d\tau = \frac{2}{13}\cos 3t + \frac{3}{13}\sin 3t - \frac{2}{13}e^{-2t}$$

Finally, the solution is

$$y(t) = y_h(t) + y_{nh}(t) = \frac{63}{13}e^{-2t} + \frac{2}{13}\cos 3t + \frac{3}{13}\sin 3t$$

Note 1: *The first term in the solution approaches zero fast, and it is often called the* transient *part of the solution. The remaining terms are then called the* steady-state *part of the solution. Note that both the initial conditions and the input contribute to the transient part of $y(t)$, through $y_h(t)$ and $y_{nh}(t)$), respectively. The steady-state part, however, comes from the input only, therefore it is often called the* forced *solution.*

Note 2: *We can solve this equation in other ways, using the Laplace transform for example. Another method is attractive too: Knowing the root of the characteristic equation and from the form of the input we can immediately write*

$$y(t) = Ae^{-2t} + B\cos 3t + C\sin 3t$$

If we substitute this into the original equation (not only its homogeneous part), the initial condition gives us one of three equations for constants A, B, and C:

$$A + B = 5$$

The other two equations are obtained by equating the coefficients next to $\cos 3t$ *and* $\sin 3t$ *terms, respectively:*

$$2B + 3C = 1 \quad \textit{and} \quad 3B - 2C = 0$$

See also Problem 3.4.2.

b) For this input

$$y_{nh}(t) = e^{-t} - e^{-2t}$$

Therefore,

$$y(t) = 4e^{-2t} + e^{-t}$$

c) For $u(t) = e^{-(2-\varepsilon)t}$ we find

$$y(t) = \left(5 - \frac{1}{\varepsilon}\right) e^{-2t} + \frac{1}{\varepsilon} e^{-(2-\varepsilon)t}$$

Note: *As* $\varepsilon \to 0$*, i.e., when the input's complex frequency approaches the system's pole, the forced output grows in magnitude. This is* resonance. *Asymptotically (as* $\varepsilon \to 0$*), the total output behaves like:*

$$\begin{aligned} \lim_{\varepsilon \to 0} \left(\left(5 - \frac{1}{\varepsilon}\right) e^{-2t} + \frac{1}{\varepsilon} e^{-(2-\varepsilon)t} \right) &= 5e^{-2t} + \lim_{\varepsilon \to 0} \frac{e^{-(2-\varepsilon)t} - e^{-2t}}{\varepsilon} \\ &= 5e^{-2t} + te^{-2t} \end{aligned}$$

d) When $u(t) = e^{-2t}$ the input's complex frequency coincides with the pole of the system. The convolution of two similar terms produces a new form:

$$y_{nh}(t) = \int_0^t e^{-2(t-\tau)} e^{-2\tau} \, d\tau = te^{-2t}$$

Therefore

$$y(t) = 5e^{-2t} + te^{-2t}$$

Problem 3.4.2 Find the output of a system described by

a) $\dddot{y}(t) + 3\ddot{y}(t) + 3\dot{y}(t) + y(t) = e^{-t}$ with $\ddot{y}(0) = 1, \dot{y}(0) = 2$, and $y(0) = 3$

b) $\dddot{y}(t) + 3\ddot{y}(t) + 4\dot{y}(t) + 12y(t) = \cos 2t$ with $\ddot{y}(0) = 1, \dot{y}(0) = 1$, and $y(0) = 1$

Solution: a) This system has a triple pole at -1, hence the homogeneous part of the solution is a linear combination of e^{-t}, te^{-t}, and t^2e^{-t} (cf. Problem 3.1.3). Since the input's complex frequency coincides with this triple pole, besides these same terms the particular solution also adds t^3e^{-t} to the solution. Hence

$$y(t) = Ae^{-t} + Bte^{-t} + Ct^2e^{-t} + Dt^3e^{-t}$$

Coefficients A, B, C, and D are found from the initial conditions for the whole equation and by substitution of this expression into the equation. Generally, such problems are easier to solve using Laplace transform method.

b) In this case poles are at $\pm 2j$ and -3, and the input coincides with a pair of poles at $\pm 2j$, therefore

$$y(t) = A\cos 2t + B\sin 2t + Ct\cos 2t + Dt\sin 2t + Ee^{-3t}$$

Problem 3.4.3 What is the output of the system described by

$$\dot{y}(t) + 3y(t) = \dot{u}(t) + 2u(t)$$

with $y(0) = 1$ and $u(t) = e^{-2t} + \cos 7t$.

Solution: If we try $y(t) = Ae^{-3t} + Be^{-2t} + C\cos 7t + D\sin 7t$ and substitute it into the equation we immediately find that $B = 0$. The input part of the equation is responsible for this. The complex frequencies for which this happens (in this case only -2) are called the *zeros* of the system. They are the roots of the characteristic equation of the input part of the equation.

Problem 3.4.4 What is the impulse response $h(t)$ of a system? What is the transfer function $T(s)$ of a system? Show that $T(s) = \mathcal{L}\{h(t)\}$.

Solution: *Impulse response.* The impulse response $h(t)$ of a system is the output of the system caused by the Dirac's delta impulse $\delta(t)$ at the input. The system is assumed to be at rest when $\delta(t)$ is applied, i.e., all initial conditions are zero.

The impulse response is important because the output of a linear time-invariant system to any given input can be determined if we know the impulse response of the system[5]: If the initial conditions are non-zero y_h is found as in Problem 3.1.3, while y_{nh} can be characterized in terms of the impulse response as follows.

By the linearity of the system and the decomposition of the input

$$u(t) = \int_0^\infty u(\tau)\delta(t-\tau)\,d\tau$$

we find (assuming the system is causal, i.e., $h(t) \equiv 0$ for $t < 0$)

$$y_{nh}(t) = \int_0^t u(\tau)h(t-\tau)\,d\tau$$

In general, if a system is given by a differential equation, the impulse response is most easily determined as the inverse Laplace transform of its transfer function. The derivation is given below.

[5]Let us mention here that if we want to measure the acoustic impulse response of a concert hall, and thus characterize its acoustic properties, firing a gun and measuring its echo is not the best thing to do, especially since the audience should be in the hall – without the audience the acoustics are completely different (cf. [49]).

Transfer function. Transfer function is the ratio of the Laplace transforms of the output and the input of the system, assuming zero initial conditions:

$$T(s) = \frac{Y(s)}{U(s)}$$

If a system is given by

$$y^{(n)}(t) + a_1 y^{(n-1)}(t) + \ldots + a_{n-1}\dot{y}(t) + a_n y(t) = b_0 u^{(m)}(t) + \ldots + b_{m-1}\dot{u}(t) + b_m u(t)$$

with zero initial conditions, then the Laplace transform yields

$$T(s) = \frac{Y(s)}{U(s)} = \frac{b(s)}{a(s)} = \frac{b_0 s^m + \ldots + b_{m-1}s + b_m}{s^n + a_1 s^{n-1} + \ldots + a_{n-1}s + a_n}$$

From this expression we see that $T(s)$ does not depend on the input $u(t)$, only on the coefficients of the differential equation.

Relation between $h(t)$ and $T(s)$. Since $T(s)$ does not depend on the particular choice of $u(t)$, we can pick $u(t) = \delta(t)$. Then $U(s) = 1$, $y(t) = h(t)$, and $Y(s) = H(s)$. Therefore

$$T(s) = \frac{Y(s)}{U(s)} = H(s) = \mathcal{L}\{h(t)\} = \int_0^\infty h(t)e^{-st}\,dt$$

Note: *This is why we often write $H(s)$ instead of $T(s)$. Another way to see this is to use the convolution property of the Laplace transform: With zero initial conditions*

$$y(t) = h(t) * u(t) \quad \Rightarrow \quad Y(s) = H(s)U(s)$$

Note also that for causal systems ($h(t) \equiv 0$ for $t < 0$) when $s = j\omega$ the transfer function $T(s)$ becomes the frequency response *$T(j\omega)$ and we find that $h(t)$ and $T(j\omega)$ are a Fourier transform pair:*

$$T(j\omega) = H(j\omega) = \mathcal{F}\{h(t)\} = \int_0^\infty h(t)e^{-j\omega t}\,dt$$

Problem 3.4.5 Find the impulse response of the system described by its transfer function

$$H(s) = \frac{s+10}{s^2 + 20s + 164}$$

Solution: The relation between the impulse response $h(t)$ of a linear system and its transfer function $H(s)$ is $H(s) = \mathcal{L}\{h(t)\}$, i.e., $h(t) = \mathcal{L}^{-1}\{H(s)\}$, therefore

$$h(t) = \mathcal{L}^{-1}\left\{\frac{s+10}{(s+10)^2 + 8^2}\right\} = e^{-10t}\cos 8t$$

Matlab note: *Here we demonstrate how simple it is to get plots in* MATLAB. *In Figure* 3.2 *we use the command* `impulse` *to plot the impulse response of a system given by a rational transfer function. It can be used to obtain the inverse Laplace transform of any rational function.*

```
num = [1 10]          % numerator
den = [1 20 164]      % denominator
impulse(num,den)      % does everything: calculations and plot
```

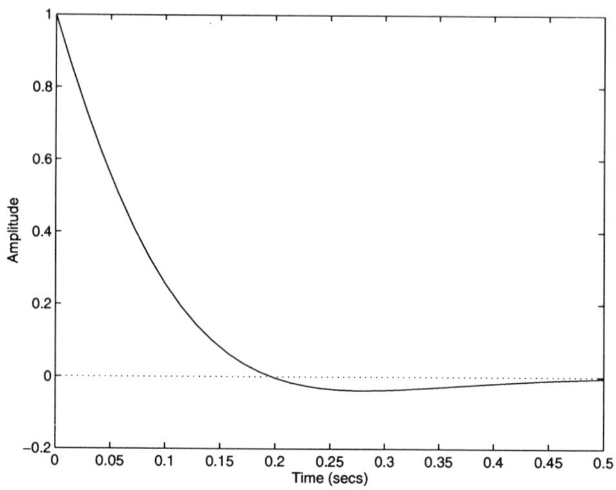

Figure 3.2: The plot produced by the MATLAB command `impulse`.

Usually, plots produced by MATLAB *are quite satisfactory, but if we want to add a "personal touch," we can do as follows (see Figure* 3.3*):*

```
num = [1 10];                     % numerator
den = [1 20 164];                 % denominator
t = 0:0.01:1;                     % time to be shown
[y,x,t] = impulse(num,den,t);     % this form of the command only does calculations
                                  % x returns the states of the system
                                  % which we shall define later
plot(t,y), axis([0 1 -0.2 1.4])
grid, xlabel('time [s]'), ylabel('impulse response')
title('Just a little bit nicer plot')
text(0.2,0.5,'you can even put some text inside the graph')
```

Problem 3.4.6 Linear time-invariant systems are described using linear differential equations with constant coefficients which relate the output $y(t)$ to the input $u(t)$:

$$y^{(n)}(t) + a_1 y^{(n-1)}(t) + \ldots + a_{n-1}\dot{y}(t) + a_n y(t) = b_0 u^{(m)}(t) + \ldots + b_{m-1}\dot{u}(t) + b_m u(t)$$

with initial conditions $y(0), \dot{y}(0), \ldots, y^{(n-1)}(0)$ given. Discuss the solution $y(t)$.

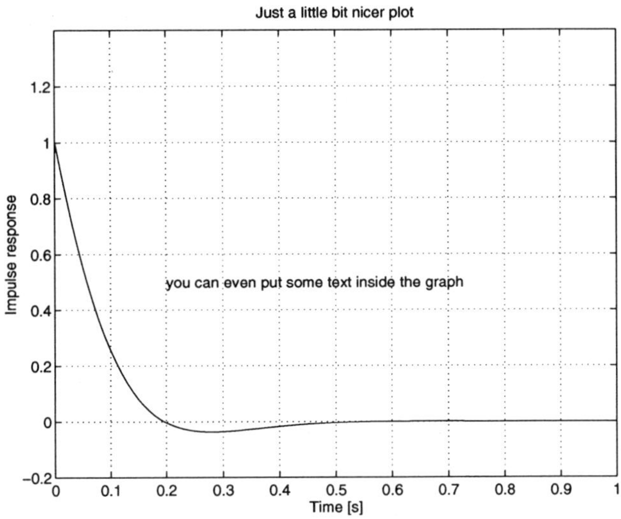

Figure 3.3: This plot shows how the user can add some comments and change the ranges on the axis.

Solution: The solution of this equation can be written as

$$y(t) = y_h(t) + y_{nh}(t)$$

where $y_h(t)$ is a homogeneous part of the solution, while $y_{nh}(t)$ is a non-homogeneous (also known as particular) solution:

- $y_h(t)$: For each multiplicity-m root a of the characteristic equation of the differential equation $y_h(t)$ contains the following term(s)

$$\alpha_0 e^{at} + \alpha_1 t e^{at} + \alpha_2 t^2 e^{at} + \ldots + \alpha_{m-1} t^{m-1} e^{at}$$

  where $\alpha_0, \ldots, \alpha_{m-1}$ are constants determined from the homogeneous part of the equation

$$y^{(n)}(t) + a_1 y^{(n-1)}(t) + \ldots + a_{n-1}\dot{y}(t) + a_n y(t) = 0$$

  and the initial conditions.

- $y_{nh}(t)$: This part of the solution is a convolution of the input $u(t)$ with $h(t)$, the impulse response of the system:

$$y_{nh}(t) = \int_0^t u(\tau)h(t-\tau)\,d\tau$$

  The impulse response is most easily determined using the inverse Laplace transform.

Note 1: *If the system is initially at rest, i.e., if all initial conditions are zero, then obviously* $y_h(t) \equiv 0$*, hence* $y(t) = y_{nh}(t)$*. On the other hand, if* $u(t) \equiv 0$*, then* $y(t) = y_h(t)$*. We say that the non-homogeneous part of the solution is a response to the input, while the homogeneous part of the solution is a response to the initial conditions.*

Note 2: *How does this convolution formula reduce to the convolution formula used in Section* 3.1*?*

Problem 3.4.7 Determine the impulse response of a system described by

$$\ddot{y}(t) + 2\dot{y}(t) + 10y(t) = \dot{u}(t) + 3u(t)$$

Solution: Obviously

$$H(s) = \frac{s+3}{s^2+2s+10} = \frac{s+1}{(s+1)^2+3^2} + \frac{2}{3}\frac{3}{(s+1)^2+3^2}$$

hence

$$h(t) = e^{-t}\cos 3t + \frac{2}{3}e^{-t}\sin 3t \quad (t > 0)$$

Matlab note: *To plot this directly from the coefficients of the differential equation do the following (see Figure* 3.4*):* `impulse([1 3],[1 2 10])`

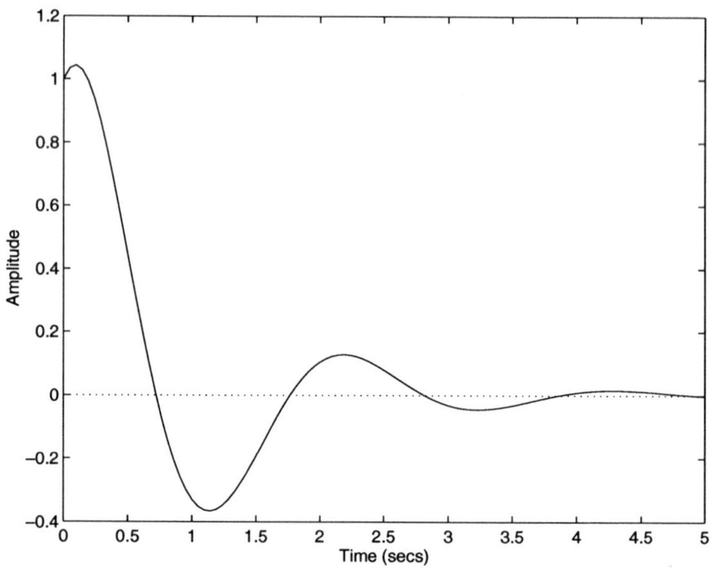

Figure 3.4: The plot produced by the MATLAB command `impulse`.

Problem 3.4.8 Determine the impulse response of a system described by

$$\ddot{y}(t) + 2\dot{y}(t) + y(t) = \dot{u}(t) + 2u(t)$$

Solution: Obviously

$$H(s) = \frac{s+2}{(s+1)^2} = \frac{1}{s+1} + \frac{1}{(s+1)^2}$$

hence

$$h(t) = e^{-t} + te^{-t} \quad (t > 0)$$

Problem 3.4.9 Show that if $u(t) = e^{s_1 t} + e^{s_2 t}$, then the output contains terms $T(s_1)e^{s_1 t}$ and $T(s_2)e^{s_2 t}$.

Solution: To simplify the analysis, assume zero initial conditions. With this input the Laplace transform of the output becomes

$$Y(s) = \frac{1}{s-s_1}\frac{b(s)}{a(s)} + \frac{1}{s-s_2}\frac{b(s)}{a(s)}$$

These two terms contribute $A/(s-s_1)$ and $B/(s-s_2)$ at complex frequencies s_1 and s_2, respectively, where

$$A = \lim_{s\to s_1}((s-s_1)Y(s)) = \frac{b(s_1)}{a(s_1)} = T(s_1)$$

and similarly $B = T(s_2)$.

Problem 3.4.10 Why do we encounter rational transfer functions so often? Why are these transfer functions always such that the degree of the numerator is $<$ than the degree of the denominator?

Solution: Linear time-invariant systems are described by linear differential equations with constant coefficients. If the corresponding equation can be written in terms of the input and output signals and their derivatives, without any of them being delayed, Laplace transform yields a rational transfer function. In general, if there are delays, we find factors e^{-sT}, and the transfer function is not rational.

If $u(t) = \cos\omega t$ and frequency ω tends to infinity, the response of any physical system at that frequency falls to zero. To reflect this general property we require that $m < n$.

Indeed, from Problem 3.4.9 and with $s_1 = j\omega$ and $s_2 = -j\omega$ we find that

$$u(t) = \cos\omega t = \frac{e^{j\omega t} + e^{-j\omega t}}{2}$$

causes the output to be

$$\begin{aligned}
y(t) &= \frac{1}{2}T(j\omega)e^{j\omega t} + \frac{1}{2}T(-j\omega)e^{-j\omega t} + \dots \\
&= \mathrm{Re}\left\{T(j\omega)e^{j\omega t}\right\} + \dots \\
&= |T(j\omega)|\cos(\omega t - \arg T(j\omega)) + \dots
\end{aligned}$$

Hence we require

$$\lim_{\omega \to \infty} |T(j\omega)| = 0$$

Since for large ω we have $|T(j\omega)| \approx \left|\frac{b_0 \omega^m}{\omega^n}\right|$ all physical systems must have $m < n$.

Problem 3.4.11 Use MATLAB to plot the amplitude of the frequency response and the locations of the poles and the zeros of the following 7th order low-pass filters with the cut-off frequency at $\omega_n = 100$ rad/s: Butterworth, Bessel, Chebyshev Type I, and Chebyshev Type II.

Solution: Figure 3.5 is easily obtained using the following MATLAB commands: `butter`, `besself`, `cheby1`, `cheby2`, `freqs`, and `tf2zp`.

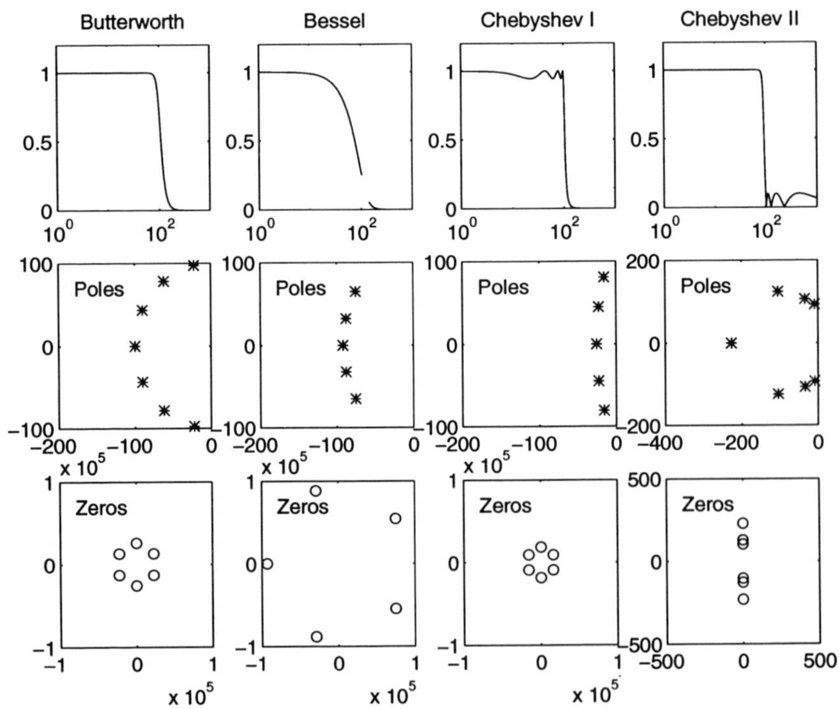

Figure 3.5: Amplitudes of the frequency responses and locations of poles and zeros of order 7 Butterworth, Bessel, Chebyshev type I, and Chebyshev type II low-pass filters with $\omega_n = 100$ rad/s.

Problem 3.4.12 Given a stable system with frequency response $H(j\omega)$ in a feedback connection with gain $-k$ as in Figure 3.6, derive the frequency response of a closed-loop system. Discuss the stability of the closed-loop system if measurements of magnitude and phase of $H(j\omega)$ are given for $0 \leq \omega < \infty$.

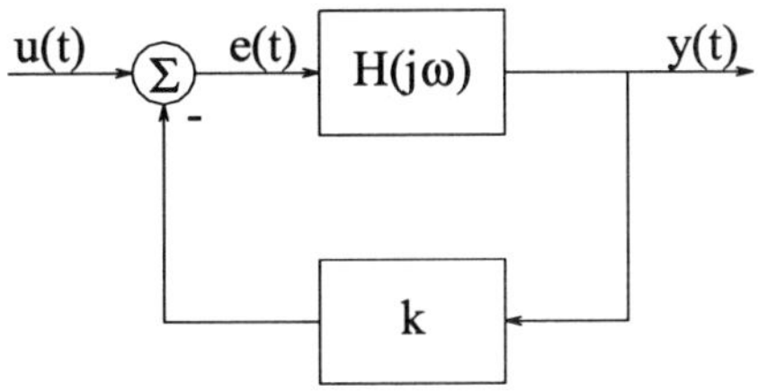

Figure 3.6: Typical feedback system.

Solution: This problem presents the *Nyquist stability criterion,* but also its history and the profound influence it had on the control theory. The following is loosely based on the seminal 1932 paper by H. Nyquist [39] and on the 1977 paper by H. S. Black [5] in which he described his 1927 invention of the negative feedback amplifier.

First, the definition of stability in Nyquist's own words:

> The circuit will be said to be stable when an impressed small disturbance, which itself dies out, results in a response which dies out. It will be said to be unstable when such a disturbance results in a response which goes on indefinitely, either staying at a relatively small value or increasing until it is limited by the non-linearity of the amplifier.

This is the so-called input-output or BIBO (bounded-input bounded-output) stability and we define it in Section 3.6.

Approach 1: Nyquist first considers the output after n "round trips" of the input disturbance:

$$Y_n(j\omega) = -kH(j\omega)Y_{n-1}(j\omega), \qquad Y_0(j\omega) = H(j\omega)U(j\omega)$$

Obviously $Y_n(j\omega) = (-kH(j\omega))^n H(j\omega)U(j\omega)$ and the total output is a sum of all $Y_n(j\omega)$:

$$Y(j\omega) = \sum_{n=0}^{\infty} Y_n(j\omega)$$

This sum exists if and only if $|kH(j\omega)| < 1$ for all ω, when

$$Y(j\omega) = \frac{H(j\omega)}{1 + kH(j\omega)} U(j\omega), \qquad |kH(j\omega)| < 1$$

Nyquist then comments on the limitations of this approach: If $|kH(j\omega)| < 1$ for all ω, then this result agrees with experimental evidence. But it incorrectly suggests that if for some frequency $|kH(j\omega)| > 1$, then there must exist a "runaway condition," i.e., the system must be unstable. For example, Black's negative feedback amplifiers[6] fall into this category, and are not necessarily unstable. In [5] H. S. Black wrote about patenting his invention:

> Although the invention had been submitted to the U.S. Patent Office on August 8, 1928, more than nine years would elapse before the patent was issued on December 21, 1937 (No. 2 102 671). One reason for the delay was that the concept was so contrary to established beliefs that the Patent Office initially did not believe it would work. The Office cited technical papers, for example, that maintained the output could not be connected back to the input unless the loop gain was less than one, whereas mine was between 40 and 50 dB. In England, our patent application was treated in the same manner as one for a perpetual-motion machine.

Thus Black's invention showed that even if for some frequency $|kH(j\omega)| > 1$ the above result may be completely valid. However, we cannot extend this result for arbitrary $kH(j\omega)$, because experiments show instability if for some frequency $kH(j\omega)$ is real and < -1. To complicate the situation further, some closed-loop systems are unstable when $kH(j\omega)$ is real and > 1. This effect too was discovered by Black and is now called *conditional stability.*

Here is what Black wrote about this phenomenon in [5]:

> Results of experiments, however, seemed to indicate something more was involved and these matters were described to Mr. H. Nyquist, who developed a more general criterion for freedom from instability applicable to an amplifier having linear positive constants.

Nyquist concluded about the "round trip" approach:

> Briefly then, the difficulty with this method is that it neglects the building-up processes.

Approach 2: In this approach Nyquist assumes that a steady state exists and writes (his notation is slightly different than our)

$$\frac{Y(j\omega)}{U(j\omega)} = \frac{H(j\omega)E(j\omega)}{E(j\omega) + kH(j\omega)E(j\omega)} = \frac{H(j\omega)}{1 + kH(j\omega)}$$

Thus we know what the closed-loop transfer function is when the steady state exists, but we know nothing about the conditions under which it does exist. Nyquist says:

> The difficulty with this method is that it does not investigate whether or not a steady state exists.

[6] After several years of attempts to reduce the nonlinearities in amplifiers used in long-distance telephony, in 1927 H. S. Black realized that the solution to this problem was not in perfecting the design of vacuum tubes, but in a new concept: negative feedback. Using a very high gain amplifier in a negative feedback configuration he could trade the high gain for a moderate gain with a very flat frequency response:

$$|kH(j\omega)| \gg 1 \quad \Rightarrow \quad \frac{H(j\omega)}{1 + kH(j\omega)} \to \frac{1}{k}$$

Approach 3: In this approach Nyquist develops his famous stability criterion and states it as follows (as we mentioned earlier, Nyquist used a slightly different notation: where he wrote $AJ(i\omega)$ we now write $-kH(j\omega)$):

> *Rule: Plot plus and minus the imaginary part of $AJ(i\omega)$ against the real part for all frequencies from 0 to ∞. If the point $1 + i0$ lies completely outside this curve the system is stable; if not it is unstable.*

In our notation (which follows Bode) this rule is the same, except that we plot $kH(j\omega)$ and examine whether the point $-1+j0$ is inside or outside the resulting curve.

The importance of the Nyquist's criterion lies very much in the fact that $H(j\omega)$ can be obtained by measurements. Hence the closed-loop stability can be verified without solving any equations or even having a mathematical model of the system.

The original proof given by Nyquist was not rigorous although it led to a very important result. Today we use Cauchy's Argument Principle (also called the Encirclement Property) to prove it. The same method is used to prove the more general version of the Nyquist stability criterion, which allows the initial system to be unstable (here we assume that $H(s)$ has neither zeros nor poles on the $j\omega$ axis; a slight modification in the formulation is necessary to include such cases as well):

If the open-loop transfer function $kH(s)$ has M unstable poles then the closed-loop system is stable if and only if the locus of $kH(j\omega)$ for $-\infty < \omega < \infty$ encircles the point $-1 + j0$ exactly M times in the counterclockwise direction.

Note: *If $H(s)$ is unstable, then $H(j\omega)$ is not defined, but is still formally used.*

Problem 3.4.13 A Nyquist (polar) plot of the amplifier frequency response $H(j\omega)$ with the feedback gain $k = 1$ is given in Figure 3.7. Determine the range of k which guarantees the stability of the closed-loop system. Explain why the conditional stability is undesirable.

Solution: For stability, the point $-\frac{1}{k} + j0$ should not be enclosed by the given curve. Since this curve intersects the real axis at approximately -0.5, -1.9, and -4.5 and from its shape we can conclude that the allowable ranges for k are approximately

$$0 < k < 0.22 \quad \text{and} \quad 0.53 < k < 2$$

In the latter range the system is conditionally stable. It is an undesirable property because for large inputs the system may get saturated, which effectively reduces its amplification and destabilizes the system.

Matlab note: *The plot in Figure 3.7 represents simulated measurements of $H(j\omega)$ of the following system*

$$H(s) = 100\frac{(s + 2e^{j2\pi/3})(s + 2e^{-j2\pi/3})}{(s + 0.1)(s + 4)(s + 6)(s + e^{j3\pi/4})(s + e^{-j3\pi/4})}$$

In Figure 3.8 we show its root locus. We see that the ranges similar to those determined above define stable operation of the system. Note again that the Nyquist criterion required neither calculations nor explicit expressions for transfer function or frequency response. The following code was used to obtain Figures 3.7 and 3.8:

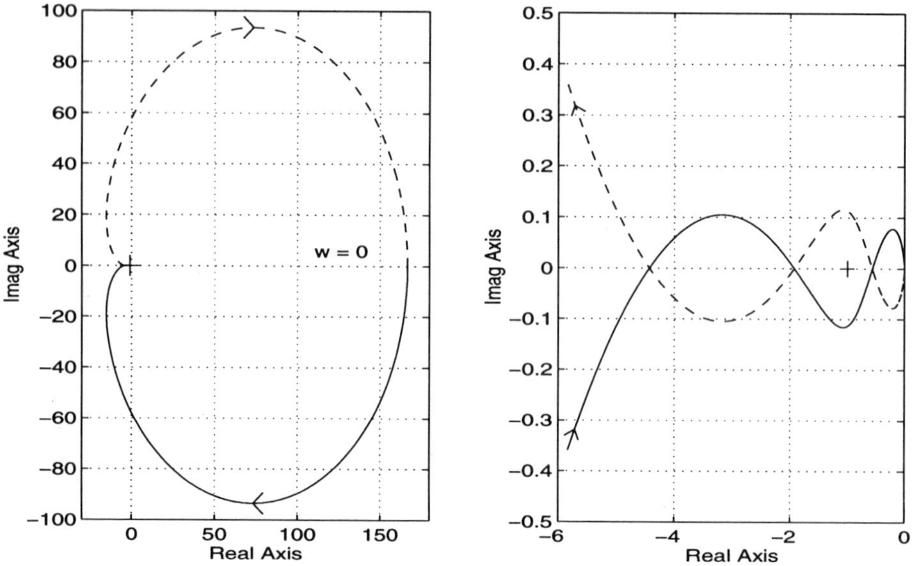

Figure 3.7: A polar plot of the measured open-loop amplifier frequency response $H(j\omega)$. The larger view around the origin is shown to the right.

```
num = 100*[1,2,4];
den = poly([-0.1,-4,-6,sqrt(2)/2*(-1+j),sqrt(2)/2*(-1-j)]);
w = 1.25:0.01:100;

figure(1), subplot(1,2,1)
nyquist(num,den)
grid, text(110,6,'w = 0'), axis([-30,180,-100,100])

subplot(1,2,2)
nyquist(num,den,w)
grid, axis([-6,0,-0.5,0.5])

figure(2), subplot(1,2,1)
rlocus(num,den)
grid

subplot(1,2,2)
rlocus(num,den)
grid, axis([-0.4,0.4,0,5])

[K,poles] = rlocfind(num,den)
s = num2str(K);
gtext(['k = ',s])
[K,poles] = rlocfind(num,den)
s = num2str(K);
gtext(['k = ',s])
[K,poles] = rlocfind(num,den)
s = num2str(K);
gtext(['k = ',s])
```

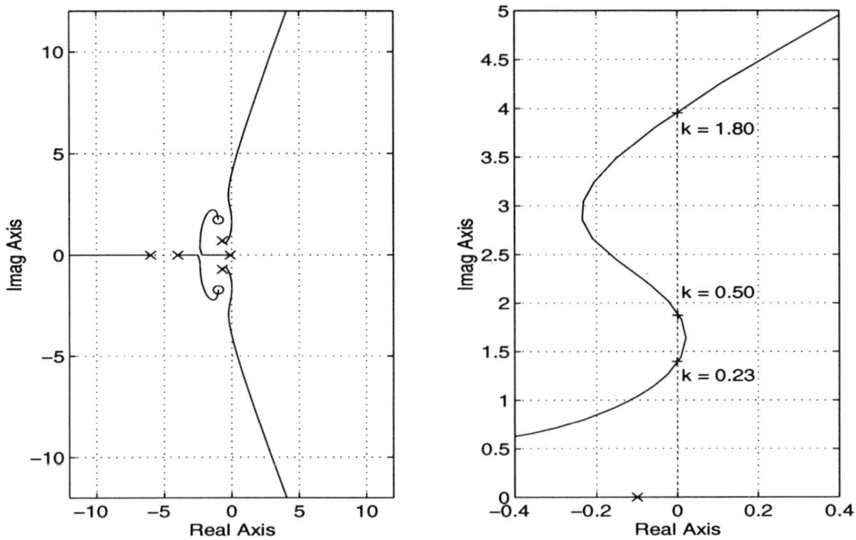

Figure 3.8: A root locus plot for $H(s)$. The larger view around the intersection points with the imaginary axis is shown to the right.

Problem 3.4.14 Given an unstable plant with

$$H(s) = \frac{1}{s-1}$$

try to stabilize it by putting a compensator in series with it. Let

$$H_c(s) = \frac{s-1}{s+1}$$

Explain why this technique is not satisfactory.

Solution: Although from a purely theoretical point of view this seems to be a satisfactory compensation, and it does not make any difference whether the compensator precedes or follows the plant, we shall see that in reality neither of the two possibilities is satisfactory, and that the reasons are completely different for each of them.

If we put the compensator between the input and the plant, ideally, the compensator eliminates the component of the input signal at the unstable complex frequency $s = 1$ because $H_c(1) = 0$. The problem here is not that the cancellation cannot be realistically achieved. Even if we could achieve the pole-zero cancellation, the slightest amount of noise entering directly to the plant will excite the unstable response of the plant.

If we put the compensator after the plant, and the cancellation is perfect, we won't see any unstable response at the output, but since the plant is unstable, its output (i.e., the input to the compensator) will grow larger and larger, and sooner or later some condition will change: either the plant will burn out, or the compensator or the plant will leave the linear mode of operation.

The conclusion to be drawn here is that the external behavior of a system does not completely describe its internal behavior. In the language of the state-space approach, the first attempt was unsuccessful because the unstable state of the plant was not controllable, while in the second attempt it was not observable.

Problem 3.4.15 Since the pole-zero cancellation is not a good approach, in order to stabilize the plant with

$$H(s) = \frac{1}{s-1}$$

try to apply negative output feedback with gain k.

Apply the same technique to stabilize a plant with

$$H(s) = \frac{1}{s(s-2)}$$

Solution: The closed-loop transfer function can be found from the differential equation of the plant $\dot{y} - y = u$ and the equation for the feedback $u = -ky + v$.

The new equation becomes $\dot{y} - (1-k)y = v$, hence the closed-loop transfer function is

$$G(s) = \frac{1}{s+k-1}$$

We see that the stabilization can be achieved by picking $k > 1$.

For a slightly more complicated plant, given by $H(s) = 1/s(s-2)$, this technique doesn't work, because the poles of the closed-loop system, i.e., the roots of $s^2 - 2s + k = 0$ are unstable for any choice of k. Indeed

$$s_{1,2} = 1 \pm \sqrt{1-k}$$

Matlab note: *To plot the root locus use* `rlocus([1],[1,-2,0])`

To stabilize this plant we would also need to feed back $\dot{y}(t)$. In general, to stabilize a plant with a characteristic polynomial of order n, we need $y(t)$, $\dot{y}(t)$, ..., $y^{(n-1)}(t)$, which is not a realistic requirement, because differentiation drastically amplifies noise. We shall see that the state-space approach offers a more elegant solution, which allows arbitrary placement of poles of the closed-loop system.

3.5 State-space representation

In this Section we shall introduce the state-space approach to analysis and design of linear systems. The state-space (linear or nonlinear) model of a system is often the most natural and the easiest description to determine. The importance of the state-space techniques is not only in that they provide solutions, or at least insights, for many problems difficult to solve by the transfer function methods. Under fairly general conditions, linearized state-space models continue to reflect the properties of nonlinear state-space models, which are many orders of magnitude more difficult to handle.

Problem 3.5.1 Assuming that ideal differentiators are available, design an analog computer to solve the following differential equation

$$\ddot{y} + 3\dot{y} + y = u$$

Solution: First rewrite the equation as follows

$$\begin{aligned} y &= -(\ddot{y} + 3\dot{y}) + u \\ &= -\frac{d}{dt}\left(\frac{d}{dt}y + 3y\right) + u \end{aligned}$$

Then we can draw the analog computer as in Figure 3.9.

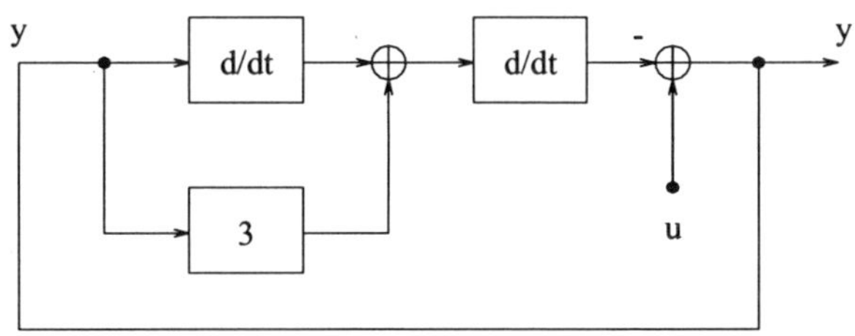

Figure 3.9: The analog computer based on differentiators. This technique is not good, because differentiation amplifies noise.

Problem 3.5.2 Assuming that ideal integrators are available, design an analog computer to solve the following differential equation

$$\ddot{y} + 9\dot{y} + 3y = u$$

Solution: First rewrite the equation as follows

$$\ddot{y} = -9\dot{y} - 3y + u$$

Then we can draw the analog computer as in Figure 3.10.

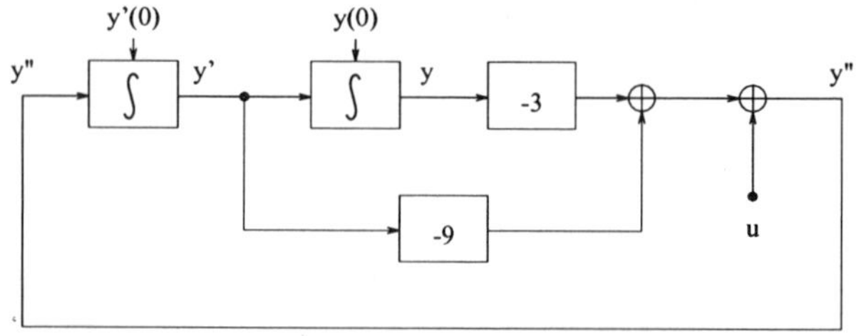

Figure 3.10: The analog computer based on integrators – the Kelvin's scheme.

Note: *This technique is called the Kelvin's scheme, after Lord Kelvin, who proposed it in 1876. First practical implementations of his ideas were made in the 1930's.*

To see how to proceed if the equation also has the derivatives of the input, let us redraw the integrator-based analog computer in a more convenient form, as in Figure 3.11.

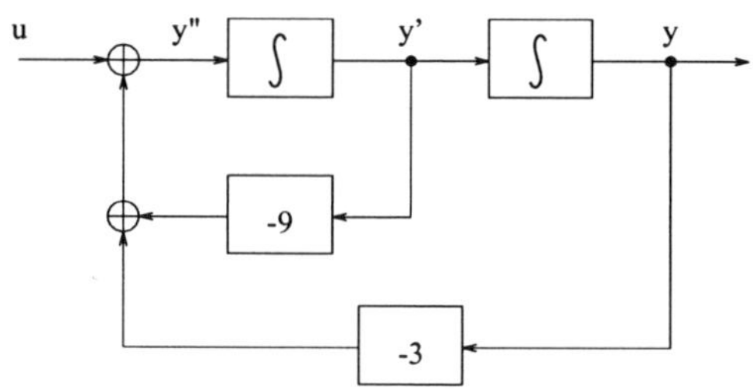

Figure 3.11: The more convenient diagram of the Kelvin's scheme.

If the equation to be solved is

$$\ddot{y} + 9\dot{y} + 3y = 2\dot{u} + 5u$$

we can use linearity of the equation, and first solve the auxiliary equation

$$\ddot{w} + 9\dot{w} + 3w = u$$

whose simulation we already have in Figures 3.10 and 3.11.

Now, from linearity, $y = 2\dot{w} + 5w$, therefore we can design the simulation as in Figure 3.12.

Problem 3.5.3 Given a diagram in Figure 3.12, write a system of two first-order differential equations, whose unknowns are the integrator outputs: $x_1 = \dot{w}$ and $x_2 = w$. Write the output y in terms of states x_1 and x_2.

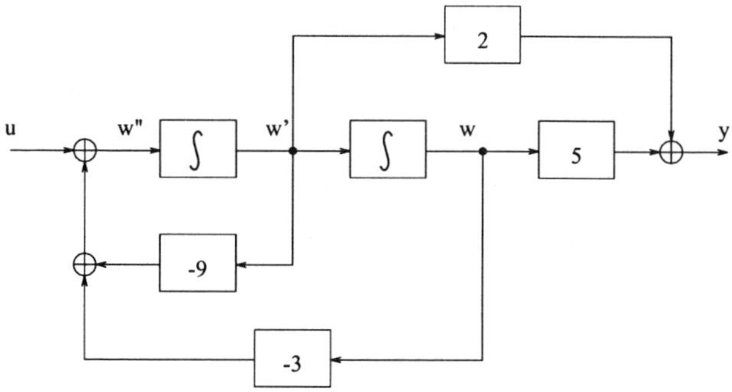

Figure 3.12: The Kelvin's scheme for $\ddot{y} + 9\dot{y} + 3y = 2\dot{u} + 5u$, which was rewritten as $\ddot{w} + 9\dot{w} + 3w = u$ and $y = 2\dot{w} + 5w$.

Solution: It is easy to see that

$$\begin{aligned} \dot{x}_1 &= -9x_1 - 3x_2 + u \\ \dot{x}_2 &= x_1 \end{aligned}$$

and that $y = 2x_1 + 5x_2$.

We usually write this in a matrix form

$$\begin{aligned} \begin{bmatrix} \dot{x}_1 \\ \dot{x}_2 \end{bmatrix} &= \begin{bmatrix} -9 & -3 \\ 1 & 0 \end{bmatrix} \begin{bmatrix} x_1 \\ x_2 \end{bmatrix} + \begin{bmatrix} 1 \\ 0 \end{bmatrix} u \\ y &= \begin{bmatrix} 2 & 5 \end{bmatrix} \begin{bmatrix} x_1 \\ x_2 \end{bmatrix} \end{aligned}$$

Note: *Compare the eigenvalues*[7] *of the differential equation* $\ddot{y}+9\dot{y}+3y = 2\dot{u}+5u$ *and the matrix* $\begin{bmatrix} -9 & -3 \\ 1 & 0 \end{bmatrix}$.

Problem 3.5.4 In the systems theory linear systems are often described using the state-space representation:

$$\begin{aligned} \dot{x}(t) &= Ax(t) + bu(t) \\ y(t) &= c'x(t) + du(t) \end{aligned}$$

where $u(t)$ is the input to the system, $y(t)$ is its output, while $x(t)$ is an $n \times 1$ vector whose components are the states of the system. A is an $n \times n$ matrix, while b and c' are $n \times 1$ and $1 \times n$ vectors, respectively. Matrix A is usually called the *system* matrix,

[7] If W. Heisenberg took a piece of advice from D. Hilbert, and looked for the differential equation with the same eigenvalues as the matrices in his matrix quantum mechanics, he would have discovered the equation now known after E. Schrödinger [49].

while vectors b and c' are called *input* and *output* vectors. For the sake of simplicity, we shall often assume that A has n distinct eigenvalues.

Express $y(t)$ in terms of $u(t)$, A, b, c', d, and the initial conditions $x(0)$. Determine the impulse response in terms of A, b, c', and d.

Solution: From Problem 3.3.2 we know that

$$x(t) = \phi(t)x(0) + \int_0^t \phi(t-\tau)\underbrace{bu(\tau)}_{f(\tau)}\,d\tau$$

where $\phi(t) = e^{At}$, therefore we can immediately write

$$y(t) = c'\phi(t)x(0) + \int_0^t c'\phi(t-\tau)bu(\tau)\,d\tau + du(t)$$

or

$$y(t) = c'\phi(t)x(0) + (c'\phi(t)b) * u(t) + du(t)$$

To determine the impulse response, put $x(0) = 0$ and $u(t) = \delta(t)$. Then

$$h(t) = y(t)|_{[x(0)=0,u(t)=\delta(t)]} = c'\phi(t)b + d\delta(t)$$

Problem 3.5.5 Solve the state-space equations

$$\begin{aligned}\dot{x}(t) &= Ax(t) + bu(t)\\ y(t) &= c'x(t) + du(t)\end{aligned}$$

in the Laplace transform domain. Determine the transfer function in terms of A, b, c', and d.

Solution: From Problem 3.3.3 we know that

$$X(s) = (sI - A)^{-1}(x(0) + bU(s))$$

therefore

$$Y(s) = c'(sI-A)^{-1}x(0) + (c'(sI-A)^{-1}b + d)U(s)$$

Hence, the transfer function is

$$H(s) = \left.\frac{Y(s)}{U(s)}\right|_{x(0)=0} = c'(sI-A)^{-1}b + d$$

Problem 3.5.6 Consider a dynamical system given by

$$\dot{x}(t) = \begin{bmatrix} -1 & -3 \\ 0 & -2 \end{bmatrix} x(t) + \begin{bmatrix} 5 \\ 6 \end{bmatrix} u(t)$$

$$y(t) = \begin{bmatrix} 3 & 4 \end{bmatrix} x(t)$$

with $x(0) = \begin{bmatrix} 4 \\ 5 \end{bmatrix}$ and $u(t) = 1$ $(t > 0)$. Find the eigenvalues of the system. Find the system response to the initial conditions, as well as the response to the input $u(t)$.

Add these two to obtain the total response of the system. Determine the transfer function and the impulse response of the system.

Solution: The Laplace transform of the response to the initial conditions is

$$c'(sI-A)^{-1}x(0) = \begin{bmatrix} 3 & 4 \end{bmatrix} \frac{\begin{bmatrix} s+2 & -3 \\ 0 & s+1 \end{bmatrix}}{(s+1)(s+2)} \begin{bmatrix} 4 \\ 5 \end{bmatrix} = \frac{32s-1}{(s+1)(s+2)}$$

Since $\frac{32s-1}{(s+1)(s+2)} = -\frac{33}{s+1} + \frac{65}{s+2}$, we find $y_{\text{init.cond.}}(t) = -33e^{-t} + 65e^{-2t}$.

Since $U(s) = \mathcal{L}\{1\} = 1/s$, the Laplace transform of the response to the input $u(t) = 1$ $(t > 0)$ is

$$c'(sI-A)^{-1}bU(s) = \frac{39}{(s+1)(s+2)} = \frac{39}{s+1} - \frac{39}{s+2}$$

therefore $y_{\text{input}}(t) = 39e^{-t} - 39e^{-2t}$.

The total response of the system is

$$y(t) = y_{\text{init.cond.}}(t) + y_{\text{input}}(t) = 6e^{-t} + 26e^{-2t} \quad (t > 0)$$

The transfer function of the system is

$$H(s) = c'(sI-A)^{-1}b = \frac{39s}{(s+1)(s+2)} = -\frac{39s}{s+1} + \frac{78}{s+2}$$

hence the impulse response is

$$h(t) = -39e^{-t} + 78e^{-2t}$$

Note: *Since the derivative of* $u(t) = 1$ $(t > 0)$, *is* $\delta(t)$, *we have*

$$h(t) = \frac{d}{dt}\left(39e^{-t} - 39e^{-2t}\right)$$

Even if $u(t)$ *was not zero for* $t < 0$, *we consider it zero, because all its influences on the system at times* $t < 0$ *are condensed in the initial conditions.*

Matlab note: MATLAB *has several very useful functions for simulation of state-space models. See, for example,* `initial`, `impulse`, `step`, *and* `lsim`.

Problem 3.5.7 Check that, in matrix notation too, the impulse response $h(t)$ and the transfer function of a system are a Laplace transform pair.

Solution: Earlier we found that (assume, without loss of generality, $d = 0$)

$$h(t) = c'\phi(t)b \quad \text{and} \quad H(s) = c'(sI-A)^{-1}b$$

and indeed we can write $H(s) = \mathcal{L}\{h(t)\}$.

Note: *Like in the scalar case, $\phi(t) = e^{At}$ and $(sI - A)^{-1}$ are a Laplace transform pair.*

Also note that

$$c'(sI - A)^{-1}b = \frac{c'\text{adj}(sI - A)b}{\det(sI - A)}$$

If the numerator $b(s) = c'\text{adj}(sI - A)b$ and the denominator $a(s) = \det(sI - A)$ are not coprime, then there are some pole-zero cancellations in the transfer function, and some of the eigenvalues of A are hidden, i.e., they do not appear as the poles of $H(s)$.

Problem 3.5.8 For $H(s) = c'(sI - A)^{-1}b$ with no cancellations show that for any v, which is not an eigenvalue of A, there are initial conditions $x(0)$ such that the response to $u(t) = e^{vt}$ is $y(t) = H(v)e^{vt}$. Use the results of Problem 3.2.14.

Solution: Here $Y(s) = H(v)/(s - v) = c'(vI - A)^{-1}b/(s - v)$, $H(s) = c'(sI - A)^{-1}b$, and $U(s) = 1/(s - v)$. Since, in general,

$$Y(s) = c'(sI - A)^{-1}(x(0) + bU(s))$$

we find

$$c'(sI - A)^{-1}(s - v)x(0) = c'((vI - A)^{-1} - (sI - A)^{-1})b$$

Using the identity (cf. Problem 3.2.14)

$$(sI - A)^{-1} - (vI - A)^{-1} = (sI - A)^{-1}(v - s)(vI - A)^{-1}$$

we can write

$$(vI - A)^{-1} - (sI - A)^{-1} = (sI - A)^{-1}(s - v)(vI - A)^{-1}$$

therefore

$$c'(sI - A)^{-1}(s - v)x(0) = c'(sI - A)^{-1}(s - v)(vI - A)^{-1}b$$

so we can pick (this solution is not unique)

$$x(0) = (vI - A)^{-1}b$$

Note: *If v is a zero of $H(v)$, than the initial conditions $x(0) = (vI - A)^{-1}b$ and the input $u(t) = e^{vt}$ cause zero output: $y(t) \equiv 0$.*

Problem 3.5.9 Assume A to have distinct eigenvalues. Let μ be one of them. Find the initial state $x(0)$ such that the response to the zero input (i.e., the response to the initial conditions only) is $e^{\mu t}$. Use the result of Problem 3.2.15.

Solution: Since $U(s) = 0$, and $Y(s) = 1/(s - \mu)$, we can write

$$c'(sI - A)^{-1}x(0) = \frac{1}{s - \mu}$$

Since (cf. Problem 3.2.15)

$$(sI - A)^{-1} = \sum_{i=1}^{n} \frac{R_i}{s - \lambda_i}$$

where $R_i = q_i p_i'$, and q_i and p_i' are the right and left eigenvectors of A, we can write

$$\sum_{i=1}^{n} \frac{c' q_i p_i' x(0)}{s - \lambda_i} = \frac{1}{s - \mu}$$

If we assume $\mu = \lambda_1$, this implies that $p_i' x(0) = \delta_{1i}$. If we recall that $p_i' q_j = \delta_{ij}$, we can pick

$$x(0) = q_1$$

i.e., the initial state should be the right eigenvector of A corresponding to μ.

Problem 3.5.10 The sources that have been switched out of the circuit shown in Figure 3.13 prior to $t = 0^-$ caused the following initial conditions: $i(0^-) = 5\,A$ and $v(0^-) = 10\,V$. An input $u(t) = 10\delta(t)\,V$ is applied.

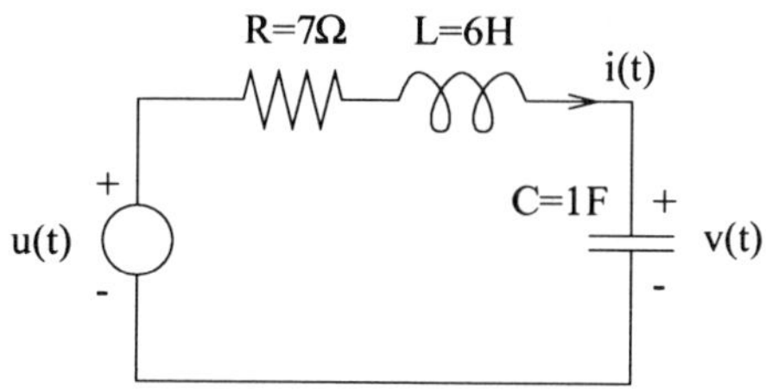

Figure 3.13: The circuit described in Problem 3.5.10.

a) Write the system equation in matrix form. Let the state variables be defined as

$$\begin{aligned} x_1(t) &= v(t) \\ x_2(t) &= i(t) \end{aligned}$$

Note: *Based on the equations*

$$\begin{aligned} v_L(t) &= L\frac{di_L(t)}{dt} \\ i_C(t) &= C\frac{dv_C(t)}{dt} \end{aligned}$$

our best choice for the states in a circuit are the currents through coils and voltages across capacitors.

b) Use Laplace transform to calculate $v(t)$.

c) Find $i(0^+)$ and $v(0^+)$.

Solution: We are given that

$$i(0^-) = 5\,A, \quad v(0^-) = 10\,V, \quad \text{and} \quad u(t) = 10\delta(t)\,V$$

a) From the circuit we can write

$$\begin{aligned} u(t) &= Ri(t) + L\frac{di(t)}{dt} + v(t) \\ i(t) &= C\frac{dv(t)}{dt} \end{aligned}$$

With $x_1(t) = v(t)$ and $x_2(t) = i(t)$, we can rewrite these equations as

$$\begin{aligned} u &= Rx_2 + L\dot{x}_2 + x_1 \\ x_2 &= C\dot{x}_1 \end{aligned}$$

i.e.,

$$\begin{aligned} \dot{x}_1 &= \frac{1}{C}x_2 \\ \dot{x}_2 &= -\frac{1}{L}x_1 - \frac{R}{L}x_2 + \frac{1}{L}u \end{aligned}$$

Therefore

$$\dot{x} = \begin{bmatrix} 0 & 1 \\ -1/6 & -7/6 \end{bmatrix} x + \begin{bmatrix} 0 \\ 1/6 \end{bmatrix} u$$

b) To use the formalism developed earlier, let $y = v = [1\ 0]\ x$. Then by taking the $\mathcal{L}_-$ Laplace transform of

$$\begin{aligned} \dot{x}(t) &= Ax(t) + bu(t) \\ y(t) &= c'x(t) \end{aligned}$$

we obtain

$$\begin{aligned} sX(s) - x(0^-) &= AX(s) + bU(s) \\ Y(s) &= c'X(s) \end{aligned}$$

Since $U(s) = \mathcal{L}_-\{10\delta(t)\} = 10$, while $x(0^-) = [10\ \ 5]'$, we see that

$$Y(s) = c'X(s) = c'(sI - A)^{-1}(bU(s) + x(0^-))$$

Finally,

$$Y(s) = [1\ \ 0]\begin{bmatrix} s & -1 \\ 1/6 & s + 7/6 \end{bmatrix}^{-1}\left(\begin{bmatrix} 0 \\ 1/6 \end{bmatrix} \cdot 10 + \begin{bmatrix} 10 \\ 5 \end{bmatrix}\right)$$

i.e.,

$$Y(s) = \frac{-10}{s+1} + \frac{20}{s+1/6}$$

which implies

$$v(t) = y(t) = (-10e^{-t} + 20e^{-t/6})\,V$$

c)

$$v(0^+) = \lim_{t\to 0+} v(t) = 10\,V$$

On the other hand

$$i(0^+) = \lim_{t\to 0+} x_2(t)$$

Since

$$x_2(t) = C\dot{x}_1(t) = (10e^{-t} - \frac{20}{6}e^{-t/6})\,A$$

obviously $i(0^+) = 6.67\,A$.

Problem 3.5.11 The exact equations which describe the dynamics of the inverted pendulum on a cart are

$$\begin{aligned} (M+m)\ddot{z} + ml\ddot{\theta}\cos\theta - ml\dot{\theta}^2\sin\theta &= f \\ m\ddot{z}\cos\theta + ml\ddot{\theta} - mg\sin\theta &= 0 \end{aligned}$$

where M and m are the masses of the cart and the bob, l is the length of the pendulum rod, z and θ are the horizontal displacement of the cart and the angle between the vertical and the pendulum rod (expressed in radians), while f is the force applied to the cart (see Figure 3.14).

Linearize these equations and write the state-space representation of the system.

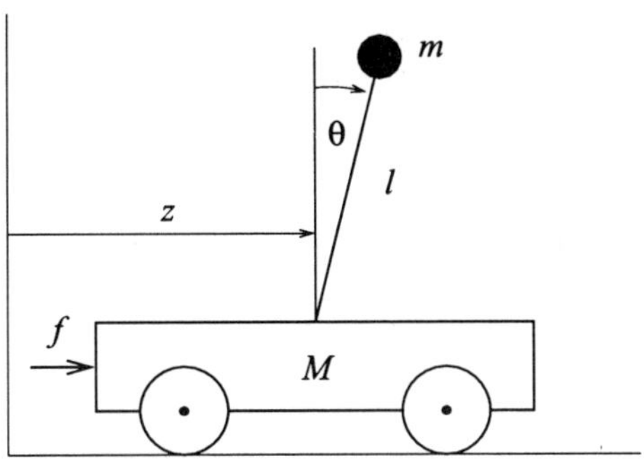

Figure 3.14: The inverted pendulum on a cart.

Solution: The above equations are nonlinear, due to the presence of trigonometric functions. In several problems scattered throughout this chapter, we shall try to stabilize the pendulum, i.e., to keep θ small, thus we can write

$$\cos\theta \approx 1 \qquad \text{and} \qquad \sin\theta \approx \theta$$

After linearization, we can write the equations in a state-space form, by first defining the vector of states:

$$x = \begin{bmatrix} z \\ \theta \\ \dot{z} \\ \dot{\theta} \end{bmatrix}$$

Now we can see that

$$\dot{x} = Ax + bu$$

where

$$A = \begin{bmatrix} 0 & 0 & 1 & 0 \\ 0 & 0 & 0 & 1 \\ 0 & -\frac{mg}{M} & 0 & 0 \\ 0 & \frac{(M+m)g}{Ml} & 0 & 0 \end{bmatrix} \qquad \text{and} \qquad b = \begin{bmatrix} 0 \\ 0 \\ 1/M \\ -1/Ml \end{bmatrix}$$

while $u = f$, the external force applied to control the cart and the inverted pendulum.

In the following problems we shall assume that the measured variables are z and θ, i.e., that

$$y = Cx$$

where

$$C = \begin{bmatrix} 1 & 0 & 0 & 0 \\ 0 & 1 & 0 & 0 \end{bmatrix}$$

Problem 3.5.12 Consider a system described by

$$\begin{aligned} \dot{x}(t) &= Ax(t) + bu(t) \\ y(t) &= c'x(t) \end{aligned}$$

where A is an $n \times n$ matrix with distinct eigenvalues, and $u(t)$ is the input to the system. The Laplace transform of the output is then

$$Y(s) = c'(sI - A)^{-1}bU(s)$$

Prove that the transfer function of the above system can be written as

$$H(s) = \frac{Y(s)}{U(s)} = \sum_{i=1}^{n} \frac{(c'q_i)(p_i'b)}{s - \lambda_i}$$

Solution: The formula for the transfer function of a system with distinct eigenvalues is a direct consequence of the result of Problem 3.2.15. This representation of the transfer function is very important, because it provides us with its rational decomposition in the transform domain, thus making the application of the inverse Laplace transform easy. Really, since

$$\mathcal{L}^{-1}\left\{\frac{1}{s-\lambda_i}\right\} = e^{\lambda_i t}$$

for the impulse response of a system with distinct eigenvalues we can write

$$h(t) = \sum_{i=1}^{n} \alpha_i e^{\lambda_i t}$$

where $\alpha_i = (c'q_i)(p_i'b)$ $(i = 1, 2, \ldots, n)$.

The general formula, for systems with multiple eigenvalues, is more complicated. In general, matrix A is not diagonalizable, hence, $\sum \lambda_i R_i \neq A$ (cf. Problem 3.2.15). Then one has to use Jordan matrices instead of diagonal matrices, when the impulse response is a linear combination of exponential functions multiplied by polynomials:

$$h(t) = \sum_{i=1}^{n} \alpha_i(t) e^{\lambda_i t}$$

The degree of each $\alpha_i(t)$ is equal to the number of generalized eigenvectors corresponding to λ_i, i.e.,

$$\deg(\alpha_i(t)) \;=\; \nu(\lambda_i I - A) - 1 \;=\; n - \rho(\lambda_i I - A) - 1$$

Problem 3.5.13 Use both the formula from Problem 3.5.12 and the Laplace transform to determine the impulse response of the system given by

$$\dot{x}(t) = \begin{bmatrix} -1 & -1 & -1 \\ 0 & -2 & -1 \\ 0 & 0 & -3 \end{bmatrix} x(t) + \begin{bmatrix} 1 \\ 0 \\ 1 \end{bmatrix} u(t)$$

$$y(t) = [1 \;\; 0 \;\; 0]\; x(t)$$

Solution: The transfer function of the system is

$$\begin{aligned}
H(s) &= c'(sI - A)^{-1} b \\
&= [1 \;\; 0 \;\; 0] \begin{bmatrix} s+1 & 1 & 1 \\ 0 & s+2 & -1 \\ 0 & 0 & s+3 \end{bmatrix}^{-1} \begin{bmatrix} 1 \\ 0 \\ 1 \end{bmatrix} \\
&= [1 \;\; 0 \;\; 0] \frac{\begin{bmatrix} (s+2)(s+3) & -(s+3) & -(s+1) \\ 0 & (s+1)(s+3) & -(s+1) \\ 0 & 0 & (s+1)(s+2) \end{bmatrix}}{(s+1)(s+2)(s+3)} \begin{bmatrix} 1 \\ 0 \\ 1 \end{bmatrix} \\
&= \frac{s^2 + 4s + 5}{(s+1)(s+2)(s+3)} \\
&= \frac{1}{s+1} - \frac{1}{s+2} + \frac{1}{s+3}
\end{aligned}$$

Since the eigenvalues of A are distinct, we can obtain the same result by applying the formula from Problem 3.5.12. First, we need to find the matrix of right eigenvectors and its inverse (the matrix of its left eigenvectors):

$$Q = \begin{bmatrix} 1 & 1 & 1 \\ 0 & 1 & 1 \\ 0 & 0 & 1 \end{bmatrix} \quad \text{and} \quad Q^{-1} = \begin{bmatrix} 1 & -1 & 0 \\ 0 & 1 & -1 \\ 0 & 0 & 1 \end{bmatrix}$$

Hence

$$\begin{aligned}
\alpha_1 &= \left([1 \ 0 \ 0] \begin{bmatrix} 1 \\ 0 \\ 0 \end{bmatrix} \right) \left([1 \ -1 \ 0] \begin{bmatrix} 1 \\ 0 \\ 1 \end{bmatrix} \right) = 1 \\
\alpha_2 &= \left([1 \ 0 \ 0] \begin{bmatrix} 1 \\ 1 \\ 0 \end{bmatrix} \right) \left([1 \ -1 \ 0] \begin{bmatrix} 1 \\ 0 \\ 1 \end{bmatrix} \right) = -1 \\
\alpha_3 &= \left([0 \ 1 \ -1] \begin{bmatrix} 1 \\ 1 \\ 1 \end{bmatrix} \right) \left([0 \ 0 \ 1] \begin{bmatrix} 1 \\ 0 \\ 1 \end{bmatrix} \right) = 1
\end{aligned}$$

Problem 3.5.14 Signal flow graphs are a useful tool in the system analysis. In the signal flow graph, each node presents a signal and is also a summing junction. The net signal at a node is the sum of all the branches coming into the node. The transmission gains are represented by a directed arrow and the gain on the branch.

The Mason's gain formula allows us to find the transfer function of a system directly from its signal flow graph:

$$H(s) = \frac{1}{\Delta(s)} \sum_{i=1}^{n} P_i(s)\Delta_i(s)$$

where

n	...	is the number of direct paths between input and output nodes
$P_i(s)$	...	is the gain of the i-th direct path
$\Delta(s)$	=	$1 - \sum_i L_i(s) + \sum_{i<j} L_i(s)L_j(s) - \sum_{i<j<k} L_i(s)L_j(s)L_k(s) + \ldots$ where each of the above summands is a product of gains of $1, 2, 3, \ldots$ non-intersecting loops, respectively
$\Delta_i(s)$	...	is defined as $\Delta(s)$, but on the graph without the i-th direct path.

Use the Mason's formula to find the transfer function of the system given by the signal flow graph in Figure 3.15.

Note: *Mason's gain formula was first derived as a method for solving systems of linear equations in the early 1950's. Its derivation can be found in* [11] *and* [65].

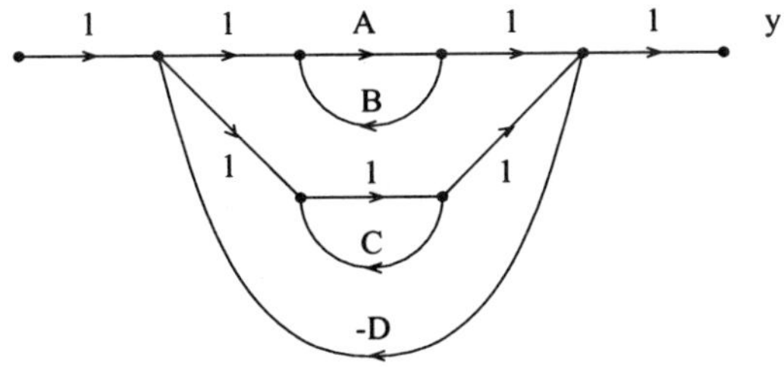

Figure 3.15: The signal flow graph for the illustration of the Mason's formula.

Solution: In this problem

$n = 2,$

$P_1 = A,\ P_2 = 1,$

$L_1 = AB,\ L_2 = C,\ L_3 = -AD,\ L_4 = -D,$

$\Delta = 1 - AB - C + AD + D + ABC - ABD - ACD,$

$\Delta_1 = 1 - C,\ \Delta_2 = 1 - AB,$

therefore

$$H(s) = \frac{A(1-C) + 1 - AB}{1 - AB - C + AD + D + ABC - ABD - ACD}$$

Problem 3.5.15 Use MATLAB to plot the phase-plane plot for the van der Pol oscillator given by the following nonlinear state-space equations:

$$\begin{aligned}\dot{x}_1 &= x_2\\ \dot{x}_2 &= -a(x_1^2 - 1)x_2 - x_1\end{aligned}$$

Solution: Put the following code into file called **vndrpol2.m**:

```
% file vndrpol2.m - simulation of the van der Pol oscillator
%
t0 = 0;
tf = 10;
for i=1:30;
  x0 = 10*(rand(2,1) - 0.5);
  [t,x] = ode45('vndrpol3',[t0,tf],x0);      % ode45 numerically solves
  hold on                                   % ordinary diff.equations (ode)
  plot(x0(1),x0(2),'b*')
  plot(x(:,1),x(:,2),'b-')
  xlabel('x1')
  ylabel('x2')
  hold off
end
title('Phase-plane plot for the van der Pol oscillator (a=0.75)')
```

and the following code into **vndrpol3.m**:

```
% file vndrpol3.m - equation for van der Pol oscillator
%
function xdot = vndrpol1(t,x)
a = 0.75;
xdot = [x(2); -a*(x(1)^2 - 1)*x(2) - x(1)];
end
```

Then run **vndrpol.m** from MATLAB to obtain the plot as in Figure 3.16. It shows the system trajectories for 30 randomly chosen initial points. The parameter a is taken to be $a = 0.75$.

Note: *For the van der Pol oscillator, no matter what the initial conditions, all trajectories converge to a curve called the* limit cycle.

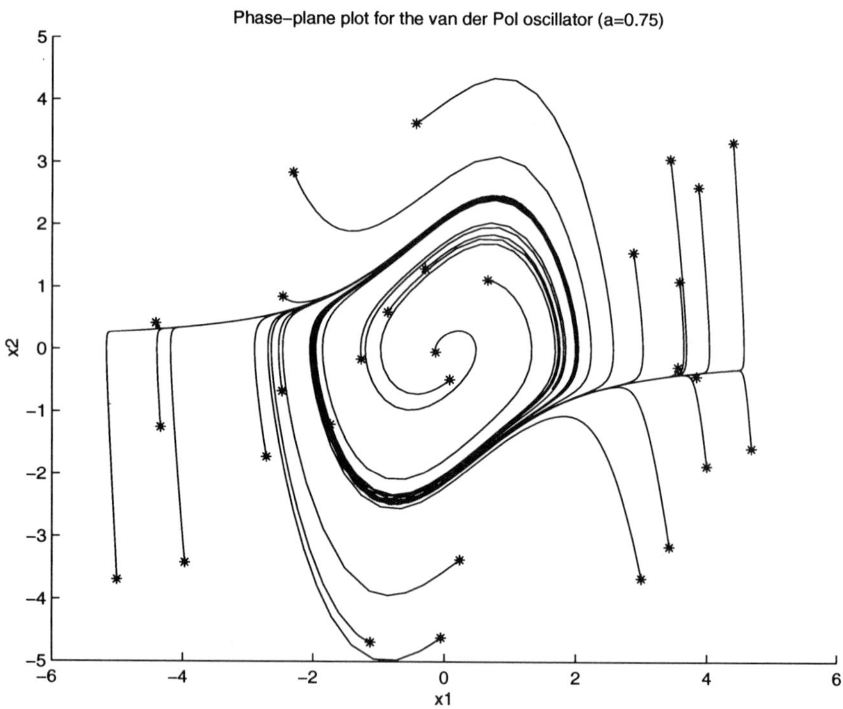

Figure 3.16: The MATLAB plot of the phase-plane for the van der Pol oscillator. In this case $a = 0.75$ was used over the time interval from 0 to $t_f = 10$s.

If $a = 0.2$ is used instead, the graph in Figure 3.17 is obtained. The convergence to the limit cycle is slower, and for some initial conditions 10s was not long enough for the corresponding trajectories to get close to it.

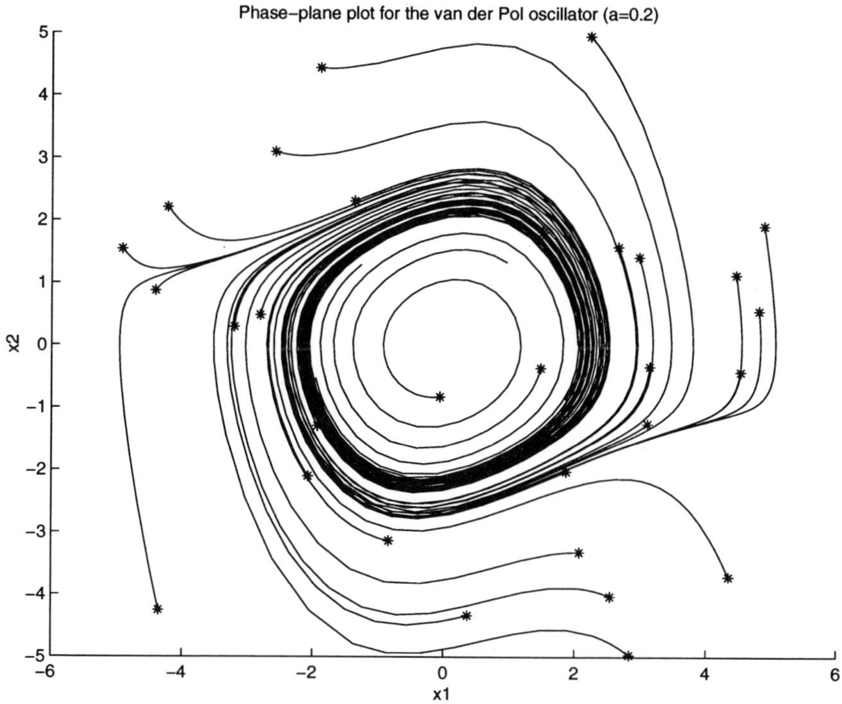

Figure 3.17: The MATLAB plot of the phase-plane for the van der Pol oscillator with $a = 0.2$.

3.6 Stability

This Section defines various types of stability and investigates their relations. It describes the stability in the sense of Lyapunov and the second method of Lyapunov, which are useful not only for linear systems, but also for the stability analysis of nonlinear systems.

Problem 3.6.1 Define BIBO (bounded-input bounded-output) stability. Give a necessary and sufficient condition for BIBO stability in terms of $h(t)$, the impulse response of the system.

Solution: A system is said to be BIBO stable if its output to any bounded input is also bounded. Since for any t, the output $y(t)$ for any input $u(t)$, $t \geq 0$, and zero initial conditions

$$y(t) = \int_0^\infty h(\tau)u(t-\tau)\,d\tau$$

the condition for BIBO stability can be written in terms of the impulse response $h(t)$:

$$\int_0^\infty |h(t)|\,dt < \infty$$

Indeed, if $h(t)$ is absolutely integrable, and $C = \max(|u(t)|)$ then

$$|y(t)| = \left|\int_0^\infty h(\tau)u(t-\tau)\,d\tau\right| \leq \int_0^\infty |h(\tau)||u(t-\tau)|\,d\tau$$

Hence

$$|y(t)| \leq C\int_0^\infty |h(\tau)|\,d\tau < \infty$$

To show that absolute integrability of $h(t)$ is also a necessary condition, suppose $h(t)$ is not absolutely integrable. Then for a particular bounded input

$$u(t) = \operatorname{sgn}(h(T-t))$$

where T is some time instance, we find

$$y(T) = \int_0^\infty h(\tau)u(T-\tau)\,d\tau = \int_0^\infty |h(\tau)|\,d\tau$$

which does not exist. Hence, if for any bounded input the system has a bounded output, the impulse response must be absolutely integrable.

Thus, a system is BIBO stable if and only if its impulse response is absolutely integrable.

Problem 3.6.2 Characterize the BIBO stability in terms of the poles of the system transfer function $H(s)$. Assume that $H(s)$ is a rational function.

Solution: If the poles $p_1, p_2, \ldots, p_n$ of $H(s)$ are distinct, then $h(t)$ is given by (cf. Problem 3.5.12)

$$h(t) = \sum_{i=1}^{n} \alpha_i e^{p_i t}$$

If $H(s)$ has repeated poles, then (again cf. Problem 3.5.12) the α_i are polynomials in t:

$$h(t) = \sum_{i=1}^{n} \alpha_i(t) e^{p_i t}$$

In both cases the condition for BIBO stability is given by the following requirement:

$$\text{Re}\{p_i\} < 0 \quad (i = 1, 2, \ldots, n)$$

Note: *A system is said to be* marginally stable *if its impulse response is bounded. If the system has distinct poles, it is marginally stable if* $Re\{p_i\} \leq 0$ $(i = 1, 2, \ldots, n)$. *If it has repeated poles, then the repeated poles must have their real parts strictly less than* 0, *while the distinct poles may lie on the imaginary axis.*

Problem 3.6.3 Any quadratic form $Q(x)$ can be expressed as

$$Q(x) = x'Px$$

where P is a symmetric matrix.

Determine P for

$$Q(x) = x_1^2 + x_2^2 + x_3^2 + 2x_1x_2 + 6x_1x_3 + 4x_2x_3$$

and determine if $Q(x)$ (and therefore P) is positive definite, positive semi-definite (nonnegative definite) or indefinite. (For more on quadratic forms and definiteness, we refer the reader to Appendix C.)

Solution: Since for

$$P = \begin{bmatrix} a & d & e \\ d & b & f \\ e & f & c \end{bmatrix}$$

the quadratic form is $Q(x) = x'Px = ax_1^2 + bx_2^2 + cx_3^2 + 2dx_1x_2 + 2ex_1x_3 + 2fx_2x_3$, in our case we easily find

$$P = \begin{bmatrix} 1 & 1 & 3 \\ 1 & 1 & 2 \\ 3 & 2 & 1 \end{bmatrix}$$

Since for $x_1 = x_2 = x_3 = 1$ we have $Q(x) = 15 > 0$, while for $x_1 = x_2 = 1$, $x_3 = -1$ we have $Q(x) = -5 < 0$, the quadratic form $Q(x)$ is indefinite. Therefore, the matrix P is also said to be indefinite.

Note: *Of course, the Rayleigh-Ritz theorem leads to the same conclusion, because* $\lambda_{min}(P) = -2.2012 < 0$, *while* $\lambda_{max}(P) = 5.1131 > 0$, *and so does the Sylvester's criterion:*

$$|1| = 1, \qquad \begin{vmatrix} 1 & 1 \\ 1 & 1 \end{vmatrix} = 0, \qquad \begin{vmatrix} 1 & 1 & 3 \\ 1 & 1 & 2 \\ 3 & 2 & 1 \end{vmatrix} = -1$$

Problem 3.6.4 Define the asymptotic Lyapunov stability for linear time-invariant systems.

Solution: A linear time-invariant system is asymptotically stable in the sense of Lyapunov if when there is no input, its states tend to 0 as $t \to \infty$, for arbitrary initial conditions $x(0)$.

Note: *The non-repeated purely imaginary eigenvalues of a system are consistent with Lyapunov stability, but not with the asymptotic Lyapunov stability. The stability defined here is the so-called stability-in-the-large, or the global stability. For nonlinear systems it is necessary to distinguish between the local and global stability, because more than one equilibrium point may exist.*

Problem 3.6.5 When do the asymptotic Lyapunov stability and the BIBO stability of a linear control system with a rational transfer function coincide.

Solution: If a linear system with rational transfer function is internally asymptotically stable, its poles are to the left from the imaginary axis of the s-plane. Therefore $y(t) = c'x(t)$ is bounded for any bounded input, hence the system is BIBO stable. But as we saw in Problem 3.4.14, if the system is not controllable or not observable, BIBO stability does not imply the internal stability, much less the asymptotic stability. It can be shown that if there are no cancellations between $c'\text{adj}(sI - A)b$ and $\det(sI - A)$, i.e., if the system is both controllable and observable, these two types of stability do coincide.

Note: *The systems which are both controllable and observable are called minimal.*

Problem 3.6.6 Describe the "second method" of Lyapunov.

Solution: Instead of solving the system equations (the "first method" of Lyapunov), which can be quite a difficulty (and for some nonlinear systems even an impossible task), we can investigate behavior of the system's energy. For many systems energy cannot be defined in the standard sense, but any positive definite function $V(x)$ of states, such that $V(0) = 0$, can be used.

If for some positive definite function $V(x)$, such that $V(0) = 0$, we find that it decreases as time goes on, then we can say that the system is asymptotically stable in the sense of Lyapunov. However, if at some times it decreases and at other times it increases, we cannot conclude anything, only try another $V(x)$. Indeed, finding the appropriate generalized energy (also called the Lyapunov function), which is positive definite and has negative definite time derivative, is often quite an art.

Problem 3.6.7 The Lyapunov stability theory applied to linear systems reduces to investigation of candidate Lyapunov functions which are quadratic (or more generally, Hermitian) forms.

Let A be a system matrix of a system, and consider a function of the state vector

$$V(x) = x'Px, \qquad \text{where } P \text{ is symmetric and positive definite}$$

The time derivative of $V(x)$ is

$$\dot{V}(x) = \dot{x}'Px + x'P\dot{x} = x'A'Px + x'PAx = x'(A'P + PA)x$$

If matrix Q defined by

$$Q = -(A'P + PA)$$

is positive definite, then the time derivative of $V(x)$ becomes zero only at the origin, therefore, the system is asymptotically stable. If Q turns out to be positive semi-definite, and $\dot{V}(x) \not\equiv 0$ along any possible system trajectory, the system is asymptotically stable. If Q is positive semi-definite, and $\dot{V}(x) \equiv 0$ along some system trajectories, the system is stable (we cannot say it is asymptotically stable, but some other choice of P may show that). If Q is indefinite, we have to try another P. If Q is negative definite, the system is unstable, and if Q is negative semi-definite, again we have to examine $\dot{V}(x)$ along the system trajectories.

Since making a good choice of P is not trivial, for simple systems we usually begin with any[8] symmetric and positive definite matrix Q, and look for the corresponding P. A is a Hurwitz matrix if and only if the solution of the Lyapunov equation

$$A'P + PA = -Q$$

is symmetric and positive definite.

However, sometimes we can come up with a natural choice of P (i.e., $V(x)$), and we can avoid solving the Lyapunov equation. For an interesting application of Lyapunov theory, see Problem 3.10.2.

Use Lyapunov equation to determine whether the matrix A given by

$$A = \begin{bmatrix} 3 & -2 \\ 1 & -1 \end{bmatrix}$$

is Hurwitz. Let Q be an identity matrix. Verify your results by checking the eigenvalues of A.

Solution: Let

$$P = \begin{bmatrix} p_1 & p_2 \\ p_2 & p_3 \end{bmatrix}$$

[8]Indeed, we can begin with *any* positive definite symmetric matrix Q. This is based on the theorem due to Lyapunov: ***A matrix A is Hurwitz if and only if for any given positive definite symmetric matrix Q there exists a positive definite symmetric matrix P such that*** $A'P+PA = -Q$. Let us mention that if A is Hurwitz, then for any positive definite symmetric Q the solution of the Lyapunov equation P is unique and

$$P = \int_0^\infty e^{A't} Q e^{At} \, dt$$

then

$$\begin{bmatrix} 3 & 1 \\ -2 & -1 \end{bmatrix} \begin{bmatrix} p_1 & p_2 \\ p_2 & p_3 \end{bmatrix} + \begin{bmatrix} p_1 & p_2 \\ p_2 & p_3 \end{bmatrix} \begin{bmatrix} 3 & -2 \\ 1 & -1 \end{bmatrix} = \begin{bmatrix} -1 & 0 \\ 0 & -1 \end{bmatrix}$$

therefore

$$P = \begin{bmatrix} 1/4 & -5/4 \\ -5/4 & 3 \end{bmatrix}$$

Using the Sylvester's criterion

$$\frac{1}{4} > 0, \qquad \frac{1}{4} \cdot 3 - \frac{5}{4} \cdot \frac{5}{4} = -\frac{13}{16} < 0$$

we see that P is not positive definite, hence A is not Hurwitz.

Indeed, the eigenvalues of A are $1 - \sqrt{2} < 0$ and $1 + \sqrt{2} > 0$.

Matlab note: MATLAB *command* `lyap` *can be used to solve for* P*:*

```
A = [3 -2;1 -1];
Q = eye(2);        % identity matrix 2x2
P = lyap(A',Q);  % A' because Matlab solves equation A*P + P*A' = -Q
```

Problem 3.6.8 For a system given by

$$\dot{x} = \begin{bmatrix} 0 & 1 \\ -6 & -5 \end{bmatrix} x$$

use the following functions to investigate stability:

$$\begin{aligned} V_1(x) &= 67x_1^2 + 10x_1x_2 + 7x_2^2 \\ V_2(x) &= 6x_1^2 + x_2^2 \\ V_3(x) &= x_1^2 + x_2^2 - x_1x_2 \end{aligned}$$

Solution: $V_1(x)$ is a good choice, because it is positive definite (to see that note $V_1(x) = (2.5x_1 + 2x_2)^2 + 60.75x_1^2 + 3x_2^2)$, and its time derivative is (we use $\dot{x}_1 = x_2$ and $\dot{x}_2 = -6x_1 - 5x_2$)

$$\dot{V}_1(x) = 134x_1\dot{x}_1 + 10x_1\dot{x}_2 + 10\dot{x}_1x_2 + 14x_2\dot{x}_2 = \ldots = -60(x_1^2 + x_2^2)$$

a negative definite function. Therefore, our system is asymptotically stable.

$V_2(x)$ is also a good choice, because it is positive definite, and its time derivative is negative semi-definite:

$$\dot{V}_2(x) = 12x_1\dot{x}_1 + 2x_2\dot{x}_2 = 12x_1x_2 - 2x_2(6x_1 + 5x_2) = -10x_2^2$$

Since $\dot{V}_2(x)$ is negative semi-definite, we need to examine $\dot{V}_2(x)$ along the trajectories of the system. $\dot{V}_2(x) \equiv 0$ only when $x_2 \equiv 0$, which (through the system equations) implies $x_1 \equiv 0$. Therefore, the generalized energy $V_2(x)$ of the system is decreasing along any trajectory of the system. Again we see that the system is asymptotically stable.

$V_3(x)$ is not a good choice, because, although it is positive definite, its time derivative is indefinite, $\dot{V}_3(x) = 6x_1^2 - 5x_1x_2 - 11x_2^2$. Hence, $V_3(x)$ does not tell us a thing about the stability of the system.

Problem 3.6.9 The dynamic equations of a simple inverted pendulum are given by

$$\dot{x} = \begin{bmatrix} 0 & 1 \\ 1 & 0 \end{bmatrix} x + \begin{bmatrix} 0 \\ -1 \end{bmatrix} u$$

$$y = \begin{bmatrix} 1 & 0 \end{bmatrix} x$$

Use the Lyapunov equation

$$A'P + P'A = -Q$$

to determine whether the matrix A for the inverted pendulum is Hurwitz (i.e., if the real parts of its eigenvalues are all negative). Let Q be an identity matrix.

Solution: The system is obviously unstable (if the description of the system as the *inverted* pendulum is not enough to convince you, look at the eigenvalues of the system), but we are asked to employ the Lyapunov function to show that A is not Hurwitz.

The solution of the Lyapunov equation

$$A'P + PA = -Q \qquad \text{(in this case } Q = I)$$

is not unique. Indeed, if

$$P = \begin{bmatrix} p_1 & p_2 \\ p_2 & p_3 \end{bmatrix}$$

we find that $p_2 = -1/2$, $p_3 = -p_1$, while p_1 is arbitrary.

Since P is not unique, A is not Hurwitz.

3.7 Controllability and observability

Illustrative examples in this Section introduce the notions of state controllability and state observability. In later sections we shall see that these two properties of dynamic systems are encountered as conditions for pole placement and state observation. Several powerful tests for controllability and observability are derived.

Problem 3.7.1 Consider a single-input single-output system given by

$$\begin{aligned} \dot{x} &= Ax + bu \\ y &= c'x + du \end{aligned}$$

where x is $n \times 1$, u, y, and d are scalars, while A is $n \times n$, b is $n \times 1$, and c' is $1 \times n$. Show that this system is state controllable if and only if

$$\rho(\mathcal{C}) = n$$

where $\mathcal{C}$ is the controllability matrix of the system

$$\mathcal{C} = [b \;\; Ab \;\; A^2b \;\; \ldots \;\; A^{n-1}b]$$

Solution: By definition, a system is state controllable if, by applying a proper input $u(t)$, we can change its state from any given state to any other given state in a finite amount of time. We know that

$$x(t) = e^{At}x(0) + \int_0^t e^{A(t-\tau)}bu(\tau)\,d\tau$$

Since e^{At} is always nonsingular, without any loss of generality we can consider a case when we wish to take vector x from its arbitrary initial state $x(0)$ to the origin when $t = t_f$. Thus we write

$$0 = e^{At_f}x(0) + \int_0^{t_f} e^{A(t_f-\tau)}bu(\tau)\,d\tau$$

i.e.,

$$x(0) = -\int_0^{t_f} e^{-A\tau}bu(\tau)\,d\tau$$

Using the Cayley-Hamilton theorem, we can write $e^{-A\tau}$ as a finite sum

$$e^{-A\tau} = \sum_{i=0}^{n-1} \alpha_i(\tau)A^i$$

Therefore

$$x(0) = -\sum_{i=0}^{n-1} A^i b\beta_i \tag{3.3}$$

where

$$\beta_i = \int_0^{t_f} \alpha_i(\tau)u(\tau)\,d\tau \qquad (i = 0, 1, \ldots, n-1)$$

Note: *The input $u(t)$ can be determined from the above set of equations because functions $\alpha_i(t)$ are linearly independent on segment $[0, t_f]$. This is so because of the way we determined them (cf. Problem 3.2.7)*

$$V_{k_1,\ldots,k_r}(\lambda_1,\ldots,\lambda_r)\,\alpha = \begin{bmatrix} e^{\lambda_1 t} & te^{\lambda_1 t} & \ldots & t^{k_1-1}e^{\lambda_1 t} & \ldots & e^{\lambda_r t} & te^{\lambda_r t} & \ldots & t^{k_r-1}e^{\lambda_r t}\end{bmatrix}'$$

Since the functions in the array on the right-hand side are linearly independent and the generalized Vandermonde matrix on the left-hand side is nonsingular, $\alpha_i(t)$ are also linearly independent. See also Problems 3.7.15 and 3.8.9.

Equation (3.3) can be viewed as a system of n equations in n unknowns β_i $(i = 0, 1, \ldots, n-1)$, which has a solution for any arbitrary $x(0)$ if and only if

$$\rho(\mathcal{C}) = n$$

where

$$\mathcal{C} = [b \;\; Ab \;\; A^2b \;\; \ldots \;\; A^{n-1}b]$$

Note: *In the case of single-input systems the controllability matrix $\mathcal{C}$ is $n \times n$, therefore we could write the above condition as $\det(\mathcal{C}) \neq 0$. The reason we didn't is that the validity of the above condition can be extended to the systems with m-dimensional inputs, when $\mathcal{C}$ is $n \times mn$.*

Thus, in general, when B is $n \times m$, the system

$$\begin{aligned} \dot{x} &= Ax + Bu \\ y &= Cx + Du \end{aligned}$$

is controllable if and only if

$$\rho(\mathcal{C}) = n$$

where

$$\mathcal{C} = [B \;\; AB \;\; A^2B \;\; \ldots \;\; A^{n-1}B]$$

Problem 3.7.2 A system is described by the transfer function

$$G(s) = \frac{Y(s)}{U(s)} = \frac{s+1}{s^2}$$

A first order negative feedback dynamic compensator $H(s)$ given by

$$H(s) = k\frac{s+p}{s+1}$$

is implemented to get a desired feedback system transfer function.

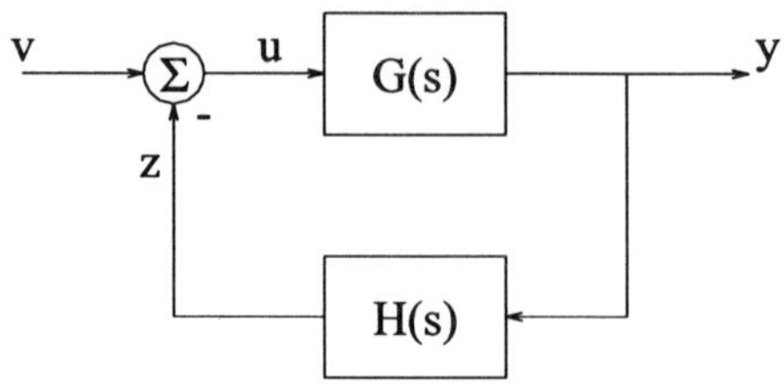

Figure 3.18: With Problem 3.7.2.

a) Find parameters k and p in the feedback compensator such that the overall transfer function from the reference input V to output Y (see Figure 3.18) is given by

$$T(s) = \frac{Y(s)}{V(s)} = \frac{s+1}{s^2+2s+4}$$

b) For the parameters chosen above, write the system state equations and discuss the controllability and observability of the system.

Solution: Given are $G(s) = \frac{Y(s)}{U(s)} = \frac{s+1}{s^2}$ and $H(s) = \frac{Z(s)}{Y(s)} = k\frac{s+p}{s+1}$.
a) In order to achieve

$$T(s) = \frac{Y(s)}{V(s)} = \frac{s+1}{s^2+2s+4}$$

we first write

$$T = \frac{Y}{V} = \frac{Y}{U+Z} = \frac{1}{U/Y + Z/Y} = \frac{1}{1/G + H}$$

i.e.,

$$T(s) = \frac{G(s)}{1+G(s)H(s)} = \frac{s+1}{s^2+ks+kp}$$

Therefore, we can immediately write

$$k = 2 \quad \text{and} \quad p = 2$$

b) Functions $G(s)$ and $H(s)$ given by

$$G(s) = \frac{s+1}{s^2} \quad \text{and} \quad H(s) = 2 + \frac{2}{s+1}$$

can be easily realized and connected, and then the state-space equation can be found:

$$\dot{x} = \begin{bmatrix} -2 & -2 & -2 \\ 1 & 0 & 0 \\ 1 & 1 & -1 \end{bmatrix} x + \begin{bmatrix} 1 \\ 0 \\ 0 \end{bmatrix} v$$

$$y = \begin{bmatrix} 1 & 1 & 0 \end{bmatrix} x$$

This realization is neither controllable nor observable.

Problem 3.7.3 Consider a multi-input multi-output system given by

$$\begin{aligned}\dot{x} &= Ax + Bu \\ y &= Cx + Du\end{aligned}$$

where B is $n \times m$.

First assume A has n distinct eigenvalues, diagonalize it and find conditions for the state controllability of the system.

Then assume A is not diagonalizable, transform it into its Jordan form, and derive the general conditions for the system to be state controllable.

Solution: If A has n distinct eigenvalues, then it has n linearly independent eigenvectors, which can be used to diagonalize A:

$$SAS^{-1} = \Lambda \qquad (\Lambda = \text{diag}(\lambda_1, \ldots, \lambda_2))$$

where the columns of S are the left eigenvectors of A (cf. Problem 3.2.1).

Since S is nonsingular, instead of states $x(t)$ we can consider an alternative state vector defined by $z(t) = Sx(t)$.

Then the equations become

$$\begin{aligned}\dot{z} &= SAS^{-1}z + SBu \\ y &= CS^{-1}z + Du\end{aligned}$$

and we can see that the system equations got uncoupled, therefore unless all rows of SB are non-zero, some state will not be controllable by the input $u(t)$. On the other hand, if all rows of SB are non-zero, then we can design $u(t)$ (cf. Problem 3.7.15) so that it takes $z(0) = Sx(0)$ into $z(t_f) = Sx(t_f) = 0$.

If A has repeated eigenvalues, the columns of the transformation matrix S are the principal vectors (eigenvectors and possibly generalized eigenvectors) of A. The similarity transformation SAS^{-1} produces a Jordan matrix similar to A. With $z(t) = Sx(t)$, the equations are again as above, but the conditions for controllability are now as follows:

1. No two Jordan blocks in $J = SAS^{-1}$ are associated with the same eigenvalue;
2. The elements of rows of SB corresponding to the last rows of Jordan blocks in J are not all zero;

Problem 3.7.4 Give a few examples of the above criteria for state controllability.

Solution: Given are only matrices after similarity transformation, and brief explanations.

$$SAS^{-1} = \begin{bmatrix} -1 & 0 & 0 \\ 0 & -2 & 0 \\ 0 & 0 & -3 \end{bmatrix}, \quad SB = \begin{bmatrix} 1 \\ 4 \\ 5 \end{bmatrix}, \qquad \text{controllable}$$

$$SAS^{-1} = \begin{bmatrix} -1 & 0 & 0 \\ 0 & -2 & 0 \\ 0 & 0 & -3 \end{bmatrix}, \quad SB = \begin{bmatrix} 1 \\ 4 \\ 0 \end{bmatrix}, \qquad \text{not controllable}$$

$$SAS^{-1} = \begin{bmatrix} -1 & 0 \\ 0 & -1 \end{bmatrix}, \quad SB = \begin{bmatrix} 1 \\ 4 \end{bmatrix}, \qquad \text{not controllable}$$

$$SAS^{-1} = \begin{bmatrix} -1 & 1 \\ 0 & -1 \end{bmatrix}, \quad SB = \begin{bmatrix} 1 \\ 4 \end{bmatrix}, \qquad \text{controllable}$$

$$SAS^{-1} = \begin{bmatrix} -1 & 1 \\ 0 & -1 \end{bmatrix}, \quad SB = \begin{bmatrix} 1 \\ 0 \end{bmatrix}, \qquad \text{not controllable}$$

$$SAS^{-1} = \begin{bmatrix} -1 & 1 \\ 0 & -1 \end{bmatrix}, \quad SB = \begin{bmatrix} 0 \\ 1 \end{bmatrix}, \qquad \text{controllable}$$

$$SAS^{-1} = \begin{bmatrix} -2 & 1 & 0 \\ 0 & -2 & 0 \\ 0 & 0 & -3 \end{bmatrix}, \quad SB = \begin{bmatrix} 1 & 2 \\ 4 & 5 \\ 0 & 0 \end{bmatrix}, \qquad \text{not controllable}$$

Problem 3.7.5 For a system given by

$$\begin{aligned} \dot{x}(t) &= Ax(t) + bu(t) \\ y(t) &= c'x(t) \end{aligned}$$

find the condition which guarantees that if i.c.'s $y(0^-)$, $\dot{y}(0^-)$, ..., $y^{(n-1)}(0^-)$, $u(0^-)$, $\dot{u}(0^-)$, ..., $u^{(n-1)}(0^-)$, are known, then the i.c.'s $x_1(0^-)$, $x_2(0^-)$, ..., $x_n(0^-)$ can be found.

Solution: If we define $\mathcal{Y}(t)$ and $\mathcal{U}(t)$ as

$$\mathcal{Y}(t) = \begin{bmatrix} y(t) \\ \dot{y}(t) \\ \vdots \\ y^{(n-1)}(t) \end{bmatrix} \quad \text{and} \quad \mathcal{U}(t) = \begin{bmatrix} u(t) \\ \dot{u}(t) \\ \vdots \\ u^{(n-1)}(t) \end{bmatrix}$$

then we can write

$$\mathcal{Y}(t) = \mathcal{O}x(t) + T\mathcal{U}(t)$$

where

$$\mathcal{O} = \begin{bmatrix} c' \\ c'A \\ \vdots \\ c'A^{n-1} \end{bmatrix}$$

and T is a lower triangular Toeplitz matrix with first column $[0 \ c'b \ \dots \ c'A^{n-2}b]'$ (numbers $h_i = c'A^ib$ are called Markov parameters, cf. Problem 3.8.4).

Therefore, we can write

$$\mathcal{O}x(0^-) = \mathcal{Y}(0^-) - T\mathcal{U}(0^-)$$

and we see that the initial conditions $x(0^-)$ can be determined from the arbitrary given initial conditions $\mathcal{Y}(0^-)$ and $\mathcal{U}(0^-)$ if and only if

$$\rho(\mathcal{O}) = n$$

If $\rho(\mathcal{O}) = n$ we say that the system is observable.

Note: *If the system is not observable, the i.c.'s $x(0^-)$ can be found only for some i.c.'s $\mathcal{Y}(0^-)$ and $\mathcal{U}(0^-)$. In that case, if i.c.'s $x(0^-)$ can be found, then they are not unique. This non-uniqueness is not important if we just need to determine any i.c.'s $x(0^-)$. But if we need the actual i.c.'s, this is a problem (see Problem 3.7.6).*

Problem 3.7.6 Describe the condition which guarantees that if we know A, b, c', $u(t)$ $(t \geq 0)$, and $y(t)$ $(t \geq 0)$, then we can determine $x(t)$.

Solution: To determine $x(t)$, besides the given matrices and signals, we only need the i.c.'s $x(0^-)$. As we found in Problem 3.7.5, this can be done if and if the observability matrix, defined by

$$\mathcal{O} = \begin{bmatrix} c' \\ c'A \\ \vdots \\ c'A^{n-1} \end{bmatrix}$$

has a full rank.

Note: *If the system is not observable, there are cases when the i.c.'s $x(0^-)$ cannot be determined, and even if they can, they are not unique. Hence if $\rho(\mathcal{O}) < n$, the states $x(t)$ either cannot be determined, or are not uniquely determined. Due to the Cayley-Hamilton theorem, further derivatives $y^{(n)}$, $y^{(n+1)}$, ... are of no help here.*

Problem 3.7.7 Derive observability conditions analogous to the controllability conditions in Problem 3.7.3. The system is given by A, B, and C matrices.

Solution: In general, when A is similar to a Jordan matrix via the similarity transformation $J = SAS^{-1}$, the conditions for observability are as follows

1. No two Jordan blocks in $J = SAS^{-1}$ are associated with the same eigenvalue;
2. The elements of columns of CS^{-1} corresponding to the first columns of Jordan blocks in J are not all zero;

Problem 3.7.8 Describe the PBH (Popov-Belevitch-Hautus) eigenvector tests for controllability and observability. Also describe the PBH rank tests.

Solution: *PBH eigenvector tests:*

- A system is controllable if and only if no left (row) eigenvector of A is orthogonal to b, i.e., if we have

$$p'b \neq 0 \qquad \text{for all left eigenvectors of } A$$

- A system is observable if and only if no right (column) eigenvector of A is orthogonal to c', i.e., if we have

$$c'q \neq 0 \qquad \text{for all right eigenvectors of } A$$

PBH rank tests:

- A system is controllable if and only if

$$\rho\,[sI - A \quad b] = n \qquad \text{(for all } s\text{)}$$

- A system is observable if and only if

$$\rho \begin{bmatrix} c' \\ sI - A \end{bmatrix} = n \qquad \text{(for all } s\text{)}$$

Problem 3.7.9 Describe the s-plane criteria for controllability and observability.

Solution: A system is controllable if and only if there are no cancellations in $(sI - A)^{-1}b$, or, more precisely, if there are no common pole-zero cancellations between the elements of $\text{adj}(sI - A)b$ and $\det(sI - A)$.

It is observable if and only if there are no cancellations in $c'(sI - A)^{-1}$, i.e., if there are no common pole-zero cancellations between the elements of $c'\text{adj}(sI - A)$ and $\det(sI - A)$.

Combining these two criteria, we can say that a system is both controllable and observable if and only if there are no pole-zero cancellations in $c'(sI - A)^{-1}b$.

Problem 3.7.10 For the system shown in Figure 3.19 write the system equations in the form

$$\begin{aligned} \dot{x}(t) &= Ax(t) + bu(t) \\ y(t) &= c'x(t) + du(t) \end{aligned}$$

and find the transfer function. Determine whether this system is controllable and/or observable. Is it stable?

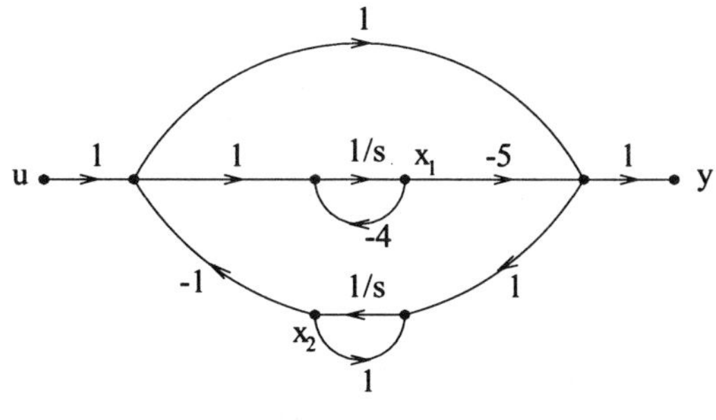

Figure 3.19: Diagram of the system described in Problem 3.7.10.

Solution: Obviously

$$\begin{bmatrix} \dot{x}_1 \\ \dot{x}_2 \end{bmatrix} = \begin{bmatrix} -4 & -1 \\ -5 & 0 \end{bmatrix} \begin{bmatrix} x_1 \\ x_2 \end{bmatrix} + \begin{bmatrix} 1 \\ 1 \end{bmatrix} u$$

$$y = \begin{bmatrix} -5 & -1 \end{bmatrix} \begin{bmatrix} x_1 \\ x_2 \end{bmatrix} + 1 \cdot u$$

The transfer function is found by taking the Laplace transform of the system equations and assuming zero initial conditions:

$$\begin{aligned} \dot{x}(t) &= Ax(t) + bu(t) \\ y(t) &= c'x(t) + du(t) \end{aligned}$$

$$\begin{aligned} sX(s) &= AX(s) + bU(s) \\ Y(s) &= c'X(s) + dU(s) \end{aligned}$$

which yields

$$Y(s) = (c'(sI - A)^{-1}b + d)U(s), \qquad \text{i.e.,} \qquad H(s) = \frac{Y(s)}{U(s)} = c'(sI - A)^{-1}b + d$$

In our case

$$\begin{aligned} H(s) &= \begin{bmatrix} -5 & -1 \end{bmatrix} \begin{bmatrix} s+4 & 1 \\ 5 & s \end{bmatrix}^{-1} \begin{bmatrix} 1 \\ 1 \end{bmatrix} + 1 \\ &= \begin{bmatrix} -5 & -1 \end{bmatrix} \frac{\begin{bmatrix} s & -1 \\ -5 & s+4 \end{bmatrix}}{s^2 + 4s - 5} \begin{bmatrix} 1 \\ 1 \end{bmatrix} + 1 \\ &= \frac{s^2 - 2s + 1}{s^2 + 4s - 5} = \frac{s-1}{s+5} \end{aligned}$$

This result can easily be obtained from the signal flow graph using the Mason formula.

The controllability matrix is

$$\mathcal{C} = [b \quad Ab] = \begin{bmatrix} 1 & -5 \\ 1 & -5 \end{bmatrix}$$

while the observability matrix is

$$\mathcal{O} = \begin{bmatrix} c' \\ c'A \end{bmatrix} = \begin{bmatrix} -5 & -1 \\ 25 & 5 \end{bmatrix}$$

Their ranks are both equal 1, which is less than the order of the system ($n = 2$), hence this system is neither controllable nor observable.

Note: *Another possibility was to observe that there is a zero-pole cancellation in $(sI - A)^{-1}b$, which is not consistent with controllability. Also there is a zero-pole cancellation in $c'(sI - A)^{-1}$, which is not consistent with observability.*

We shall see later some special types of realizations. Using these realizations the transfer function from this problem can be realized as controllable (but not observable) or observable (but not controllable) system. Due to the pole-zero cancellation it can never be realized as both controllable and observable system.

Problem 3.7.11 Consider the two feedback systems in Figure 3.20. Write a state-space representation in each case. What can you say about the controllability, observability, and stability of these two configurations?

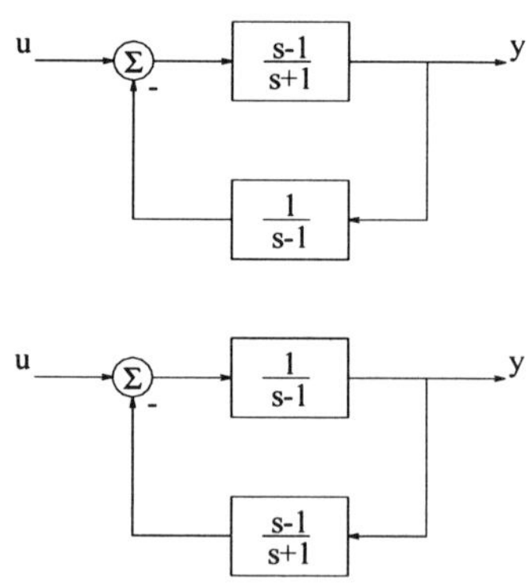

Figure 3.20: With Problem 3.7.11.

Solution: The two "black boxes" can be realized as in Figure 3.21. Therefore, for the first configuration we can write:

$$\begin{bmatrix} \dot{x}_1 \\ \dot{x}_2 \end{bmatrix} = \begin{bmatrix} -1 & -1 \\ -2 & 0 \end{bmatrix} \begin{bmatrix} x_1 \\ x_2 \end{bmatrix} + \begin{bmatrix} 1 \\ 1 \end{bmatrix} u$$

$$y = \begin{bmatrix} -2 & -1 \end{bmatrix} \begin{bmatrix} x_1 \\ x_2 \end{bmatrix} + u$$

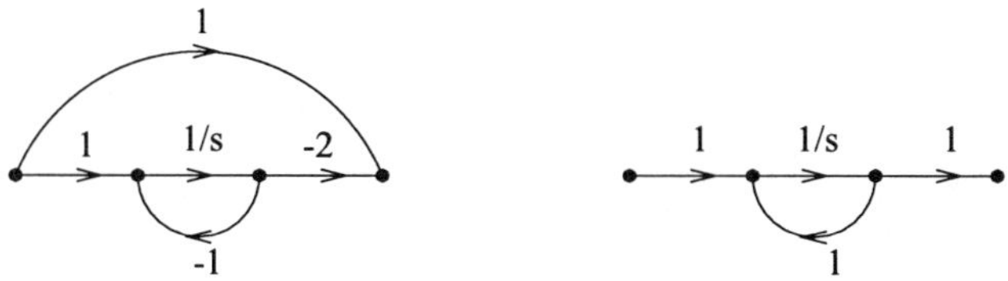

Figure 3.21: Possible realizations of the two "black boxes" in Problem 3.7.11.

Hence

$$\det(\mathcal{C}) = \begin{vmatrix} 1 & -2 \\ 1 & -2 \end{vmatrix} = 0 \quad \Rightarrow \quad \text{not controllable}$$

$$\det(\mathcal{O}) = \begin{vmatrix} -2 & -1 \\ 4 & 2 \end{vmatrix} = 0 \quad \Rightarrow \quad \text{not observable.}$$

Although the transfer function is

$$H(s) = c'(sI - A)^{-1}b + d = [-2 \ -1]\frac{\begin{bmatrix} s & -1 \\ -2 & s+1 \end{bmatrix}}{s^2+s-2}\begin{bmatrix} 1 \\ 1 \end{bmatrix} + 1 = \frac{(s-1)^2}{(s-1)(s+2)} = \frac{s-1}{s+2}$$

the system is not stable, because we had a zero-pole cancellation of an unstable pole $s = 1$.

We could better examine the stability by finding the eigenvalues of the system:

$$\det(\lambda I - A) = 0 \quad \Rightarrow \quad \lambda^2 + \lambda - 2 = 0 \quad \Rightarrow \quad \lambda_1 = -2, \ \lambda_2 = 1$$

The second configuration can be realized as follows:

$$\begin{bmatrix} \dot{x}_1 \\ \dot{x}_2 \end{bmatrix} = \begin{bmatrix} 0 & 2 \\ 1 & -1 \end{bmatrix}\begin{bmatrix} x_1 \\ x_2 \end{bmatrix} + \begin{bmatrix} 1 \\ 0 \end{bmatrix} u$$

$$y = \begin{bmatrix} 1 & 0 \end{bmatrix}\begin{bmatrix} x_1 \\ x_2 \end{bmatrix}$$

Hence

$$\det(\mathcal{C}) = \begin{vmatrix} 1 & 0 \\ 0 & 1 \end{vmatrix} = 1 \neq 0 \quad \Rightarrow \quad \text{controllable}$$

$$\det(\mathcal{O}) = \begin{vmatrix} 1 & 0 \\ 0 & 2 \end{vmatrix} = 2 \neq 0 \quad \Rightarrow \quad \text{observable.}$$

This time the transfer function is

$$H(s) = c'(sI - A)^{-1}b = \begin{bmatrix} 1 & 0 \end{bmatrix}\frac{\begin{bmatrix} s+1 & 2 \\ 1 & s \end{bmatrix}}{s^2+s-2}\begin{bmatrix} 1 \\ 0 \end{bmatrix} = \frac{s+1}{(s-1)(s+2)}$$

hence the system is not stable.

Problem 3.7.12 Let the parameters of the inverted pendulum (see Problem 3.5.11) on a cart be as follows: $m = 0.102\,kg$, $g = 9.81\,m/s^2$, $M = 1\,kg$, and $l = 0.5\,m$.

Write the corresponding state-space equations, and find the eigenvalues of the system. Is the system stable, controllable, and/or observable? Simulate the system using the MATLAB package, first for the unit step input of magnitude 0.1 and zero initial conditions, and then for zero input and initial conditions: $x'(0) = [0 \;\; 0.1 \;\; 0 \;\; 0]'$.

Solution: The system equations are now

$$\dot{x} = \begin{bmatrix} 0 & 0 & 1 & 0 \\ 0 & 0 & 0 & 1 \\ 0 & -1 & 0 & 0 \\ 0 & 21.6 & 0 & 0 \end{bmatrix} x + \begin{bmatrix} 0 \\ 0 \\ 1 \\ -2 \end{bmatrix} u$$

$$y = \begin{bmatrix} 1 & 0 & 0 & 0 \\ 0 & 1 & 0 & 0 \end{bmatrix} x$$

The eigenvalues are found from

$$\begin{vmatrix} \lambda & 0 & -1 & 0 \\ 0 & \lambda & 0 & -1 \\ 0 & 1 & \lambda & 0 \\ 0 & -21.6 & 0 & \lambda \end{vmatrix} = 0$$

$$\lambda^2 \begin{vmatrix} \lambda & -1 \\ -21.6 & \lambda \end{vmatrix} = 0$$

$$\lambda^2(\lambda^2 - 21.6) = 0$$

therefore

$$\lambda_1 = 0, \quad \lambda_2 = 0, \quad \lambda_3 = 4.65, \quad \lambda_4 = -4.65$$

Since $\lambda_3 > 0$, the system is not stable.

The system is controllable because

$$\mathcal{C} = [b \;\; Ab \;\; A^2b \;\; A^3b] = \begin{bmatrix} 0 & 1 & 0 & 2 \\ 0 & -2 & 0 & -43.2 \\ 1 & 0 & 2 & 0 \\ -2 & 0 & -43.2 & 0 \end{bmatrix}$$

is a full rank matrix. Really,

$$\begin{vmatrix} 2 & 1 & 0 & 0 \\ -43.2 & -2 & 0 & 0 \\ 0 & 0 & 2 & 1 \\ 0 & 0 & -43.2 & -2 \end{vmatrix} = -(-1)^{1+2+1+2} \begin{vmatrix} 2 & 1 \\ -43.2 & -2 \end{vmatrix} = -(39.2)^2 \neq 0$$

The system is also observable, because

$$\mathcal{O} = \begin{bmatrix} C \\ CA \\ CA^2 \\ CA^3 \end{bmatrix}$$

is a full rank matrix. Indeed

$$\rho(\mathcal{O}) = \rho \begin{bmatrix} 1 & 0 & 0 & 0 \\ 0 & 1 & 0 & 0 \\ 0 & 0 & 1 & 0 \\ 0 & 0 & 0 & 1 \\ 0 & -1 & 0 & 0 \\ 0 & 21.6 & 0 & 0 \\ 0 & 0 & 0 & -1 \\ 0 & 0 & 0 & 21.6 \end{bmatrix} = \rho(I_4) = 4$$

Note: *Examine the observability of this system if the only state available is $\theta(t)$. What if only $z(t)$ is available?*

Simulation for the unit step input of magnitude 0.1 and zero initial conditions:

First create the file called **pend1a.m** containing the following lines:

```
% file pend1a.m
%
function xdot = pend1a(t,x)
u = 0.1;
xdot(1) =       x(3);
xdot(2) =       x(4);
xdot(3) =      -x(2) + 1*u;
xdot(4) = 21.6*x(2) - 2*u;
end
```

and then create and run the following file in MATLAB to obtain Figure 3.22:

```
% file simul1a.m
%
t0 = 0;
tf = 0.8;
x0 = [0;0;0;0];
[t,x] = ode45('pend1a',t0,tf,x0);   % ode45 numerically solves ord.diff.equations (ode)
subplot(2,1,1)
plot(t,x(:,1))
title('Inverted Pendulum Simulation 1a')
xlabel('t [s]'), ylabel('z(t) [m]'), grid
subplot(2,1,2)
plot(t,x(:,2))
xlabel('t [s]'), ylabel('theta(t) [rad]'), grid
```

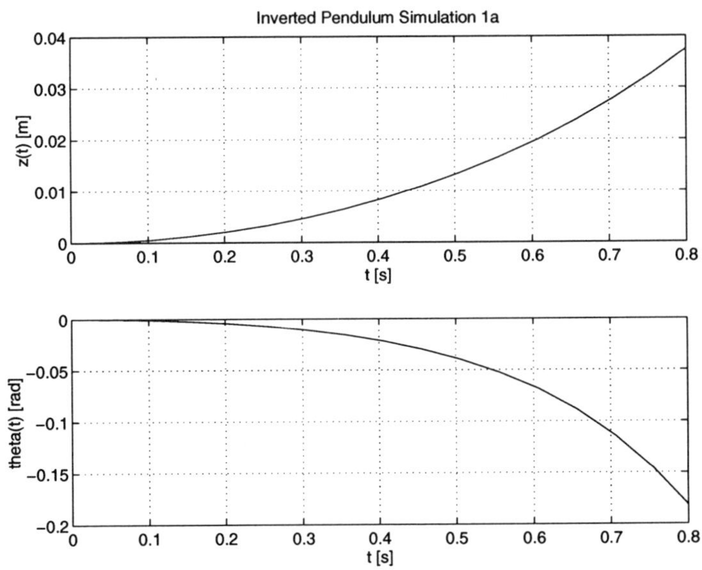

Figure 3.22: The results of the MATLAB Simulation 1a. The results are expected because the system is unstable.

Simulation for the zero input and initial conditions: $x'(0) = [0 \;\; 0.1 \;\; 0 \;\; 0]'$.

Now begin by creating the file called `pend1b.m` containing the following lines

```
% file pend1b.m
%
function xdot = pend1b(t,x)
xdot(1) =        x(3);
xdot(2) =        x(4);
xdot(3) =       -x(2);
xdot(4) = 21.6*x(2);
end
```

and then create and run the following file in MATLAB to obtain Figure 3.23:

```
% file simul1b.m
%
t0 = 0;
tf = 0.8;
x0 = [0;0.1;0;0];
[t,x] = ode45('pend1b',t0,tf,x0);
subplot(2,1,1)
plot(t,x(:,1))
title('Inverted Pendulum Simulation 1b')
xlabel('t [s]'), ylabel('z(t) [m]'), grid
subplot(2,1,2)
plot(t,x(:,2))
xlabel('t [s]'), ylabel('theta(t) [rad]'), grid
```

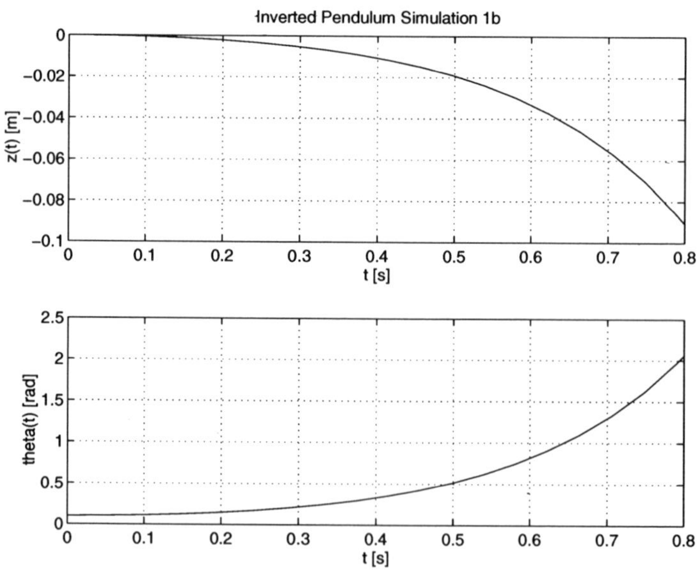

Figure 3.23: The results of the MATLAB Simulation 1b. Again, the results are as expected.

Problem 3.7.13 Consider the following dynamical system equation

$$\dot{x} = \begin{bmatrix} -1 & 1 & 0 & 0 \\ 0 & -1 & 0 & 0 \\ 0 & 0 & -2 & 1 \\ 0 & 0 & 0 & -2 \end{bmatrix} x + \begin{bmatrix} 1 \\ 0 \\ 0 \\ 1 \end{bmatrix} u$$

$$y = \begin{bmatrix} 1 & 1 & 1 & 0 \end{bmatrix} x$$

Is this system controllable and/or observable? Is it stable? Draw a signal flow graph and find a transfer function, both using the Mason's formula, and the formula in terms of system matrices.

Solution: Since $\det(\mathcal{C}) = 0$ and $\det(\mathcal{O}) = 1$, this system is not controllable, but is observable. Its eigenvalues are obviously -1, -1, -2, and -2, so the system is stable. The signal flow graph is shown in Figure 3.24.

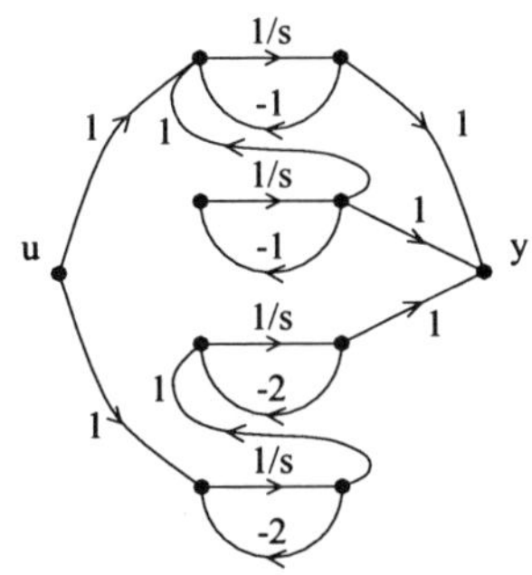

Figure 3.24: With Problem 3.7.13. If we were allowed to add one new connection in the graph, which nodes should we connect in order to make the system controllable? Which output connection we can remove without destroying observability?

From the Mason's rule, we find

$$\Delta(s) = 1 + \frac{6}{s} + \frac{13}{s^2} + \frac{12}{s^3} + \frac{4}{s^4}$$

and

$$P_1(s) = 1/s \qquad \Delta_1(s) = 1 + 5/s + 8/s^2 + 4/s^3$$

$$P_2(s) = 1/s^2 \qquad \Delta_2(s) = 1 + 2/s + 1/s^2$$

so we have

$$H(s) = \frac{\sum_i P_i(s)\Delta_i(s)}{\Delta(s)} = \ldots = \frac{s^2 + 5s + 5}{(s+1)(s+2)^2}$$

Of course,

$$H(s) = c'(sI - A)^{-1}b = \ldots = \frac{1}{(s+1)} + \frac{1}{(s+2)^2} = \frac{s^2 + 5s + 5}{(s+1)(s+2)^2}$$

Problem 3.7.14 For the system described in Problem 3.7.13

$$\dot{x} = \begin{bmatrix} -1 & 1 & 0 & 0 \\ 0 & -1 & 0 & 0 \\ 0 & 0 & -2 & 1 \\ 0 & 0 & 0 & -2 \end{bmatrix} x + \begin{bmatrix} 1 \\ 0 \\ 0 \\ 1 \end{bmatrix} u$$

$$y = \begin{bmatrix} 1 & 1 & 1 & 0 \end{bmatrix} x$$

determine the controllability and observability matrices using the MATLAB commands `ctrb` and `obsv`. Also, determine its transfer function using the command `ss2tf`.

Use the following code to obtain the controllability and observability matrices and the transfer function:

```
A = [ -1  1  0  0;
       0 -1  0  0;
       0  0 -2  1;
       0  0  0 -2 ];
b = [1; 0; 0; 1];
c = [1; 1; 1; 0];

C = ctrb(A,b)
det(C)
O = obsv(A,c')
det(O)
[num,den] = ss2tf(A,b,c',0,1)
[R,P,K] = residue(num,den)
```

The transfer function results are as follows:

```
num =     0     1.0000     6.0000    10.0000     5.0000


den =     1       6      13      12      4


R = 0.0000  1.0000  1.0000  0.0000


P = -2.0000  -2.0000  -1.0000  -1.0000


K = []
```

From the results of `ss2tf` and `residue` we can write

$$H(s) = \frac{s^2+5s+5}{(s+1)(s+2)^2} = \frac{1}{(s+1)} + \frac{1}{(s+2)^2}$$

Note: *Use* MATLAB *to answer the questions asked in the caption of Figure* 3.24.

Problem 3.7.15 For the system with

$$A = \begin{bmatrix} 1 & 2 & 3 \\ 0 & 1 & 5 \\ 2 & 1 & 0 \end{bmatrix} \quad \text{and} \quad b = \begin{bmatrix} 2 \\ 3 \\ 1 \end{bmatrix}$$

investigate its stability and controllability. Determine any input $u(t)$ which can take the system from $x(0) = [10 \;\; 5 \;\; 3]'$ to the origin in $t_f = 1.2\,s$.

Solution: The system is unstable because one of its eigenvalues has a positive real part:

$$\lambda_1 = 4.5995, \quad \lambda_2 = -1.2998 + 0.5170j, \quad \lambda_3 = -1.2998 - 0.5170j$$

If the system is to be stabilized, it should be state controllable, or at least the unstable mode of the system should be controllable by the appropriate input $u(t)$.

We shall see later how to stabilize systems by the appropriate state feedback. At the moment we are interested only in its controllability. Since

$$\det(\mathcal{C}) = |b \quad Ab \quad A^2b| = \begin{vmatrix} 2 & 11 & 48 \\ 3 & 8 & 43 \\ 1 & 7 & 30 \end{vmatrix} \neq 0$$

this system is state controllable.

Now let us find some input $u(t)$ which in finite time ($t_f = 1.2\,s$) takes the system from $x(0) = [10 \quad 5 \quad 3]'$ to the origin. From what we know, such an input must satisfy the following three equations:

$$\int_0^{t_f} \alpha_0(t)u(t)\,dt = \beta_0 \qquad \int_0^{t_f} \alpha_1(t)u(t)\,dt = \beta_1 \qquad \int_0^{t_f} \alpha_2(t)u(t)\,dt = \beta_2$$

where β_0, β_1, and β_2 are the solutions (cf. Equation 3.3) of

$$\begin{bmatrix} x_1(0) \\ x_2(0) \\ x_3(0) \end{bmatrix} = -\mathcal{C} \begin{bmatrix} \beta_0 \\ \beta_1 \\ \beta_2 \end{bmatrix}$$

In our case

$$\begin{bmatrix} \beta_0 \\ \beta_1 \\ \beta_2 \end{bmatrix} = -\mathcal{C}^{-1} \begin{bmatrix} x_1(0) \\ x_2(0) \\ x_3(0) \end{bmatrix} = -\frac{-1}{15} \begin{bmatrix} -61 & 6 & 89 \\ -47 & 12 & 58 \\ 13 & -3 & -17 \end{bmatrix} \begin{bmatrix} 10 \\ 5 \\ 3 \end{bmatrix} = \begin{bmatrix} -20.87 \\ -15.73 \\ 4.27 \end{bmatrix}$$

There are many different inputs which can do that. The simplest such input is a piece-wise constant function of time. Let $R_1(t)$, $R_2(t)$, and $R_3(t)$ be defined by the graphs in Figure 3.25, and let

$$u(t) = \gamma_1 R_1(t) + \gamma_2 R_2(t) + \gamma_3 R_3(t)$$

This choice for $u(t)$ greatly simplifies the above integral equations – now they become a system of three linear equations in unknowns γ_1, γ_2, and γ_3:

$$\begin{aligned} \int_0^{t_f} \alpha_i(t)u(t)\,dt = \beta_i \quad &\Rightarrow \quad \int_0^{t_f} \alpha_i(t) \left(\sum_{j=1}^{3} \gamma_j R_j(t) \right) dt = \beta_i \\ &\Rightarrow \quad \sum_{j=1}^{3} \gamma_j \left(\int_0^{t_f} \alpha_i(t) R_j(t)\,dt \right) = \beta_i \\ &\Rightarrow \quad \sum_{j=1}^{3} \gamma_j \left(\int_{t_{j-1}}^{t_j} \alpha_i(t)\,dt \right) = \beta_i \qquad (i = 0, 1, 2) \end{aligned}$$

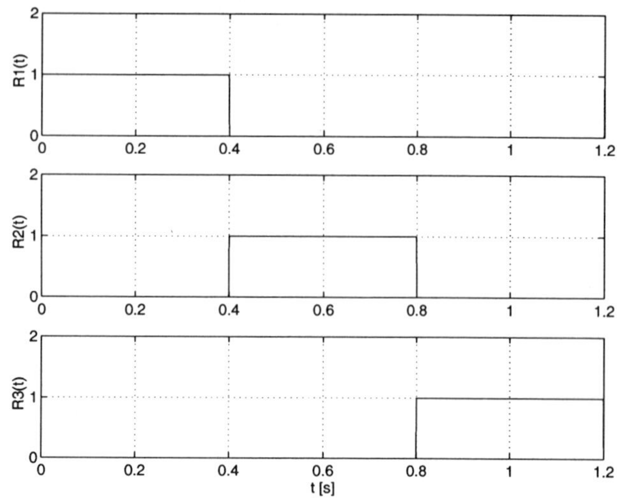

Figure 3.25: Graphs defining $R_1(t)$, $R_2(t)$, and $R_3(t)$. Each is non-zero for $t_f/3$ seconds.

Using the Cayley-Hamilton theorem (as in Problem 3.2.6) we can find the $\alpha_i(t)$'s. Note that $\alpha_i(t)$'s were defined as coefficients next to A^i (rather than next to $(-A)^i$) in the expansion of e^{-At}, hence we have to be careful with the signs.

$$\begin{bmatrix}\alpha_0(t)\\ \alpha_1(t)\\ \alpha_2(t)\end{bmatrix} = \begin{bmatrix}1 & \lambda_1 & \lambda_1^2\\ 1 & \lambda_2 & \lambda_2^2\\ 1 & \lambda_3 & \lambda_3^2\end{bmatrix}^{-1}\begin{bmatrix}e^{-\lambda_1 t}\\ e^{-\lambda_2 t}\\ e^{-\lambda_3 t}\end{bmatrix}$$

therefore

$$\int_{t_{j-1}}^{t_j}\begin{bmatrix}\alpha_0(t)\\ \alpha_1(t)\\ \alpha_2(t)\end{bmatrix} dt = \begin{bmatrix}1 & \lambda_1 & \lambda_1^2\\ 1 & \lambda_2 & \lambda_2^2\\ 1 & \lambda_3 & \lambda_3^2\end{bmatrix}^{-1}\begin{bmatrix}-(e^{-\lambda_1 t_j} - e^{-\lambda_1 t_{j-1}})/\lambda_1\\ -(e^{-\lambda_2 t_j} - e^{-\lambda_2 t_{j-1}})/\lambda_2\\ -(e^{-\lambda_3 t_j} - e^{-\lambda_3 t_{j-1}})/\lambda_3\end{bmatrix} \qquad (j = 1, 2, 3)$$

Denote these values for different values of j by g_{0j}, g_{1j}, and g_{2j}, then with

$$G = \begin{bmatrix} g_{01} & g_{02} & g_{03}\\ g_{11} & g_{12} & g_{13}\\ g_{21} & g_{22} & g_{23}\end{bmatrix} = \begin{bmatrix} 0.3914 & 0.2725 & -0.1862\\ -0.0889 & -0.3606 & -0.9040\\ 0.0095 & 0.0669 & 0.2056\end{bmatrix}$$

we can write

$$G\gamma = \beta \quad \text{i.e.,} \quad \gamma = G^{-1}\beta$$

and finally

$$\begin{bmatrix}\gamma_1\\ \gamma_2\\ \gamma_3\end{bmatrix} = \begin{bmatrix}-11.08\\ -37.74\\ 33.55\end{bmatrix}$$

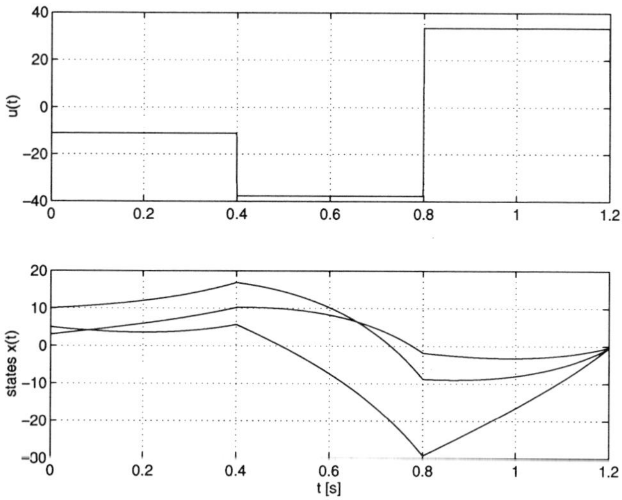

Figure 3.26: Shown are the input $u(t)$ and the states $x_1(t)$, $x_2(t)$, and $x_3(t)$, which are driven by the input to the origin.

Therefore we can pick $u(t)$ as in Figure 3.26.

Note: *Since the system is unstable, if greater t_f were chosen, the accumulation of numerical errors and noise would make it very improbable that the origin would be reached. That is why it is much better to design the state feedback, which uses the current information (on-line or real-time calculations) than to use the predesigned inputs, which do not account for unpredictable changes.*

Problem 3.7.16 If a pair (A, b) is not controllable, can c' be always chosen so that (A, c') is observable?

Solution: No, such c' cannot be found when A has two or more Jordan blocks associated with the same eigenvalue (see Problems 3.7.3 and 3.7.7). The simplest example is

$$A = \begin{bmatrix} 1 & 0 \\ 0 & 1 \end{bmatrix}$$

Problem 3.7.17 If a pair (A, b) is controllable, can c' be always chosen so that (A, c') is observable?

Solution: Yes, any c' without any zeros makes a good choice.

3.8 Canonical realizations

This Section provides an overview of five most commonly used realizations of systems: controller, observer, controllability, observability, and modal.

Problem 3.8.1 Show that the complete state controllability (observability) or a lack of it is preserved under a nonsingular similarity transformation.

Solution: We will show that $\rho(\mathcal{C}_{\text{new}}) = \rho(\mathcal{C}_{\text{old}})$, where

$$\mathcal{C}_{\text{new}} = [b_{\text{new}} \quad A_{\text{new}}b_{\text{new}} \quad \ldots \quad A_{\text{new}}^{n-1}b_{\text{new}}]$$

$$\mathcal{C}_{\text{old}} = [b_{\text{old}} \quad A_{\text{old}}b_{\text{old}} \quad \ldots \quad A_{\text{old}}^{n-1}b_{\text{old}}]$$

while

$$A_{\text{new}} = SA_{\text{old}}S^{-1}, \quad b_{\text{new}} = Sb_{\text{old}}, \quad \text{and} \quad c'_{\text{new}} = c'_{\text{old}}S^{-1}$$

Indeed, since

$$A_{\text{new}}^k b_{\text{new}} = SA_{\text{old}}^k b_{\text{old}} \qquad (k = 0, 1, \ldots, n-1)$$

we have

$$\mathcal{C}_{\text{new}} = S\mathcal{C}_{\text{old}}$$

Since S is nonsingular, $\rho(\mathcal{C}_{\text{new}}) = \rho(\mathcal{C}_{\text{old}})$.

The proof of $\rho(\mathcal{O}_{\text{new}}) = \rho(\mathcal{O}_{\text{old}})$ is completely analogous to the above derivation.

Problem 3.8.2 Given a transfer function

$$H(s) = \frac{s^2 + 2}{(s+1)(s+2)(s+4)}$$

develop controller, observer, controllability, and observability canonical forms of realization. For each canonical realization write the system equations and draw a signal flow graph.

After that write $H(s)$ in partial fractions form, and draw the corresponding signal flow graph. This is the modal canonical representation of this system.

Solution: Directly from the definitions of the controller, observer, controllability, and observability forms, and from

$$H(s) = \frac{s^2 + 2}{(s+1)(s+2)(s+4)} = \frac{\frac{1}{s} + \frac{2}{s^3}}{1 + \frac{7}{s} + \frac{14}{s^2} + \frac{8}{s^3}}$$

we find

Controller form (Figure 3.27):

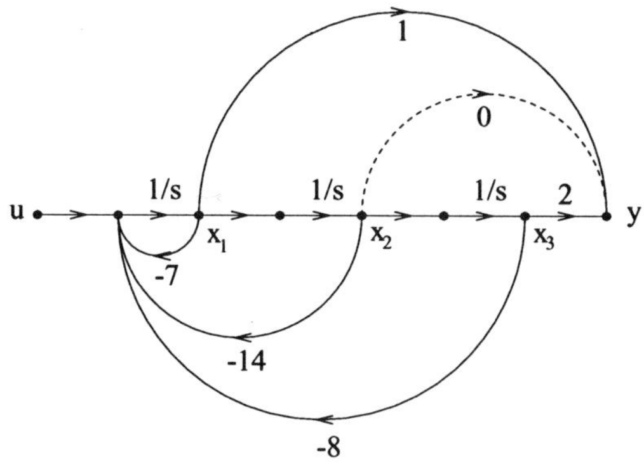

Figure 3.27: Realization of $H(s)$ in the controller form. Controller form is helpful in pole placement, because the Bass-Gura formula becomes very simple (see Problem 3.9.4).

$$\begin{bmatrix} \dot{x}_1 \\ \dot{x}_2 \\ \dot{x}_3 \end{bmatrix} = \begin{bmatrix} -7 & -14 & -8 \\ 1 & 0 & 0 \\ 0 & 1 & 0 \end{bmatrix} \begin{bmatrix} x_1 \\ x_2 \\ x_3 \end{bmatrix} + \begin{bmatrix} 1 \\ 0 \\ 0 \end{bmatrix} u$$

$$y = \begin{bmatrix} 1 & 0 & 2 \end{bmatrix} \begin{bmatrix} x_1 \\ x_2 \\ x_3 \end{bmatrix}$$

In general, with

$$H(s) = \frac{b(s)}{a(s)} = \frac{b_1 s^2 + b_2 s + b_3}{s^3 + a_1 s^2 + a_2 s + a_3}$$

we can write

$$A_c = \begin{bmatrix} -a_1 & -a_2 & -a_3 \\ 1 & 0 & 0 \\ 0 & 1 & 0 \end{bmatrix} \quad b_c = \begin{bmatrix} 1 \\ 0 \\ 0 \end{bmatrix} \quad c_c' = \begin{bmatrix} b_1 & b_2 & b_3 \end{bmatrix}$$

In Section 2.2 we derived the following results (Equations 2.5 and 2.9),

$$\mathcal{C}_c = \mathsf{a}_-^{-T} \quad \text{and} \quad \mathcal{O}_c = \tilde{I} b(A_c)$$

Since a_- is lower triangular with 1's on its main diagonal (hence $\det(\mathcal{C}_c) = 1$), this realization is always controllable.

Observer form (Figure 3.28):

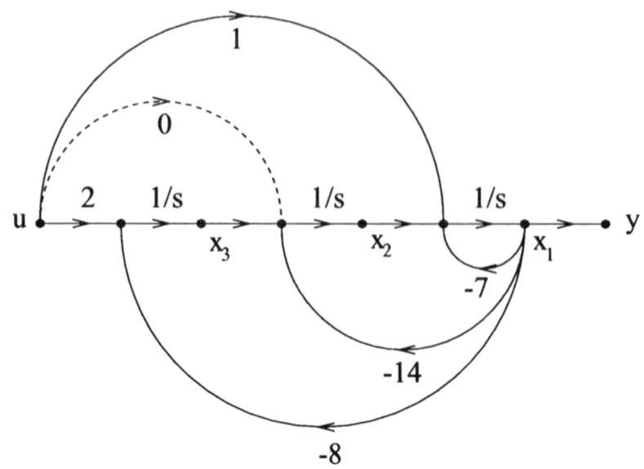

Figure 3.28: Realization of $H(s)$ in the observer form. It is useful in design of the state observer, because the dual of the Bass-Gura formula becomes very simple (see Problem 3.11.1).

$$\begin{bmatrix} \dot{x}_1 \\ \dot{x}_2 \\ \dot{x}_3 \end{bmatrix} = \begin{bmatrix} -7 & 1 & 0 \\ -14 & 0 & 1 \\ -8 & 0 & 0 \end{bmatrix} \begin{bmatrix} x_1 \\ x_2 \\ x_3 \end{bmatrix} + \begin{bmatrix} 1 \\ 0 \\ 2 \end{bmatrix} u$$

$$y = \begin{bmatrix} 1 & 0 & 0 \end{bmatrix} \begin{bmatrix} x_1 \\ x_2 \\ x_3 \end{bmatrix}$$

For this realization, in general,

$$A_o = \begin{bmatrix} -a_1 & 1 & 0 \\ -a_2 & 0 & 1 \\ -a_3 & 0 & 0 \end{bmatrix} \qquad b_o = \begin{bmatrix} b_1 \\ b_2 \\ b_3 \end{bmatrix} \qquad c_o' = \begin{bmatrix} 1 & 0 & 0 \end{bmatrix}$$

By duality to the controller realization

$$\mathcal{O}_o = \mathsf{a}_{-}^{-1} \quad \text{and} \quad \mathcal{C}_o = b(A_o)\tilde{I}$$

therefore this realization is always observable.

Controllability form (Figure 3.29):

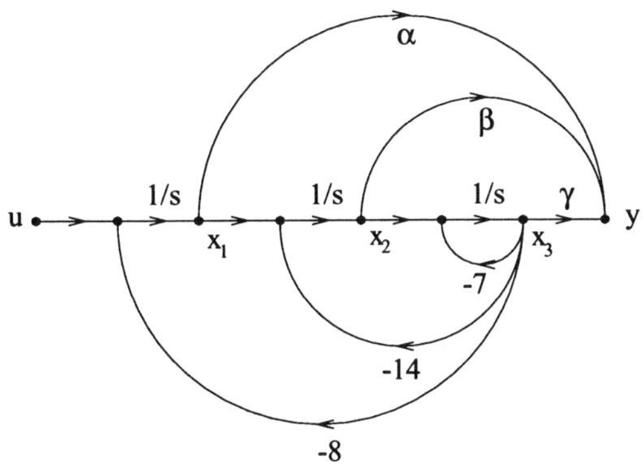

Figure 3.29: Realization of $H(s)$ in the controllability form. It is useful in partitioning a system into controllable and uncontrollable parts. The controllable part is usually written in the controllability form. It is also useful in determining the input necessary to set desirable initial conditions (see Problem 3.8.9).

$$\begin{bmatrix} \dot{x}_1 \\ \dot{x}_2 \\ \dot{x}_3 \end{bmatrix} = \begin{bmatrix} 0 & 0 & -8 \\ 1 & 0 & -14 \\ 0 & 1 & -7 \end{bmatrix} \begin{bmatrix} x_1 \\ x_2 \\ x_3 \end{bmatrix} + \begin{bmatrix} 1 \\ 0 \\ 0 \end{bmatrix} u$$

$$y = \begin{bmatrix} \alpha & \beta & \gamma \end{bmatrix} \begin{bmatrix} x_1 \\ x_2 \\ x_3 \end{bmatrix}$$

where

$$\begin{bmatrix} \alpha & \beta & \gamma \end{bmatrix} = \begin{bmatrix} 1 & 0 & 2 \end{bmatrix} \begin{bmatrix} 1 & 7 & 14 \\ 0 & 1 & 7 \\ 0 & 0 & 1 \end{bmatrix}^{-1} = \begin{bmatrix} 1 & -7 & 37 \end{bmatrix}$$

For this realization, in general,

$$A_{co} = \begin{bmatrix} 0 & 0 & -a_3 \\ 1 & 0 & -a_2 \\ 0 & 1 & -a_1 \end{bmatrix} \qquad b_{co} = \begin{bmatrix} 1 \\ 0 \\ 0 \end{bmatrix} \qquad c'_{co} = \begin{bmatrix} \alpha & \beta & \gamma \end{bmatrix}$$

where

$$\begin{bmatrix} \alpha & \beta & \gamma \end{bmatrix} = \begin{bmatrix} b_1 & b_2 & b_3 \end{bmatrix} \underline{a}_{-}^{-T}$$

It is interesting to note that α, β, and γ are the first three Markov parameters (see Equation 2.2 and Problem 3.8.4). For this realization (cf. Equations 2.1 and 2.4)

$$\mathcal{C}_{co} = I \quad \text{and} \quad \mathcal{O}_{co} = \mathcal{M}$$

therefore this realization is always controllable.

Observability form (Figure 3.30):

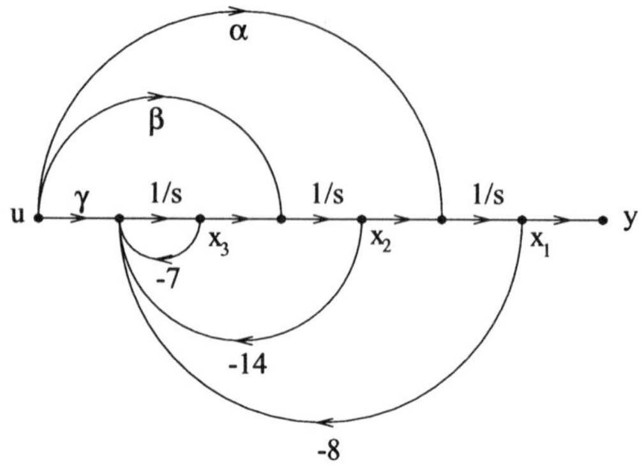

Figure 3.30: Realization of $H(s)$ in the observability form. It is useful in partitioning a system into observable and unobservable parts. The observable part is usually written in the observability form. Since $\mathcal{O}_{ob} = I$, it is also useful in determining the initial conditions $x(0^-)$ from $\mathcal{O}x(0^-) = \mathcal{Y}(0^-) - T\mathcal{U}(0^-)$. See also Problem 3.8.10.

$$\begin{bmatrix}\dot{x}_1\\ \dot{x}_2\\ \dot{x}_3\end{bmatrix} = \begin{bmatrix} 0 & 1 & 0\\ 0 & 0 & 1\\ -8 & -14 & -7\end{bmatrix}\begin{bmatrix}x_1\\ x_2\\ x_3\end{bmatrix} + \begin{bmatrix}\alpha\\ \beta\\ \gamma\end{bmatrix} u$$

$$y = \begin{bmatrix} 1 & 0 & 0 \end{bmatrix}\begin{bmatrix}x_1\\ x_2\\ x_3\end{bmatrix}$$

where again

$$\begin{bmatrix}\alpha\\ \beta\\ \gamma\end{bmatrix} = \begin{bmatrix} 1 & 0 & 0\\ 7 & 1 & 0\\ 14 & 7 & 1\end{bmatrix}^{-1}\begin{bmatrix}1\\ 0\\ 2\end{bmatrix} = \begin{bmatrix}1\\ -7\\ 37\end{bmatrix}$$

For this realization, in general,

$$A_{ob} = \begin{bmatrix} 0 & 1 & 0\\ 0 & 0 & 1\\ -a_3 & -a_2 & -a_1\end{bmatrix} \qquad b_{ob} = \begin{bmatrix}\alpha\\ \beta\\ \gamma\end{bmatrix} \qquad c'_{ob} = \begin{bmatrix} 1 & 0 & 0 \end{bmatrix}$$

where again

$$\begin{bmatrix}\alpha\\ \beta\\ \gamma\end{bmatrix} = \underline{a}_-^{-1}\begin{bmatrix}b_1\\ b_2\\ b_3\end{bmatrix}$$

Here

$$\mathcal{O}_{ob} = I \quad \text{and} \quad \mathcal{C}_{ob} = \mathcal{M}$$

therefore this realization is always observable.

Modal form (Figure 3.31):

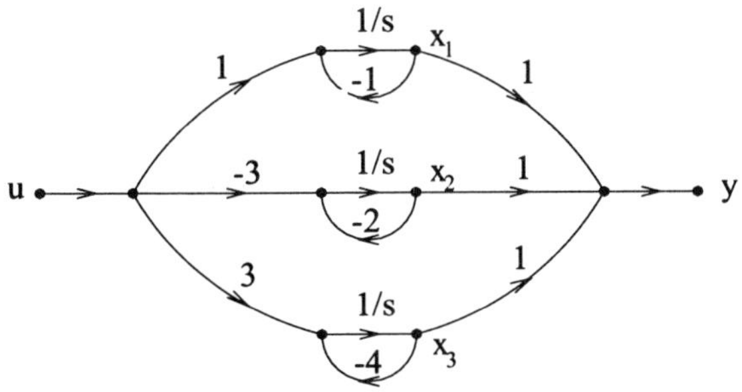

Figure 3.31: Realization of $H(s)$ in the modal canonical form. This form corresponds to the Jordan (diagonal) canonical form of the matrix. It is helpful in determining which modes are uncontrollable and/or unobservable (see Problems 3.7.3, 3.7.7, 3.7.13, 3.7.16, 3.7.17, and 3.9.10).

To find the modal canonical form of the system, we first need the partial fraction representation of $H(s)$:

$$H(s) = \frac{s^2+2}{(s+1)(s+2)(s+4)} = \frac{A}{s+1} + \frac{B}{s+2} + \frac{C}{s+4}$$

$$s^2+2 = A(s+2)(s+4) + B(s+1)(s+4) + C(s+1)(s+2)$$

$$\left.\begin{array}{lcc} s=-1 & \Rightarrow & A=1 \\ s=-2 & \Rightarrow & B=-3 \\ s=-4 & \Rightarrow & C=3 \end{array}\right\} \quad \Rightarrow \quad H(s) = \frac{1}{s+1} - \frac{3}{s+2} + \frac{3}{s+4}$$

The choice of vectors b and c' is not unique, it is just necessary to pick them so that $b_1c_1 = 1$, $b_2c_2 = -3$, and $b_3c_3 = 3$.

Note: *As an exercise, the reader may try to determine the eigenvalues of the system, its impulse response, and the transfer function, both using the formula* $H(s) = c'(sI - A)^{-1}b + d$ *and the Mason's rule, for some, or for all five realizations.*

Problem 3.8.3 Show that the product $\mathcal{OC}$ is invariant under a nonsingular similarity transformation.

Solution: If an "old" system is transformed into a "new" system by a nonsingular matrix S, then from $x_{\text{new}} = Sx_{\text{old}}$, we see that

$$\dot{x}_{\text{new}} = \underbrace{SA_{\text{old}}S^{-1}}_{A_{\text{new}}} x_{\text{new}} + \underbrace{Sb_{\text{old}}}_{b_{\text{new}}} u$$

$$y = \underbrace{c'_{\text{old}}S^{-1}}_{c'_{\text{new}}} x_{\text{new}}$$

Therefore $\mathcal{O}_{\text{new}} = \mathcal{O}_{\text{old}}S$ and $\mathcal{C}_{\text{new}} = S^{-1}\mathcal{C}_{\text{old}}$, and finally

$$\mathcal{O}_{\text{new}}\mathcal{C}_{\text{new}} = \mathcal{O}_{\text{old}}SS^{-1}\mathcal{C}_{\text{old}} = \mathcal{O}_{\text{old}}\mathcal{C}_{\text{old}}$$

Note: *This product is equal to the Hankel matrix of Markov coefficients $\mathcal{M}$, which is defined in Problem* 3.8.4. *See also Problem* 3.8.11.

Problem 3.8.4 For the transfer function

$$H(s) = \frac{s^2+2}{(s+1)(s+2)(s+4)} = \frac{s^2+2}{s^3+7s^2+14s+8}$$

five different canonical forms were examined in Problem 3.8.2. Find the transformations to convert between different canonical forms.

Solution: If the transformation is given by matrix S, than with $x_{\text{new}} = Sx_{\text{old}}$, we have

$$\dot{x}_{\text{new}} = \underbrace{SA_{\text{old}}S^{-1}}_{A_{\text{new}}} x_{\text{new}} + \underbrace{Sb_{\text{old}}}_{b_{\text{new}}} u$$

$$y = \underbrace{c'_{\text{old}}S^{-1}}_{c'_{\text{new}}} x_{\text{new}}$$

Most of the following identities can be derived from Equations 2.3 and 2.6 and Problems 3.8.2 and 5.1.24:

- any controllable form $\to$ controller form: $S = \mathcal{C}_c\mathcal{C}^{-1} = \mathsf{a}_-^{-T}\mathcal{C}^{-1}$
- any controllable form $\to$ controllability form: $S = \mathcal{C}^{-1}$
- any observable form $\to$ observer form: $S = \mathcal{O}_o^{-1}\mathcal{O} = \mathsf{a}_-\mathcal{O}$
- any observable form $\to$ observability form: $S = \mathcal{O}$
- controller $\to$ observer form: $S = -\tilde{I}\mathcal{B}\tilde{I}$
- controller $\to$ controllability form: $S = \mathsf{a}_-^T$
- controller $\to$ observability form: $S = \tilde{I}b(A_c)$
- controllability $\to$ observer form: $S = b(A_o)\tilde{I}$
- controllability $\to$ observability form: $S = \mathcal{M}$
- observer $\to$ observability form: $S = \mathsf{a}_-^{-1}$
- any of the above $\to$ modal form: see Example C.3.3

The following matrices appear in the list above:

$$\tilde{I} = \begin{bmatrix} 0 & \dots & 0 & 1 \\ 0 & \dots & 1 & 0 \\ & \dots & & \\ 1 & \dots & 0 & 0 \end{bmatrix}$$

$$\mathcal{B} = \tilde{I}(\mathsf{a}_+\mathsf{b}_- - \mathsf{b}_+\mathsf{a}_-), \qquad \text{the so-called Bezoutian}$$

where with

$$H(s) = \frac{b(s)}{a(s)} = \frac{b_1 s^{n-1} + b_2 s^{n-2} + \ldots + b_{n-1}s + b_n}{s^n + a_1 s^{n-1} + a_2 s^{n-2} + \ldots + a_{n-1}s + a_n}$$

the matrices a_+, a_-, b_+, and b_- are defined by

$$\mathsf{a}_+ = \begin{bmatrix} a_n & a_{n-1} & \dots & a_2 & a_1 \\ 0 & a_n & \dots & a_3 & a_2 \\ \vdots & \vdots & \ddots & \vdots & \vdots \\ 0 & 0 & \dots & a_n & a_{n-1} \\ 0 & 0 & \dots & 0 & a_n \end{bmatrix} \quad \mathsf{a}_- = \begin{bmatrix} 1 & 0 & \dots & 0 & 0 \\ a_1 & 1 & \dots & 0 & 0 \\ \vdots & \vdots & \ddots & \vdots & \vdots \\ a_{n-2} & a_{n-3} & \dots & 1 & 0 \\ a_{n-1} & a_{n-2} & \dots & a_1 & 1 \end{bmatrix}$$

$$\mathsf{b}_+ = \begin{bmatrix} b_n & b_{n-1} & \dots & b_2 & b_1 \\ 0 & b_n & \dots & b_3 & b_2 \\ \vdots & \vdots & \ddots & \vdots & \vdots \\ 0 & 0 & \dots & b_n & b_{n-1} \\ 0 & 0 & \dots & 0 & b_n \end{bmatrix} \quad \mathsf{b}_- = \begin{bmatrix} 0 & 0 & \dots & 0 & 0 \\ b_1 & 0 & \dots & 0 & 0 \\ \vdots & \vdots & \ddots & \vdots & \vdots \\ b_{n-2} & b_{n-3} & \dots & 0 & 0 \\ b_{n-1} & b_{n-2} & \dots & b_1 & 0 \end{bmatrix}$$

The matrices A_o and A_c are the system matrices in observer and controller form, respectively, while

$$\mathcal{M} = \begin{bmatrix} h_1 & h_2 & \dots & h_n \\ h_2 & h_3 & \dots & h_{n+1} \\ \vdots & \vdots & \ddots & \vdots \\ h_n & h_{n+1} & \dots & h_{2n-1} \end{bmatrix}$$

is the Hankel matrix of Markov parameters, h_i, defined by

$$H(s) = c'(sI - A)^{-1}b = \sum_{i=1}^{\infty} h_i s^{-i}$$

Note: *Since*

$$(sI - A)^{-1} = \frac{1}{s}\left(I - \frac{A}{s}\right)^{-1} = \frac{1}{s}\left(I + \frac{A}{s} + \frac{A^2}{s^2} + \ldots\right)$$

we have

$$h_i = c'A^{i-1}b \quad (i = 1, 2, \ldots)$$

Since the impulse response can be written as $h(t) = c'e^{At}b$ *and since* $H(s) = \mathcal{L}\{h(t)\}$ *and* $H(s) = c'(sI - A)^{-1}b$, *we see that*

$$h_i = \left.\frac{d^{i-1}}{dt^{i-1}}h(t)\right|_{t=0} \qquad (i = 1, 2, \ldots)$$

Let us see some of these transformations at work:

- Controller → observer form: $S = -\tilde{I}\mathcal{B}\tilde{I}$

$$S = -\tilde{I}\mathcal{B}\tilde{I} = -\tilde{I}\tilde{I}(a_+b_- - b_+a_-)\tilde{I} = -(a_+b_- - b_+a_-)\tilde{I}$$

$$S = -\left(\begin{bmatrix} 8 & 14 & 7 \\ 0 & 8 & 14 \\ 0 & 0 & 8 \end{bmatrix}\begin{bmatrix} 0 & 0 & 0 \\ 1 & 0 & 0 \\ 0 & 1 & 0 \end{bmatrix} - \begin{bmatrix} 2 & 0 & 1 \\ 0 & 2 & 0 \\ 0 & 0 & 2 \end{bmatrix}\begin{bmatrix} 1 & 0 & 0 \\ 7 & 1 & 0 \\ 14 & 7 & 1 \end{bmatrix}\right)\begin{bmatrix} 0 & 0 & 1 \\ 0 & 1 & 0 \\ 1 & 0 & 0 \end{bmatrix}$$

$$S = \begin{bmatrix} 1 & 0 & 2 \\ 0 & -12 & 6 \\ 2 & 6 & 28 \end{bmatrix}$$

Now we see that

$$\begin{aligned} SA_cS^{-1} &= \begin{bmatrix} 1 & 0 & 2 \\ 0 & -12 & 6 \\ 2 & 6 & 28 \end{bmatrix}\begin{bmatrix} -7 & -14 & -8 \\ 1 & 0 & 0 \\ 0 & 1 & 0 \end{bmatrix}\frac{1}{324}\begin{bmatrix} 372 & -12 & -24 \\ -12 & -24 & 6 \\ -24 & 6 & 12 \end{bmatrix} \\ &= \begin{bmatrix} -7 & 1 & 0 \\ -14 & 0 & 1 \\ -8 & 0 & 0 \end{bmatrix} = A_o \end{aligned}$$

- Controllability → observer form: $S = b(A_o)\tilde{I}$
In this case

$$S = b(A_o)\tilde{I} = (b_1A_o^2 + b_2A_o + b_3I)\tilde{I} = \begin{bmatrix} 1 & -7 & 37 \\ 0 & -12 & 90 \\ 2 & -8 & 56 \end{bmatrix}$$

Problem 3.8.5 Show that if a system given by $\{A, b, c'\}$ is controllable, it can be transformed into the controllability form using the following transformation matrix:

$$S_{co} = \mathcal{C}^{-1}$$

where $\mathcal{C} = [b \;\; Ab \;\; \ldots \;\; A^{n-1}b]$ is the controllability matrix of the original system.

Solution: We need to show that $S_{co}AS_{co}^{-1} = A_{co}$, $S_{co}b = b_{co}$, and $c'S_{co}^{-1} = c'_{co}$. To show that $S_{co}AS_{co}^{-1} = A_{co}$ we will prove that $A\mathcal{C} = \mathcal{C}A_{co}$. Indeed

$$A\mathcal{C} = A[b \;\; Ab \;\; \ldots \;\; A^{n-1}b] = [Ab \;\; A^2b \;\; \ldots \;\; A^nb]$$

while

$$\mathcal{C}A_{co} = [Ab \;\; A^2b \;\; \ldots \;\; (-a_nI - a_{n-1}A - \ldots - a_1A^{n-1})b]$$

From the Cayley-Hamilton theorem $-a_nI - a_{n-1}A - \ldots - a_1A^{n-1} = A^n$ and therefore $A\mathcal{C} = \mathcal{C}A_{co}$.

It is obvious that $S_{co}b = b_{co}$, because

$$\mathcal{C}b_{co} = [b \;\; Ab \;\; \ldots \;\; A^{n-1}b]\begin{bmatrix} 1 \\ 0 \\ \vdots \\ 0 \end{bmatrix} = b$$

To show $c'S_{co}^{-1} = [h_1 \ h_2 \ \dots \ h_n]$ observe that

$$c'S_{co}^{-1} = [c'b \ c'Ab \ \dots \ c'A^{n-1}b] = [h_1 \ h_2 \ \dots \ h_n]$$

where h_i's are Markov parameters.

Note: *Recall from Section* 2.2 *that*

$$[h_1 \ h_2 \ \dots \ h_n]\mathsf{a}_-^T = [b_1 \ b_2 \ \dots \ b_n]$$

is just another way of writing the definition of Markov parameters

$$\sum_{i=1}^{\infty} h_i s^{-i} = \frac{b(s)}{a(s)}$$

Problem 3.8.6 Show that a controllable system given by $\{A, b, c'\}$ can be transformed into the controller form using

$$S_c = \mathcal{C}_c\mathcal{C}^{-1} \qquad (\mathcal{C}_c = \mathsf{a}_-^{-T})$$

Solution: Using the result of Problem 3.8.5 we know that the transformation from controller into the controllability form is given by $S = \mathcal{C}_c^{-1} = \mathsf{a}_-^T$. Thus, to go from any controllable form into the controller form we can go via the controllability form, when we find

$$S_c = \mathsf{a}_-^{-T}\mathcal{C}^{-1}$$

Problem 3.8.7 For the transfer function given by

$$H(s) = \frac{s+3}{s^3 + 9s^2 + 24s + 18}$$

find a controllability form of system realization.

a) Find the controllability matrix $\mathcal{C}_{co}$. Is the system controllable?

b) Find the observability matrix $\mathcal{O}_{co}$. Is the system observable?

Solution: If we want to use the controllability canonical form, we need to determine the coefficients α, β, and γ in Figure 3.32.

To do that, we can use the Mason's formula which yields

$$H(s) = \frac{\frac{\alpha}{s}\left(1 + \frac{9}{s} + \frac{24}{s^2}\right) + \frac{\beta}{s^2}\left(1 + \frac{9}{s}\right) + \frac{\gamma}{s^3} \cdot 1}{1 + \frac{9}{s} + \frac{24}{s^2} + \frac{18}{s^3}}$$

which, when equated with the given expression for $H(s)$, implies

$$\alpha = 0, \quad \beta = 1, \quad \text{and} \quad \gamma = -6$$

Of course, the same result is obtained using

$$\begin{bmatrix} \alpha \\ \beta \\ \gamma \end{bmatrix} = \begin{bmatrix} 1 & 0 & 0 \\ 9. & 1 & 0 \\ 24 & 9 & 1 \end{bmatrix}^{-1} \begin{bmatrix} 0 \\ 1 \\ 3 \end{bmatrix} = \begin{bmatrix} 0 \\ 1 \\ -6 \end{bmatrix}$$

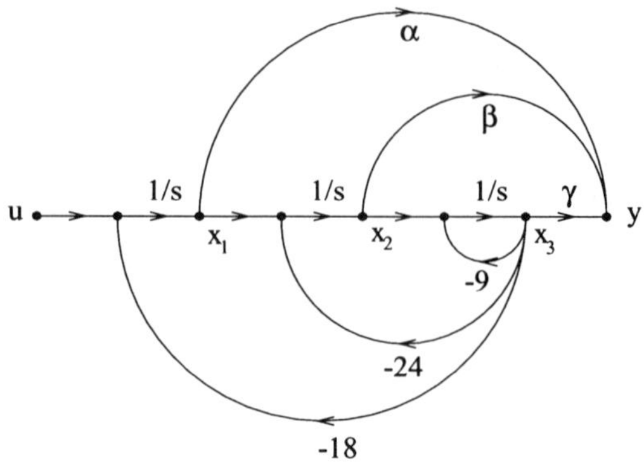

Figure 3.32: Realization of $H(s)$ from Problem 3.8.7 in the controllability form.

Therefore

$$\dot{x} = \begin{bmatrix} 0 & 0 & -18 \\ 1 & 0 & -24 \\ 0 & 1 & -9 \end{bmatrix} x + \begin{bmatrix} 1 \\ 0 \\ 0 \end{bmatrix} u$$

$$y = \begin{bmatrix} 0 & 1 & -6 \end{bmatrix} x$$

while

$$\mathcal{C}_{co} = \begin{bmatrix} 1 & 0 & 0 \\ 0 & 1 & 0 \\ 0 & 0 & 1 \end{bmatrix} \quad \text{and} \quad \mathcal{O}_{co} = \begin{bmatrix} 0 & 1 & -6 \\ 1 & -6 & 30 \\ -6 & 30 & -154 \end{bmatrix}$$

Since $\det(\mathcal{C}_{co}) \neq 0$ and $\det(\mathcal{O}_{co}) = 0$, we see this realization is controllable but it is not observable. If we chose to use the observer or observability form, the realization would be observable but not controllable.

As we saw in Problem 3.7.10 there are realizations that can be neither controllable nor observable.

Problem 3.8.8 Show that the controller (or any controllable) realization of $H(s) = b(s)/a(s)$ is observable if and only if $a(s)$ and $b(s)$ are coprime polynomials. Similarly, the observer (or any observable) realization of $H(s) = b(s)/a(s)$ is controllable if and only if $a(s)$ and $b(s)$ are coprime.

Solution: Use the theorem from Problem 3.7.9.

Problem 3.8.9 A system is given by

$$H(s) = \frac{s^2+2}{(s+1)(s+2)(s+4)} = \frac{\frac{1}{s} + \frac{2}{s^3}}{1 + \frac{7}{s} + \frac{14}{s^2} + \frac{8}{s^3}}$$

Write it in any state-space representation and determine the coefficients in the following impulsive input

$$u(t) = \mu_0\delta(t) + \mu_1\dot{\delta}(t) + \mu_2\ddot{\delta}(t)$$

so that the initial conditions are "instantaneously" changed from $x(0^-)$ to $x(0^+)$.

What canonical form is the most convenient for these calculations? What condition guarantees that any desired change in initial conditions can be made in this way?

Solution: In general, regardless of the form the system is given in, from Problem 3.5.4 we know

$$x(t) = e^{At}x(0) + \int_0^t e^{A(t-\tau)}bu(\tau)\,d\tau$$

Using

$$\int_{0^-}^{0^+} f(t)\delta^{(i)}(t)\,dt = (-1)^i f^{(i)}(0)$$

with

$$u(t) = \sum_{i=0}^{n-1} \mu_i\delta^{(i)}(t)$$

we obtain

$$x(0^+) = x(0^-) + \mathcal{C}\mu$$

where $\mu = [\mu_0 \;\; \mu_1 \;\; \dots \;\; \mu_{n-1}]'$, while $\mathcal{C}$ is the controllability matrix.

Note 1: *In the controllability canonical form $\mathcal{C}_{co} = I$, hence the calculations become trivial:*

$$\mu = x_{co}(0^+) - x_{co}(0^-)$$

Note 2: *We can achieve any change in initial conditions in zero time if and only if $\rho(\mathcal{C}) = n$, i.e., under the same conditions as with the finite time (Problem 3.7.1).*

Problem 3.8.10 Consider again Problems 3.7.5 and 3.7.6 and determine the canonical form which is most suitable for determining initial conditions and observing states.

Solution: In Problems 3.7.5 and 3.7.6 we found that

$$\mathcal{Y}(t) = \mathcal{O}x(t) + \mathcal{T}\mathcal{U}(t)$$

In order to calculate $x(t)$ or $x(0)$ it is necessary to invert $\mathcal{O}$, the observability matrix. Since in the observability canonical form $\mathcal{O}_{ob} = I$ that form is the most suitable for such calculations. Even with this simplification these calculations are not a good solution, because they involve differentiation, which amplifies any error due to noise or round-off errors. In discrete-time systems there is no differentiation involved and this formula can be practical.

Problem 3.8.11 In Problem 3.8.3 we showed that the product $\mathcal{OC}$ is invariant under similarity transformations. Prove that

$$\mathcal{OC} = \mathcal{M}$$

where

$$\mathcal{M} = \begin{bmatrix} h_1 & h_2 & \dots & h_n \\ h_2 & h_3 & \dots & h_{n+1} \\ \vdots & \vdots & \ddots & \vdots \\ h_n & h_{n+1} & \dots & h_{2n-1} \end{bmatrix}$$

is the Hankel matrix of Markov parameters, h_i, defined by

$$H(s) = c'(sI - A)^{-1}b = \sum_{i=1}^{\infty} h_i s^{-i}$$

Solution: It is easy to see this because

$$\begin{aligned} \mathcal{OC} &= \begin{bmatrix} c' \\ c'A \\ \vdots \\ c'A^{n-1} \end{bmatrix} [b \;\; Ab \;\; A^2b \;\; \dots \;\; A^{n-1}b] \\ &= \begin{bmatrix} c'b & c'Ab & \dots & c'A^{n-1}b \\ c'Ab & c'A^2b & \dots & c'A^nb \\ \vdots & \vdots & \ddots & \vdots \\ c'A^{n-1}b & c'A^nb & \dots & c'A^{2n-2}b \end{bmatrix} \end{aligned}$$

Note: *In the controllability canonical realization $\mathcal{C}_{co} = I$, therefore*

$$\mathcal{O}_{co} = \mathcal{M}$$

Similarly, in the observability canonical $\mathcal{O}_{ob} = I$, hence

$$\mathcal{C}_{ob} = \mathcal{M}$$

Problem 3.8.12 A system is given by

$$\begin{bmatrix} \dot{x}_1 \\ \dot{x}_2 \\ \dot{x}_3 \end{bmatrix} = \begin{bmatrix} -1 & 0 & 1 \\ 1 & -2 & 0 \\ 0 & 2 & -1 \end{bmatrix} \begin{bmatrix} x_1 \\ x_2 \\ x_3 \end{bmatrix} + \begin{bmatrix} 1 \\ 0 \\ 1 \end{bmatrix} u$$

$$y = \begin{bmatrix} 0 & 0 & 1 \end{bmatrix} \begin{bmatrix} x_1 \\ x_2 \\ x_3 \end{bmatrix}$$

Use MATLAB commands to calculate Markov parameters h_1, h_2, $\dots h_5$. Compare that with the result of long division of the numerator by denominator in $H(s)$.

Solution: Type in the following in MATLAB prompt: `M = obsv(A,c')*ctrb(A,b)` to obtain

$$\mathcal{M} = \begin{bmatrix} 1 & -1 & 3 \\ -1 & 3 & -7 \\ 3 & -7 & 13 \end{bmatrix}$$

The transfer function can be obtained by typing the following line in MATLAB: `[num,den] = ss2tf(A,b,c',d,1)` which yields

$$H(s) = \frac{s^2 + 3s + 4}{s^3 + 4s^2 + 5s}$$

The result of long division agrees with elements of $\mathcal{M}$ above:

$$\frac{s^2 + 3s + 4}{s^3 + 4s^2 + 5s} = s^{-1} - s^{-2} + 3s^{-3} - 7s^{-4} + 13s^{-5} - 17s^{-6} + \dots$$

3.9 State feedback

In this Section we shall see how to move the poles to any desired position, something that was not possible by a simple output feedback (cf. Problem 3.4.15). We shall see that the condition for this so-called modal controllability is the state controllability and observability.

Problem 3.9.1 Consider a system given by

$$\dot{x} = Ax + bu$$

Let the system be controllable, i.e.,

$$\rho\,[b\ Ab\ A^2b\ \ldots\ A^{n-1}b] = n$$

If the state vector is observable and a state feedback is implemented using an arbitrarily chosen gain vector as in Figure 3.33, i.e.,

$$u = -k'x + v$$

write the new system equation of the feedback system, and show that the new system is controllable for any feedback gain vector k'.

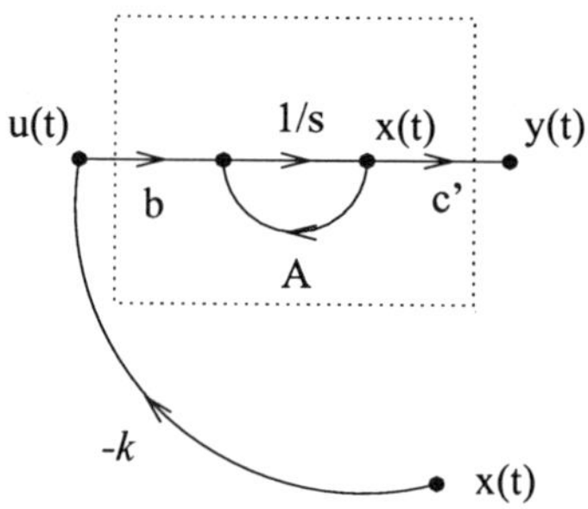

Figure 3.33: State feedback feeds back the state vector $x(t)$ to the input. If the state vector is not directly available, it has to be determined using the observability of the system.

Solution: The system equation of the feedback system is found from

$$\left.\begin{array}{rcl} \dot{x} & = & Ax + bu \\ u & = & -k'x + v \end{array}\right\} \quad \Rightarrow \quad \dot{x} = (\underbrace{A - bk'}_{A_f})x + bv$$

Since

$$\underbrace{[b\ Ab\ A^2b\ \ldots\ A^{n-1}b]}_{C} = \underbrace{[b\ A_fb\ A_f^2b\ \ldots\ A_f^{n-1}b]}_{C_f} \underbrace{\begin{bmatrix} 1 & k'b & k'Ab & \ldots & k'A^{n-2}b \\ 0 & 1 & k'b & \ldots & k'A^{n-3}b \\ 0 & 0 & 1 & \ldots & k'A^{n-4}b \\ \vdots & \vdots & \vdots & \ddots & \vdots \\ 0 & 0 & 0 & \ldots & 1 \end{bmatrix}}_{D}$$

i.e.,

$$\mathcal{C} = \mathcal{C}_f D$$

we have

$$\det(\mathcal{C}) = \det(\mathcal{C}_f D) = \det(\mathcal{C}_f) \underbrace{\det(D)}_{1} = \det(\mathcal{C}_f)$$

Since the system is single-input-single-output,

$$\rho(\mathcal{C}) = n \quad \Rightarrow \quad \det(\mathcal{C}) \neq 0 \quad \Rightarrow \quad \det(\mathcal{C}_f) \neq 0 \quad \Rightarrow \quad \rho(\mathcal{C}_f) = n$$

Therefore, the controllability of the system is not affected by the implementation of the state feedback. Also, if the initial system was not controllable, the state feedback cannot make it controllable, because

$$\rho(\mathcal{C}) < n \quad \Rightarrow \quad \det(\mathcal{C}) = 0 \quad \Rightarrow \quad \det(\mathcal{C}_f) = 0 \quad \Rightarrow \quad \rho(\mathcal{C}_f) < n$$

Note: *In Problem* 3.9.2 *we shall see that the state feedback can affect the observability of the system.*

Problem 3.9.2 Given a continuous-time system

$$\dot{x} = Ax + bu$$

$$y = c'x$$

where

$$A = \begin{bmatrix} 1 & 2 \\ 3 & 1 \end{bmatrix}, \quad b = \begin{bmatrix} 0 \\ 1 \end{bmatrix}, \quad c' = [1 \;\; 2]$$

discuss its controllability and observability.

A state feedback controller is used such that

$$u = -[3 \;\; 1]\, x + v$$

Find the system equation of the feedback system, and discuss its controllability and observability.

Solution: Although the initial system is both controllable and observable:

$$\det(\mathcal{C}) = -2 \neq 0 \quad \text{and} \quad \det(\mathcal{O}) = -19 \neq 0$$

the feedback system is controllable (by the previous problem it remains controllable for any choice of k'), but is not observable:

$$A_f = A - bk' = \begin{bmatrix} 1 & 2 \\ 0 & 0 \end{bmatrix}, \quad b_f = b = \begin{bmatrix} 0 \\ 1 \end{bmatrix}, \quad c_f' = c' = [1 \;\; 2]$$

$$\det(\mathcal{C}_f) = -2 \neq 0 \quad \text{but} \quad \det(\mathcal{O}_f) = 0$$

Problem 3.9.3 For the system given by

$$\dot{x} = \begin{bmatrix} 2 & 1 \\ -1 & 1 \end{bmatrix} x + \begin{bmatrix} 1 \\ 2 \end{bmatrix} u$$

$$y = \begin{bmatrix} 1 & 1 \end{bmatrix} x$$

draw a signal flow graph. Find the transfer function using the Mason's formula. Check this result by using the formula $H(s) = c'(sI - A)^{-1}b + d$. Find the eigenvalues of the system. Is the system stable? Is it controllable? Is it observable? Find a gain vector such that the state feedback system defined by

$$u(t) = -k'x(t) + v(t)$$

has eigenvalues at -1 and -2.

Solution: The signal flow graph looks as in Figure 3.34. To apply the Mason's formula, first write

$$\left.\begin{array}{ll} P_1(s) = 1/s, & \Delta_1(s) = 1 - 1/s, \\ P_2(s) = 2/s, & \Delta_2(s) = 1 - 2/s, \\ P_3(s) = -1/s^2, & \Delta_3(s) = 1, \\ P_4(s) = 2/s^2, & \Delta_4(s) = 1, \end{array}\right\} \Rightarrow \sum_{i=1}^{4} P_i(s)\Delta_i(s) = \frac{3s-4}{s^2}$$

and

$$\Delta(s) = 1 - \left(\frac{2}{s} + \frac{1}{s} - \frac{1}{s^2}\right) + \frac{2}{s^2} = \frac{s^2 - 3s + 3}{s^2}$$

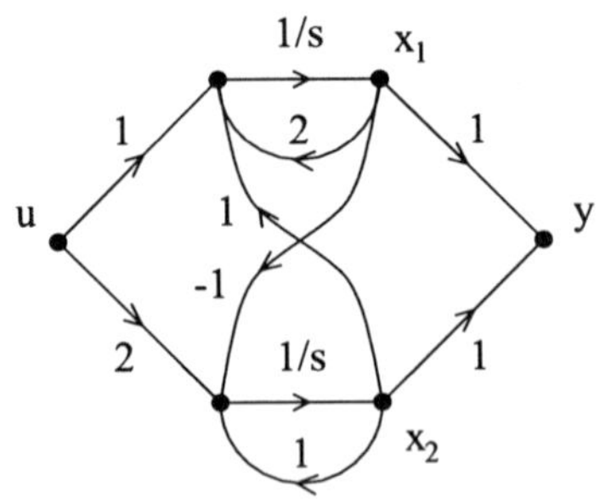

Figure 3.34: The signal flow graph of the system in Problem 3.9.3.

Finally, with $n = 4$,

$$H(s) = \frac{1}{\Delta(s)} \sum_{i=1}^{n} P_i(s)\Delta_i(s) = \frac{3s-4}{s^2 - 3s + 3}$$

The eigenvalues are found from

$$\det(\lambda I - A) = 0$$

hence

$$\lambda_{1,2} = \frac{3}{2} \pm j\frac{\sqrt{3}}{2}$$

therefore the system is not stable.

Since $\det(\mathcal{C}) \neq 0$ and $\det(\mathcal{O}) \neq 0$, the system is both controllable and observable.

To find the feedback gain vector k' which moves the eigenvalues to -1 and -2, write

$$\det(\lambda I - (A - bk')) = (\lambda + 1)(\lambda + 2)$$

Thus

$$\begin{vmatrix} \lambda + k_1 - 2 & k_2 - 1 \\ 2k_1 + 1 & \lambda + 2k_2 - 1 \end{vmatrix} = (\lambda + 1)(\lambda + 2)$$

i.e.,

$$\lambda^2 + (k_1 + 2k_2 - 3)\lambda + k_1 - 5k_2 + 3 = \lambda^2 + 3\lambda + 2$$

which finally implies

$$k' = [k_1 \; k_2] = [4 \;\; 1]$$

Problem 3.9.4 In general, if the system is controllable, we can arbitrarily change its eigenvalues by a proper choice of the feedback gain vector. If the characteristic polynomial of the initial system is

$$a(s) = s^n + a_1 s^{n-1} + \ldots + a_{n-1}s + a_n$$

while the characteristic polynomial of the desired closed-loop system is

$$\alpha(s) = s^n + \alpha_1 s^{n-1} + \ldots + \alpha_{n-1}s + \alpha_n$$

then with

$$a' = [a_1 \; \ldots \; a_n] \quad \text{and} \quad \alpha' = [\alpha_1 \; \ldots \; \alpha_n]$$

we can use the Bass-Gura formula

$$k' = (\alpha' - a')\mathcal{C}_c\mathcal{C}^{-1}$$

or the Ackermann formula

$$k' = [0 \; \ldots \; 0 \; 1]\mathcal{C}^{-1}\alpha(A)$$

Note that always $\mathcal{C}_c = a_-^{-T}$, where a_- is as defined in Problem 3.8.4.

Prove the Bass-Gura formula by transforming the original system into its controller form, designing the feedback for that form, and transforming back to the original form.

Solution: In Problem 3.8.6 we found that if a system given by

$$\dot{x} = Ax + bu$$

$$y = c'x$$

is controllable then it can be transformed into a controller form by a nonsingular transformation. The similarity transformation is given by

$$S = \mathcal{C}_c\mathcal{C}^{-1} \qquad (\mathcal{C}_c = \mathsf{a}_-^{-T})$$

When the system is written in the controller form, it is easy to find k_c' such that

$$\det(sI - (A_c - b_c k_c')) = \alpha(s)$$

Indeed, due to the special forms of A_c and b_c, matrix $A_c - b_c k_c'$ is also a companion matrix with the following characteristic polynomial

$$\det(sI - (A_c - b_c k_c')) = s^n + (a_1 + k_1^c)s^{n-1} + \ldots + (a_{n-1} + k_{n-1}^c)s + a_n + k_n^c$$

Hence,

$$k_c' = \alpha' - a'$$

and, back to the original form,

$$k' = k_c' S \qquad (S = \mathcal{C}\mathcal{C}_c^{-1})$$

Note: *In Problem* 3.9.7 *we show that controllability is also a necessary condition. It is quite remarkable that the condition for arbitrary pole placement (also called modal controllability) is the same as for state controllability:* $\rho(\mathcal{C}) = n$. *The reader should try to apply these formulas to the Problem* 3.9.3.

Problem 3.9.5 Consider a system with the transfer function

$$H(s) = \frac{(s-1)(s+2)}{(s+1)(s-2)(s+3)}$$

Note that $H(s)$ is irreducible (i.e., there are no pole-zero cancellations). Is it possible to change $H(s)$ into

$$G(s) = \frac{s-1}{(s+2)(s+3)}$$

by state feedback? If it is, calculate the corresponding feedback gain vector.

Solution: Yes, $H(s)$ can be transformed into $G(s)$ by the state feedback, because the irreducibility of $H(s)$ implies any of its representations is controllable, so we can apply the Bass-Gura formula to design a feedback such that the closed-loop eigenvalues are $\mu_1 = -2$, $\mu_2 = -2$, and $\mu_3 = -3$. Since this technique does not affect the zeros of the system (cf. Problem 3.9.6), we see that the new transfer function is going to be equal to $G(s)$.

If we write the system in the controller form

$$A = \begin{bmatrix} -2 & 5 & 6 \\ 1 & 0 & 0 \\ 0 & 1 & 0 \end{bmatrix} \qquad b = \begin{bmatrix} 1 \\ 0 \\ 0 \end{bmatrix} \qquad c' = [1 \;\; 1 \; -2]$$

we don't need the Bass-Gura formula. Indeed, the closed-loop system will have

$$A_f = A - bk' = \begin{bmatrix} -2-k_1 & 5-k_2 & 6-k_3 \\ 1 & 0 & 0 \\ 0 & 1 & 0 \end{bmatrix}$$

On the other hand, we want the first row of A_f to be $[-7 \; -16 \; -12]$, because

$$G(s) = \frac{s-1}{(s+2)(s+3)} = \frac{(s-1)(s+2)}{(s+2)^2(s+3)} = \frac{s^2+s-2}{s^3+7s^2+16s+12}$$

Therefore, we just need to pick $k' = [5 \;\; 21 \;\; 18]$.

Problem 3.9.6 Prove that the application of the state feedback does not affect the zeros of the system.

Solution: Consider a system in the controller form given by $\{A_c, b_c, c_c'\}$, where c_c' is made up of the coefficients of $b(s)$, the numerator of the transfer function. After the feedback, the system is still in the controller form, now given by $\{A_c - bk', b_c, c_c'\}$. Obviously, the coefficients of the numerator of the transfer function have not changed. Thus, the zeros are invariant under the state feedback.

Problem 3.9.7 Show that the pole placement by the state feedback is possible if and only if the system is controllable.

Solution: This is the proof originally given by Bass and Gura:

By definition

$$\begin{aligned} \alpha(s) &= \det(sI - A + bk') \\ &= \det((sI-A)(I + (sI-A)^{-1}bk')) \\ &= \det(sI-A)\det(I + (sI-A)^{-1}bk') \\ &= a(s)(1 + k'(sI-A)^{-1}b) \end{aligned}$$

therefore

$$\alpha(s) - a(s) = a(s)k'(sI-A)^{-1}b$$

By equating the coefficients of the powers on both sides we find

$$\begin{aligned} \alpha_1 - a_1 &= k'b \\ \alpha_2 - a_2 &= k'Ab + a_1k'b \\ \alpha_3 - a_3 &= k'A^2b + a_1k'Ab + a_2k'b \\ &\cdots \end{aligned}$$

i.e.,

$$\alpha' - a' = k'\mathcal{C}\mathsf{a}_-^T$$

Since a_- is always nonsingular, arbitrary pole placement is possible if and only if $\mathcal{C}$ has a full rank.

Problem 3.9.8 The dynamic equations of a simple inverted pendulum are given by

$$\dot{x} = \begin{bmatrix} 0 & 1 \\ 1 & 0 \end{bmatrix} x + \begin{bmatrix} 0 \\ -1 \end{bmatrix} u$$
$$y = \begin{bmatrix} 1 & 0 \end{bmatrix} x$$

Design a state feedback gain vector, i.e., $u(t) = -k'x + v$, to move the poles of the system to $-0.5 \pm 0.5j$, assuming that both state variables are available.

Solution: To make sure we can change both poles of the system, we first check if the system is controllable. Since $\det(\mathcal{C}) \neq 0$, we can continue:

original characteristic polynomial:	$\lambda^2 - 1$	$\Rightarrow$	$a' = [0\ -1]$
desired characteristic polynomial:	$\lambda^2 + \lambda + \frac{1}{2}$	$\Rightarrow$	$\alpha' = [1\ \ 0.5]$

Using the Bass-Gura formula we find

$$k' = (\alpha' - a')\mathcal{C}_c\mathcal{C}^{-1} = [-1.5\ \ -1]$$

Problem 3.9.9 For the inverted pendulum on a cart problem (analyzed in Problems 3.5.11 and 3.7.12), assume that all the system states are available. A feedback system is to be designed to obtain the desired eigenvalues of -1, -2, $-1+j$, and $-1-j$.

Let $u(t) = -[k_1\ k_2\ k_3\ k_4]\ x(t) + v(t)$. Find the gain vector $k' = [k_1\ k_2\ k_3\ k_4]$ to get the desired eigenvalues. (Since the system equation is quite simple with many zero elements, it is not necessary to convert this system to a controller form to design the feedback gain vector.)

Using this feedback gain vector repeat the simulations as in Problem 3.7.12.

Solution: The characteristic polynomial of the system is

$$a(s) = s^4 - 21.6s^2$$

therefore

$$a' = [0\ -21.6\ \ 0\ \ 0]$$

The characteristic polynomial of the desired closed-loop system is

$$\alpha(s) = (s+1)(s+2)(s+1-j)(s+1+j) = s^4 + 5s^3 + 10s^2 + 10s + 4$$

hence

$$\alpha' = [5\ \ 10\ \ 10\ \ 4]$$

Since

$$\mathcal{C}_c = a_-^{-T} = \begin{bmatrix} 1 & 0 & 0 & 0 \\ 0 & 1 & 0 & 0 \\ -21.6 & 0 & 1 & 0 \\ 0 & -21.6 & 0 & 1 \end{bmatrix}^{-T} = \begin{bmatrix} 1 & 0 & 21.6 & 0 \\ 0 & 1 & 0 & 21.6 \\ 0 & 0 & 1 & 0 \\ 0 & 0 & 0 & 1 \end{bmatrix}$$

and

$$\mathcal{C} = [b\ Ab\ A^2b\ A^3b] = \begin{bmatrix} 0 & 1 & 0 & 2 \\ 0 & -2 & 0 & -43.2 \\ 1 & 0 & 2 & 0 \\ -2 & 0 & -43.2 & 0 \end{bmatrix}$$

according to the Bass-Gura formula

$$k' = (\alpha' - a')\mathcal{C}_c\mathcal{C}^{-1} = [-0.2041\ -15.902\ -0.5102\ -2.7551]$$

Simulation for the unit step input of magnitude 0.1 and zero initial conditions:

First create the file called **pend2a.m** containing the following lines:

```
% file pend2a.m
%
function xdot = pend2a(t,x)
u = 0.2041*x(1) + 15.902*x(2) + 0.5102*x(3) + 2.7551*x(4) + 0.1;
xdot(1) =        x(3);
xdot(2) =        x(4);
xdot(3) =       -x(2) + 1*u;
xdot(4) = 21.6*x(2) - 2*u;
end
```

and then create and run the following file in MATLAB to obtain Figure 3.35:

```
% file simul2a.m
%
t0 = 0;
tf = 8;
x0 = [0;0;0;0];
[t,x] = ode45('pend2a',t0,tf,x0);
subplot(2,1,1)
plot(t,x(:,1))
title('Inverted Pendulum Simulation 2a')
xlabel('t [s]'), ylabel('z(t) [m]'), grid
subplot(2,1,2)
plot(t,x(:,2))
xlabel('t [s]'), ylabel('theta(t) [rad]'), grid
```

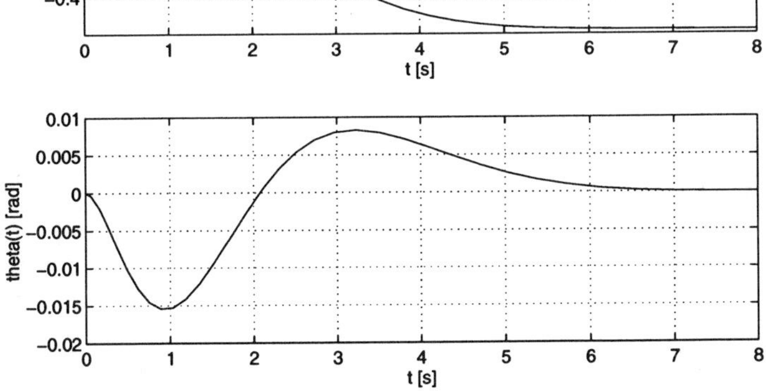

Figure 3.35: The results of the MATLAB Simulation 2a.

Simulation for the zero input and initial conditions: $x'(0) = [0\ 0.1\ 0\ 0]'$.

Now begin by creating the file called `pend2b.m` containing the following lines

```
% file pend2b.m
%
function xdot = pend2b(t,x)
u = 0.2041*x(1) + 15.902*x(2) + 0.5102*x(3) + 2.7551*x(4);
xdot(1) =        x(3);
xdot(2) =        x(4);
xdot(3) =       -x(2) + 1*u;
xdot(4) = 21.6*x(2) - 2*u;
end
```

and then create and run the following file in MATLAB to obtain Figure 3.36:

```
% file simul2b.m
%
t0 = 0;
tf = 8;
x0 = [0;0.1;0;0];
[t,x] = ode45('pend2b',t0,tf,x0);
subplot(2,1,1)
plot(t,x(:,1))
title('Inverted Pendulum Simulation 2b')
xlabel('t [s]'), ylabel('z(t) [m]'), grid
subplot(2,1,2)
plot(t,x(:,2))
xlabel('t [s]'), ylabel('theta(t) [rad]'), grid
```

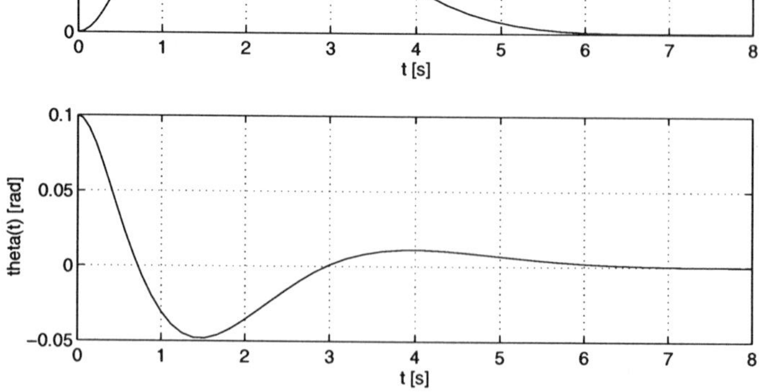

Figure 3.36: The results of the MATLAB Simulation 2b.

If we use the commands from the MATLAB CONTROLS TOOLBOX such as **initial** and **lsim**, we can simulate continuous linear systems. These two commands give us the response of the system to initial conditions and the controlling input, respectively.

The following program can be used instead of the previous two programs:

```
% file simul2c.m
%
A = [0    0 1 0;
     0    0 0 1;
     0   -1 0 0;
     0 21.6 0 0];
B = [0;0;1;-2];
C = [1 0 0 0;
     0 1 0 0];
D = [0;0];
K = [-0.2041 -15.902 -0.5102 -2.7551];  % picked so that eig(A-B*K)
                                         % are -1, -2, -1+j, -1-j
t0 = 0;
tf = 8;
dt = 0.05;
t = (t0:dt:tf)';

x0 = [0;0;0;0];              % simulation
u = 0.1*ones(size(t));       % 2a

% x0 = [0;0.1;0;0];          % simulation
% u = zeros(size(t));        % 2b

[Yinit,Xinit] = initial(A-B*K,B,C,D,x0,t);
[Yinp,Xinp] = lsim(A-B*K,B,C,D,u,t);
x = Xinit+Xinp;
y = Yinit+Yinp;

subplot(2,1,1)
plot(t,x(:,1))
title('Inverted Pendulum Simulation 2c')
xlabel('t [s]'), ylabel('z(t) [m]'), grid
subplot(2,1,2)
plot(t,x(:,2))
xlabel('t [s]'), ylabel('theta(t) [rad]'), grid
```

Problem 3.9.10 Consider the state equation of a system given by

$$\dot{x} = \begin{bmatrix} 2 & 1 & 0 & 0 \\ 0 & 2 & 0 & 0 \\ 0 & 0 & -1 & 0 \\ 0 & 0 & 0 & -1 \end{bmatrix} x + \begin{bmatrix} 0 \\ 1 \\ 1 \\ 1 \end{bmatrix} u$$

$$y = \begin{bmatrix} 1 & 1 & 1 & 1 \end{bmatrix} x$$

Is this system controllable? Is it stable? Can it be stabilized by a state feedback $u = -k'x$?

Solution: Since $\det(\mathcal{C}) = 0$, this system is not controllable. It is also not stable, because its eigenvalues are -1, -1, 2, and 2. But if we draw its signal flow graph as in Figure 3.37, we can see that the unstable modes are controllable, therefore the system is stabilizable.

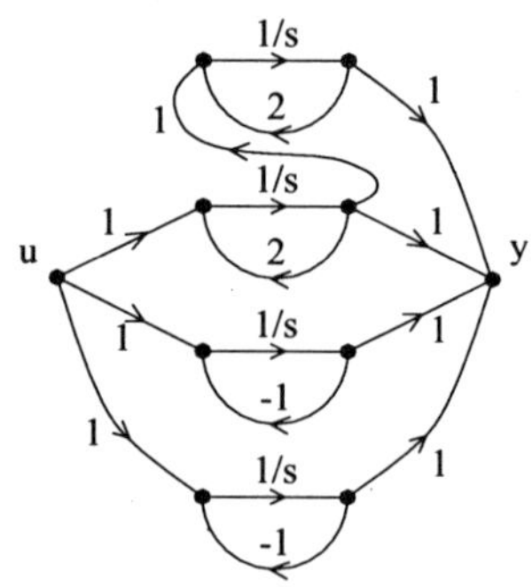

Figure 3.37: Although this system is not controllable, it can be stabilized. This is because all unstable modes are controllable. This system is said to be *stabilizable.*

If we wish to move the unstable eigenvalues from 2 to -2, and leave the stable eigenvalues where they are, we can write $k' = [k_1 \;\; k_2 \;\; 0 \;\; 0]$. Then from

$$\det(sI - (A - bk')) = (s+1)^2(s+2)^2$$

we find

$$k' = [16 \;\; 8 \;\; 0 \;\; 0]$$

Note 1: *Can we move all eigenvalues of the system to -2? Can we move the eigenvalues from -1 at all?*

Note 2: *Based on the PBH controllability criterion (cf. Problem 3.7.8) we can say that the system is stabilizable if and only if no left eigenvectors corresponding to the unstable eigenvalues of A are orthogonal to vector b. In other words the system is stabilizable if and only if*

$$p'b \neq 0 \qquad \textit{for all unstable eigenvalues of } A$$

Problem 3.9.11 Why is system observability important in the state feedback design? Define and explain the system property dual to stabilizability (Problem 3.9.10), the so-called detectability.

Solution: Throughout this Section we assumed that the states of the system were directly available for the state feedback. In Sections 3.11, 3.12, and 3.13 we shall learn how to design state observers. They determine (in noisy environments we say *estimate*) the states $x(t)$ from the output $y(t)$ and the input $u(t)$. This is why observability is so critical in the state feedback design.

If not all modes of the system are unstable, it is not necessary to observe all states of the system. If all unstable modes are observable, we say that the system is *detectable.*

Note 1: *Using the PBH observability criterion (cf. Problem* 3.7.8*) we see that the system is detectable if and only if no right eigenvectors corresponding to the unstable eigenvalues of* A *are orthogonal to vector* c'. *In other words the system is detectable if and only if*

$$c'q \neq 0 \qquad \textit{for all unstable eigenvalues of } A$$

3.10 Optimal control

In the process of stabilizing an unstable system, we must move all right-hand side poles to the left-hand side. How far should these poles be moved? One possible solution is to define a cost function indicating the relative cost of error versus the cost of control. In this Section we derive the formula for the optimal linear state feedback gain using a quadratic cost function (linear-quadratic-regulator — LQR). These optimality requirements reduce to the algebraic Riccati equation, whose solution is used in determination of the optimal feedback gain.

Problem 3.10.1 Let the system described by

$$\dot{x}(t) = Ax(t) + Bu(t)$$

be disturbed at $t = 0$, and let us consider the problem of finding the input $u(t)$ which will return the system to the equilibrium at the origin.

If the system is not asymptotically stable, we need to design a negative feedback $u(t) = -Kx(t)$ to make sure that the system will return to the origin. If the system is asymptotically stable, it will go back to the origin by itself, but even in that case it is useful to design a feedback, to make the return to the origin faster, or to satisfy some other optimality criteria.

The trade-off between the speed of return and the cost of control[9] is usually described by the following index function:

$$J = \int_0^\infty (x'(t)Qx(t) + u'(t)Ru(t))\, dt$$

in which matrices Q and R are chosen so that they reflect the prices (also called penalties) associated with values of states and control. Q is positive semi-definite, while R is positive definite. Our final goal will be to minimize the total cost of returning the system to the origin.

Derive the formula for the feedback gain K that minimizes the total cost J.

Solution: One possible approach to solving this problem is via the Lyapunov equation. In that approach it is shown that $J = x'(0)Px(0)$, where P is a positive definite solution of the algebraic Riccati equation

$$PA + A'P - PBR^{-1}B'P + Q = 0$$

The minimization of J with respect to K requires that

$$K = R^{-1}B'P$$

Here we shall use the calculus of variations approach, in which we wish to find $u(t)$ such that

$$J = \int_0^\infty (x'(t)Qx(t) + u'(t)Ru(t))\, dt$$

[9]Fast return to the origin requires the closed-loop eigenvalues far to the left, but that implies large values of the feedback gains in K, i.e., high cost of control.

is minimized under the constraints of the system equations

$$\dot{x}(t) = Ax(t) + Bu(t)$$

$$x(0) = x_0$$

To do that we write

$$J = \int_0^\infty (\underbrace{x'(t)Qx(t) + u'(t)Ru(t)}_{L} + \lambda'(Ax(t) + Bu(t) - \dot{x}(t)))\, dt$$

or

$$J = \int_0^\infty (H - \lambda'\dot{x})\, dt$$

where $H = L + \lambda'(Ax(t) + Bu(t))$ is the Hamiltonian.

Hence

$$J = -\lambda' x\Big|_0^\infty + \int_0^\infty (H + \dot{\lambda}' x)\, dt$$

If $u_{opt}(t)$ exists, then infinitesimal variations δu cause no change in J, i.e., $\delta J = 0$. Since

$$J = -\lambda'\delta x\Big|_0^\infty + \int_0^\infty \left(\left(\frac{\partial H}{\partial x} + \dot{\lambda}'\right)\delta x + \frac{\partial H}{\partial u}\delta u\right)\, dt$$

and $\delta x|_0 = 0$ (because x_0 is specified), with a convenient choice

$$\dot{\lambda}' = -\frac{\partial H}{\partial x} \qquad \lambda'(\infty) = 0$$

this reduces to

$$\delta J = \int_0^\infty \frac{\partial H}{\partial u}\delta u\, dt = 0$$

which implies

$$\frac{\partial H}{\partial u} = 0$$

Equations

$$\dot{\lambda}' = -\frac{\partial H}{\partial x} \qquad \lambda'(\infty) = 0 \qquad \frac{\partial H}{\partial u} = 0$$

are called the Euler-Lagrange equations.

Since

$$H = x'Qx + u'Ru + \lambda'(Ax + Bu)$$

the Euler-Lagrange equations become

$$\dot{\lambda}(t) = -A'\lambda(t) - Qx(t) \qquad \lambda(\infty) = 0 \qquad u(t) = -R^{-1}B'\lambda(t)$$

We can write these equations as

$$\begin{bmatrix}\dot{x}(t)\\ \dot{\lambda}(t)\end{bmatrix} = \begin{bmatrix} A & -BR^{-1}B' \\ -Q & -A' \end{bmatrix} \begin{bmatrix}x(t)\\ \lambda(t)\end{bmatrix}$$

with the two-point boundary condition

$$x(0) = x_0 \qquad \lambda(\infty) = 0$$

It can be shown that if (A, B) is controllable (or at least stabilizable) then

$$\lambda(t) = Px(t)$$

Then obviously $u(t) = -Kx(t)$, where $K = R^{-1}B'P$.

To learn more about P, write

$$\left.\begin{aligned} \dot{\lambda} &= -Qx - A'\lambda = -(Q + A'P)x \\ \dot{\lambda} &= P\dot{x} = (PA - PBR^{-1}B'P)x \end{aligned}\right\} \Rightarrow \quad A'P + PA - PBR^{-1}B'P = -Q$$

i.e., P must be a solution of the algebraic Riccati equation.

Note: *The Riccati equation may not have a solution, or if it does, the solution may not be unique. It can be shown that if the system is controllable and observable, then the Riccati equation has a solution. In addition, if the solution is not unique, then there is only one solution which corresponds to the optimal* K. *This solution is the only symmetric positive definite solution of the Riccati equation. If it is too complicated to find all solutions and check which one of them is positive definite, we use the following MacFarlane-Potter-Fath method:*
The Hamiltonian matrix

$$\mathcal{H} = \begin{bmatrix} A & -BR^{-1}B' \\ -Q & -A' \end{bmatrix}$$

has $2n$ *eigenvalues symmetric with respect to the imaginary axis in the complex plane. For the* n *eigenvalues in the left half-plane we can write*

$$\mathcal{H}\begin{bmatrix} f_i \\ g_i \end{bmatrix} = \lambda_i \begin{bmatrix} f_i \\ g_i \end{bmatrix}$$

and finally

$$P = GF^{-1}$$

Problem 3.10.2 Show that the application of the optimal input, i.e., the input $u(t) = -Kx(t)$ which minimizes the cost function

$$J(x(0), u(t)) = \int_0^\infty (x'(\tau)Qx(\tau) + Ru^2(\tau))\, d\tau$$

stabilizes the system.

Solution: In a way it is obvious that the optimal feedback stabilizes the system, because the cost J for any stabilized system is certainly less than the cost for an unstable system. We can give a formal proof using the following Lyapunov function:

$$V(x) = x'Px$$

where P is the only symmetric positive definite solution of the algebraic Riccati equation

$$A'P + PA - PBR^{-1}B'P = -Q$$

Obviously, since $P > 0$, also $V(x) > 0$. In addition to that

$$\dot{V}(x) = \dot{x}'Px + x'P\dot{x} = x'(A'P + PA - 2PBR^{-1}B'P)x = x'(-Q - PBR^{-1}B'P)x \leq 0$$

The inequality at the end is true because $Q \geq 0$ and (note that $P > 0 \Rightarrow \det(P) \neq 0$)

$$\mathrm{sgn}_{(\text{any } x)}(x'PBR^{-1}B'Px) = \mathrm{sgn}_{(\text{any } y)}(y'BR^{-1}B'y) = \mathrm{sgn}_{(z\,=\,B'y)}(z'R^{-1}z) \geq 0$$

Therefore, the system is stable in the sense of Lyapunov.

Note: *To guarantee the asymptotic stability, we require $V(x) \not\equiv 0$ for any system trajectory, but this part depends on the specifics of the system. If $Q > 0$, the optimal system is guaranteed to be asymptotically stable.*

Problem 3.10.3 Show that the minimum cost is given by

$$J_{min}(x_0) = x_0'Px_0$$

where P is a solution of the algebraic Riccati equation

$$A'P + PA - PBR^{-1}B'P = -Q$$

Solution: The cost is minimized by $u = -Kx$, where $K = R^{-1}B'P$. Therefore

$$J_{min}(x_0) = \int_0^\infty x'(t)(Q + K'RK)x(t)\,dt$$

Since $x(t) = e^{(A-BK)t}x_0$, and $A - BK$ is stable, we can write

$$J_{min}(x_0) = x_0' \underbrace{\left(\int_0^\infty e^{(A-BK)'t}(Q + K'RK)e^{(A-BK)t}\,dt\right)}_{Z} x_0$$

From the Lyapunov theory (cf. Problem 3.6.7) we know that since $A - BK$ is stable, matrix Z is a unique solution of the following Lyapunov equation (we also use $K = R^{-1}B'P$):

$$(A - BK)'Z + Z(A - BK) = -(Q + PBR^{-1}B'P)$$

Since the Riccati equation $A'P + PA - PBR^{-1}B'P = -Q$ can be rewritten as

$$(A - BK)'P + P(A - BK) = -(Q + PBR^{-1}B'P)$$

and because of the uniqueness of Z, we can write $Z = P$, i.e.,

$$J_{min}(x_0) = x_0'Px_0$$

Problem 3.10.4 Consider a linear time-invariant system given by

$$\dot{x} = Ax + Bu$$

with the cost function defined by

$$J(x(0), u(t)) = \int_0^\infty (x'(\tau)Qx(\tau) + Ru^2(\tau))\, d\tau$$

Let

$$A = \begin{bmatrix} 0 & 1 \\ 2 & -1 \end{bmatrix}, \quad B = \begin{bmatrix} 0 \\ 1 \end{bmatrix}, \quad \text{and} \quad x(0) = \begin{bmatrix} 1 \\ 1 \end{bmatrix}$$

and let $Q = I_{2\times 2}$, while R is a positive scalar. Value of R is the relative cost of control with respect to state error.

The optimal control input is given by

$$u(t) = -Kx(t)$$

with $K = R^{-1}B'P$, where P is a positive definite symmetric solution of the following equation:

$$PA + A'P - PBR^{-1}B'P + Q = 0$$

The optimal cost is then

$$J(x(0)) = x'(0)Px(0)$$

a) For $R = 1$ solve for P. Find K and calculate the optimal cost. Calculate the open-loop and the closed-loop eigenvalues.

b) Repeat the above calculations for $R = 0.1$.

c) Repeat the above calculations for $R = 10$.

Solution: In all three cases we put

$$P = \begin{bmatrix} p_1 & p_2 \\ p_2 & p_3 \end{bmatrix}$$

and find exactly four solutions of the equation $PA + A'P - PBR^{-1}B'P + Q = 0$. In each case only one of four solutions is positive definite (of course, the symmetry is insured by the initial choice of the elements of P).

a) $P = \begin{bmatrix} 7+\sqrt{5} & 2+\sqrt{5} \\ 2+\sqrt{5} & \sqrt{5} \end{bmatrix} = \begin{bmatrix} 9.24 & 4.24 \\ 4.24 & 2.24 \end{bmatrix}, \quad K = [4.24 \;\; 2.24], \quad J = 19.9.$

The open-loop eigenvalues are $\lambda_1 = 1$ and $\lambda_2 = -2$, while the closed-loop eigenvalues are $\mu_1 = -1$ and $\mu_2 = -2.24$.

b) $P = \begin{bmatrix} 1.97 & 0.57 \\ 0.57 & 0.37 \end{bmatrix}$, $K = [5.74 \quad 3.74]$, $J = 3.5$. The open-loop eigenvalues are $\lambda_1 = 1$ and $\lambda_2 = -2$, while the closed-loop eigenvalues are $\mu_1 = -1$ and $\mu_2 = -3.74$.

c) $P = \begin{bmatrix} 81.25 & 40.25 \\ 40.25 & 20.25 \end{bmatrix}$, $K = [4.02 \quad 2.02]$, $J = 182$. The open-loop eigenvalues are $\lambda_1 = 1$ and $\lambda_2 = -2$, while the closed-loop eigenvalues are $\mu_1 = -1$ and $\mu_2 = -2.02$.

Note: *Note that for large values of R unstable eigenvalues are not just moved from the right half-plane to the left half-plane, but are moved into their mirror images in the left half-plane. Also, note that the optimal feedback has stabilized the system. See also Problem* 3.10.2.

Problem 3.10.5 Consider a linear quadratic optimization problem

$$\dot{x}(t) = Ax(t) + Bu(t) \qquad x(0) = [1 \quad 1]'$$

$$J = \frac{1}{2}\int_0^\infty (x'Qx + u'Ru)\,dt$$

In particular, consider the case when

$$A = \begin{bmatrix} 0 & 1 \\ -4 & 0 \end{bmatrix}, \quad B = \begin{bmatrix} 0 \\ 1 \end{bmatrix}, \quad Q = \begin{bmatrix} 1 & 0 \\ 0 & q_v \end{bmatrix}, \quad R = 1$$

where q_v is a positive number.

The optimal feedback is given by

$$K = R^{-1}B'P$$

where P is a positive definite symmetric solution of

$$A'P + PA - PBR^{-1}B'P = -Q$$

Find the open-loop eigenvalues. Let $q_v = 1$, i.e., let the penalties for position and velocity be equal. Find P, K, the closed-loop eigenvalues, and the cost function. Do the same for $q_v = 4$.

Solution: The open-loop eigenvalues are found from $\det(\lambda I - A) = 0$:

$$\lambda_{1,2} = \pm 2j$$

To solve the Riccati equation, let

$$P = \begin{bmatrix} a & b \\ b & c \end{bmatrix}$$

Then the Riccati equation implies that

$$b^2 + 8b - 1 = 0$$

$$c^2 = 2b + 1$$

$$a = (b+4)c$$

The only solution of the above system which yields a positive definite matrix P is $a = 4.60$, $b = 0.12$, $c = 1.12$, i.e.,

$$P = \begin{bmatrix} 4.60 & 0.12 \\ 0.12 & 1.12 \end{bmatrix}$$

The feedback gain is then $K = R^{-1}B'P = [0.12 \;\; 1.12]$. The closed-loop eigenvalues are found from $\det(A - BK) = 0$:

$$\lambda_{1,2} = -0.56 \pm 1.95j$$

The cost of the optimal control is

$$J_{opt}(x(0)) = \frac{1}{2}x'(0)Px(0) = 2.98$$

For $q_v = 4$ we find

$$P = \begin{bmatrix} 8.50 & 0.12 \\ 0.12 & 2.06 \end{bmatrix}, \qquad K = [0.12 \;\; 2.06]$$

while

$$\lambda_{1,2} = -1.03 \pm 1.75j, \qquad \text{and} \qquad J_{opt}(x(0)) = 5.40$$

Note: *Note that when the penalty for velocity is high, the cost of control is higher, and the eigenvalues are moved further to the left. Again, note that the optimal feedback has stabilized the system.*

Problem 3.10.6 Consider a linear quadratic optimization problem

$$\dot{x}(t) = Ax(t) + Bu(t)$$

$$y(t) = Cx(t)$$

$$J = \frac{1}{2}\int_0^\infty (x'Qx + u'Ru)\,dt$$

Consider a particular system

$$A = \begin{bmatrix} -1 & 2 \\ 1 & 0 \end{bmatrix}, \quad B = \begin{bmatrix} 1 \\ 0 \end{bmatrix}, \quad C = [0 \;\; 1], \quad Q = \begin{bmatrix} 2 & 0 \\ 0 & 1 \end{bmatrix}, \quad R = 4$$

The optimal feedback is given by

$$K = R^{-1}B'P$$

where P is a positive definite symmetric solution of

$$A'P + PA - PBR^{-1}B'P = -Q$$

Find the open-loop eigenvalues. Check controllability and observability of this system. Assuming that both state variables are available, find positive definite and symmetric P from the algebraic Riccati equation. Find the optimal feedback gains and the closed-loop eigenvalues.

Solution: The following MATLAB commands can be used to find a solution:

```
A = [-1 2 ; 1 0 ];
B = [ 1 ; 0 ];
C = [ 0   1 ];
Q = [ 2 0 ; 0 1 ];
R = 4;

Eopen = eig(A)
   Eopen =
    -2
     1

rank(ctrb(A,B))
   ans =
     2

rank(obsv(A,C))
   ans =
     2

[K,P,Eclosed] = lqr(A,B,Q,R)
  K =
    2.1021    4.0616
  P =
    8.4085   16.2462
   16.2462   33.5807
  Eclosed =
   -2.1378
   -0.9644
```

3.11 State observers

In this Section the design of state observers is explained. They are used to estimate the states for the state feedback when the original states are not directly available or measurable.

Problem 3.11.1 A system is given by

$$\begin{bmatrix}\dot{x}_1\\ \dot{x}_2\\ \dot{x}_3\end{bmatrix} = \begin{bmatrix} 0 & 3 & 0\\ 0 & 2 & 1\\ 1 & 1 & 0\end{bmatrix}\begin{bmatrix}x_1\\ x_2\\ x_3\end{bmatrix} + \begin{bmatrix}2\\ 1\\ -1\end{bmatrix} u$$

$$y = \begin{bmatrix} 0 & 2 & 1\end{bmatrix}\begin{bmatrix}x_1\\ x_2\\ x_3\end{bmatrix}$$

Is it observable? Design a state observer such that the eigenvalues of the observer are all at -2.

A state observer is a simulator of the original system. Usually it is designed when the states $x(t)$ of the original system are not directly measurable, but can be estimated from the knowledge of the parameters of the system (A, b, and c'), the input $u(t)$, and the output $y(t)$. From Problems 3.7.5 and 3.7.6 we know that this can be done if and only if the original system is observable.

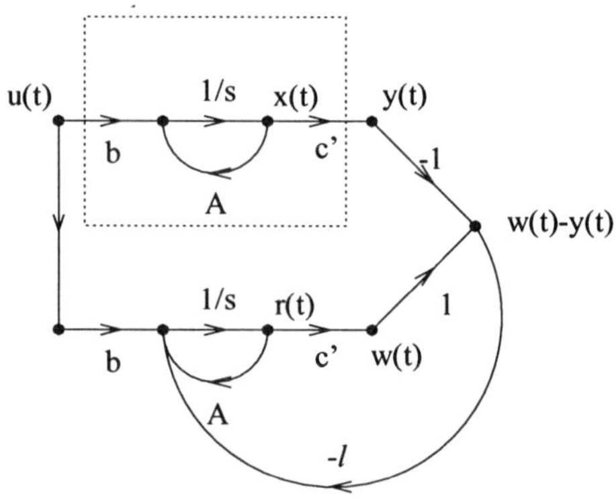

Figure 3.38: If the system is observable, than we can calculate the unmeasurable states $x(t)$ from the available information: system parameters A, b, and c', the input $u(t)$, and the output $y(t)$. The result of this estimation are the states $r(t)$, which soon after the beginning of the observation closely follow the actual states $x(t)$.

Let us call the estimated states $r(t)$ (see Figure 3.38). To improve the reliability of the estimated states $r(t)$, and their convergence to the real states $x(t)$, we use the difference between the estimated output $w(t) = c'r(t)$ and the actual output $y(t) = c'x(t)$. It is particularly important to use this difference if the system is unstable, because for unstable systems any discrepancy between the actual initial conditions

$x(0)$ and the supposed initial conditions $r(0)$ causes the estimation error $r(t) - x(t)$ to diverge. Another benefit from using this difference is the reduction of errors due to our imperfect knowledge of the system parameters A, b, and c'.

The gain vector l which weighs the influence of components of this difference on the estimated states determines the eigenvalues of the observer.

If we decide to design l so that the characteristic polynomial of the observer is $\alpha(s)$ (we should pick the eigenvalues of the observer to be much "faster" than those of the system), whereas the characteristic polynomial of the original system is $a(s)$, we can use the dual of the Bass-Gura formula to calculate l:

$$l = \mathcal{O}^{-1}\mathcal{O}_o(\alpha - a)$$

Solution: The system is observable because

$$\det(\mathcal{O}) = \begin{vmatrix} c' \\ c'A \\ c'A^2 \end{vmatrix} = \begin{vmatrix} 0 & 2 & 1 \\ 1 & 5 & 2 \\ 2 & 15 & 5 \end{vmatrix} = 3 \neq 0$$

Since $a(s) = \det(sI - A) = s^3 - 2s^2 - s - 3$ and $\alpha(s) = (s+2)^3 = s^3 + 6s^2 + 12s + 8$, we find the observer gain to be

$$l = \mathcal{O}^{-1}\mathcal{O}_o(\alpha - a) = (a_{-}\mathcal{O})^{-1}(\alpha - a) = \begin{bmatrix} 0 & 1 & -1/3 \\ 0 & 0 & 1/3 \\ 1 & 0 & -2/3 \end{bmatrix} \begin{bmatrix} 8 \\ 13 \\ 11 \end{bmatrix} = \begin{bmatrix} 9.33 \\ 3.67 \\ 0.67 \end{bmatrix}$$

Problem 3.11.2 A standard application of the state observers is in the state feedback design (see Figure 3.39). Assume that the system is both state controllable and observable. Assume also that the state feedback is designed as $u(t) = -k'x(t)$, where k' is such that the closed-loop system has the desired eigenvalues $\mu_1, \ldots, \mu_n$. Since the actual states are unavailable, use the observer-estimated states $r(t)$ instead of $x(t)$ and show that this substitution does not affect the designed closed-loop eigenvalues.

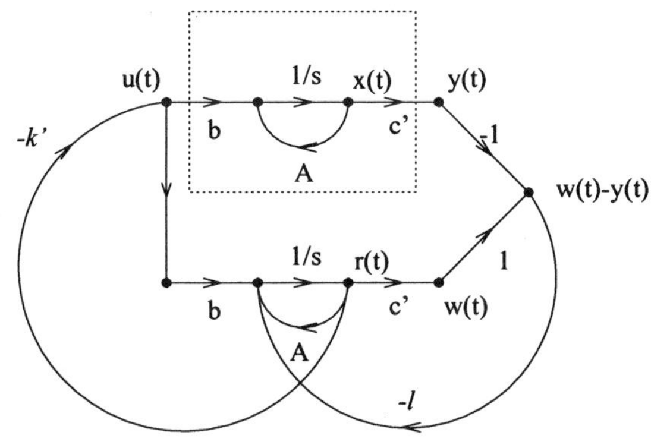

Figure 3.39: If the actual states are not measurable, the state feedback is implemented with the states estimated by the observer.

Solution: If instead of the actual state vector $x(t)$ we use the estimated state vector $r(t)$, we can write

$$\dot{x}(t) = Ax(t) - bk'r(t) = (A - bk')x(t) + bk'e(t)$$

where $e(t) = x(t) - r(t)$.

The dynamics of the estimation error vector $e(t)$ are described by

$$\dot{e}(t) = \dot{x}(t) - \dot{r}(t) = Ax(t) - bk'r(t) - ((A - lc' - bk')r(t) + lc'x(t)) = (A - lc')e(t)$$

The complete system, which includes both the original system and the observer, can be described by

$$\begin{bmatrix} \dot{x}(t) \\ \dot{e}(t) \end{bmatrix} = \begin{bmatrix} A - bk' & bk' \\ 0 & A - lc' \end{bmatrix} \begin{bmatrix} x(t) \\ e(t) \end{bmatrix}$$

The characteristic equation of the complete system is then

$$\det \begin{bmatrix} A - bk' & bk' \\ 0 & A - lc' \end{bmatrix} = 0$$

i.e.,

$$\det(A - bk')\det(A - lc') = 0$$

Due to this uncoupling of equations the eigenvalues of the original states are as desired, even though we used the feedback based on $r(t)$ instead of $x(t)$.

Note: *It is also interesting that the observer eigenvalues do not depend on k'. They are determined from $\det(A - lc') = 0$. This allows a complete separation of observer and controller design processes.*

Problem 3.11.3 For the system given by

$$\dot{x} = \begin{bmatrix} -1 & -2 & -2 \\ 0 & -1 & 1 \\ 1 & 0 & -1 \end{bmatrix} x + \begin{bmatrix} 2 \\ 0 \\ 1 \end{bmatrix} u$$

$$y = \begin{bmatrix} 1 & 1 & 0 \end{bmatrix} x$$

check controllability and observability, and design the state feedback such that the closed-loop eigenvalues are all at -2. The state feedback should be based on the observer with all eigenvalues at -4.

Solution: It is easy to check that $\det(\mathcal{C}) = -10 \neq 0$ and $\det(\mathcal{O}) = 5 \neq 0$, therefore we can proceed with the controller and observer design.

To find k', write $a(s) = \det(sI - A) = s^3 + 3s^2 + 5s + 5$, and $\alpha(s) = (s+2)^3 = s^3 + 6s^2 + 12s + 8$.

Then, by the Bass-Gura formula,

$$k' = (\alpha' - a')\mathcal{C}_c\mathcal{C}^{-1}$$

In this case

$$\mathcal{C}_c = \underline{a}^{-T} = \begin{bmatrix} 1 & 0 & 0 \\ 3 & 1 & 0 \\ 5 & 3 & 1 \end{bmatrix}^{-T} = \begin{bmatrix} 1 & -3 & 4 \\ 0 & 1 & -3 \\ 0 & 0 & 1 \end{bmatrix}$$

and

$$\mathcal{C}^{-1} = \begin{bmatrix} 2 & -4 & 0 \\ 0 & 1 & 0 \\ 1 & 1 & -5 \end{bmatrix}^{-1} = \begin{bmatrix} 0.5 & 2 & 0 \\ 0 & 1 & 0 \\ 0.1 & 0.6 & -0.2 \end{bmatrix}$$

hence

$$k' = [0.9 \;\; 0.4 \;\; 1.2]$$

To determine l, the feedback vector gain for the observer, write $a(s) = \det(sI - A) = s^3 + 3s^2 + 5s + 5$, and $\alpha(s) = (s+4)^3 = s^3 + 12s^2 + 48s + 64$.

Then, by the dual of the Bass-Gura formula,

$$l = \mathcal{O}^{-1}\mathcal{O}_o(\alpha - a)$$

In this case

$$\mathcal{O}_o = \underline{a}^{-1} = \mathcal{C}_c' = \begin{bmatrix} 1 & 0 & 0 \\ -3 & 1 & 0 \\ 4 & -3 & 1 \end{bmatrix}$$

and

$$\mathcal{O}^{-1} = \begin{bmatrix} 1 & 1 & 0 \\ -1 & -3 & -1 \\ 0 & 5 & 0 \end{bmatrix}^{-1} = \begin{bmatrix} 1 & 0 & -0.2 \\ 0 & 0 & 0.2 \\ -1 & -1 & -0.4 \end{bmatrix}$$

hence

$$l = \begin{bmatrix} 15.8 \\ -6.8 \\ -11.4 \end{bmatrix}$$

Problem 3.11.4 For the inverted pendulum on a cart problem (Problems 3.5.11, 3.7.12, and 3.9.9), design a controller as in Problem 3.9.9, but based on the states estimated by a state observer whose eigenvalues are -4, -4.5, -5, and -5.5.

Using MATLAB commands `reg`, `parallel`, and `cloop`, repeat the simulations as in Problem 3.9.9. To do that, first augment the system matrices so that the "control" and the "known" inputs are separated.

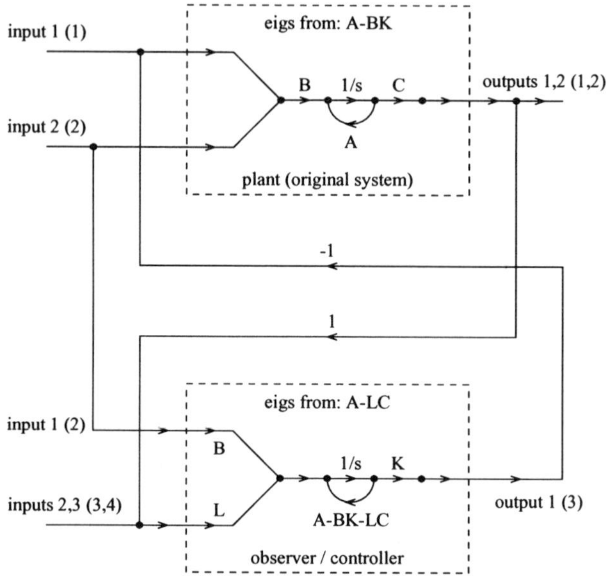

Figure 3.40: MATLAB commands for building systems are very easy to use when this diagram is kept in mind. The parenthesized numbers are those applicable after the command `parallel`. Note that the command `reg` provides the BK and the LC parts of feedback. Hence we write `reg(AA,BB,CC,DD,K,L,sensors,known,control)`, rather than `reg(AA-BB*K-L*CC,BB,CC,DD,K,L,sensors,known,control)`. Also note that the command `cloop` provides a feedback gain equal to 1. This is changed to -1 by putting a minus sign in front of the input label (`cloop(Abig,Bbig,Cbig,Dbig,[1 2 3],[3 4 -1])`).

Solution: Figure 3.40 shows the standard diagram used in MATLAB simulations. As we found earlier,

$$K = (\alpha' - a')\mathcal{C}_c\mathcal{C}^{-1} = [-0.2041 \ -15.902 \ -0.5102 \ -2.7551]$$

MATLAB provides two functions to do this, `place` and `acker`. For example, `K = place(A,B,P)` finds the state feedback gain K such that the eigenvalues of $A - BK$ are those specified in P.

The observer gain L should be found such that the eigenvalues of $A - LC$ are -4, -4.5, -5, and -5.5. We can use the MATLAB function `place` again, but first we have to adjust the input matrices to the problem – we can ask for L such that the eigenvalues of $A' - C'L'$ are as required: `L = (place(A',C',[-4, -4.5, -5, -5.5]))'`, which yields

$$L = \begin{bmatrix} 8.5 & 0 \\ 0 & 10.5 \\ 18 & -1 \\ 0 & 49.1 \end{bmatrix}$$

The MATLAB program should look as follows (see its output in Figure 3.41):

```
% file simul3.m
%

A = [0    0 1 0;
     0    0 0 1;
     0   -1 0 0;
     0 21.6 0 0];
B = [0; 0; 1; -2];
C = [1 0 0 0;
     0 1 0 0];
D = [0;0];

K = place(A,B,[-1, -2, -1+j, -1-j]);
L = (place(A',C',[-4, -4.5, -5, -5.5]))';

AA = A;              % augment B and D to separate feedback control input
BB = [B,B];          % control input from the external forcing function
CC = C;
DD = [D,D];

sensors = [1 2];
known = [2];           % info on control u1 is internally fed back
control = [1];
[Ac,Bc,Cc,Dc] = reg(AA,BB,CC,DD,K,L,sensors,known,control);

[Abig,Bbig,Cbig,Dbig] = parallel(AA,BB,CC,DD,Ac,Bc,Cc,Dc,[2],[1],[],[]);
[Abig,Bbig,Cbig,Dbig] = cloop(Abig,Bbig,Cbig,Dbig,[1 2 3],[3 4 -1]);

t0 = 0;
tf = 8;
dt = 0.0001;
t=(t0:dt:tf)';

% x0 = [0;0;0;0;0;0;0;0];   % simul 3a
% u2 = 0.1*ones(size(t));

x0 = [0;0.1;0;0;0;0;0;0];   % simul 3b
u2 = zeros(size(t));

u1 = zeros(size(t));
u3 = zeros(size(t));
u4 = zeros(size(t));
u = [u1 u2 u3 u4];

[y,x] = lsim(Abig,Bbig,Cbig,Dbig,u,t,x0); % both initial and forced response

y = y(1:300:size(t),:);     % reduce the amount of data by a factor of 300
x = x(1:300:size(t),:);
t = t(1:300:size(t),:);

subplot(2,1,1)
plot(t,y(:,1),'y-',t,x(:,5),'w.')
title('Inverted Pendulum Simulation 3b')
ylabel('z(t) [m]'), grid
subplot(2,1,2)
plot(t,y(:,2),'y-',t,x(:,6),'w.')
ylabel('theta(t) [rad]'), grid
```

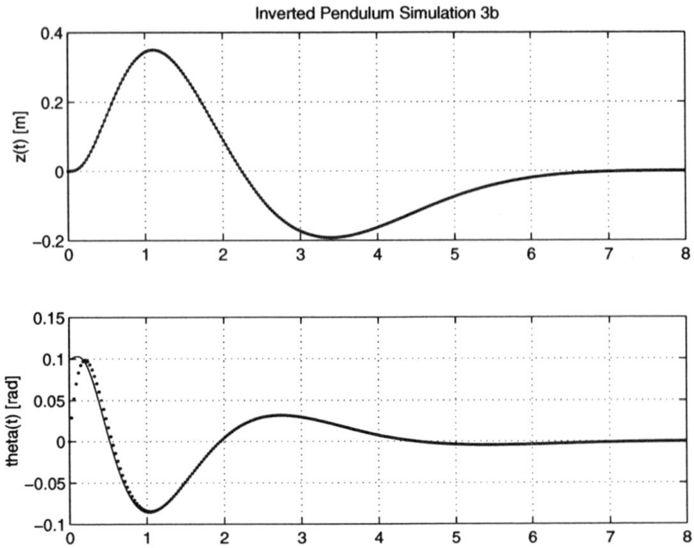

Figure 3.41: Results of MATLAB simulations of combined observer-controller applied to the inverted pendulum on a cart problem.

Problem 3.11.5 Show that the states can be estimated by the observer if and only if the system is state observable.

Solution: See Problems 3.7.5 and 3.7.6.

Problem 3.11.6 In the direct transfer function design procedures let the original system transfer function be

$$H(s) = \frac{b(s)}{a(s)} = \frac{s-1}{s(s-2)}$$

Show that a simple feedback with gain k cannot stabilize this system.

Using the diagram as in Figure 3.42 derive the equation which relates the desired characteristic polynomial $\alpha(s)$ to the given characteristic polynomial $a(s)$, and the feedback transfer functions

$$F(s) = \frac{p_u(s)}{\delta(s)} \qquad G(s) = \frac{p_y(s)}{\delta(s)}$$

Note that $\delta(s)$ is determined by the desired eigenvalues of the observer error.

Find $p_u(s)$, $p_y(s)$, and $\delta(s)$, so that the poles of the system are moved to -1 and -2, while the observer error eigenvalues are both at -4.

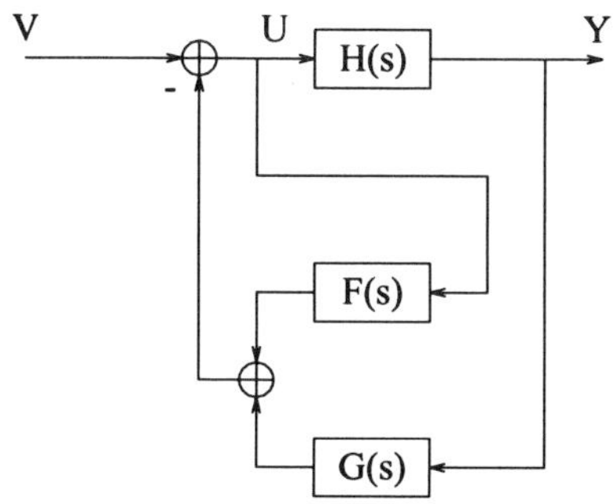

Figure 3.42: State-space approach provided the background for this transfer function design procedure.

Solution: If a feedback with gain k is applied, the new characteristic polynomial is $a(s)+kb(s)$, which in our case is $s^2+(k-2)s-k$. To satisfy the necessary condition (this is not a sufficient condition) for stability, we would need to have $k<0$ and $k>2$, which is, of course, impossible.

But with ideas and insights from the state-space approach, we can form the system as in Figure 3.42. From

$$U(s)=V(s)-\frac{p_u(s)}{\delta(s)}U(s)-\frac{p_y(s)}{\delta(s)}Y(s) \quad \text{and} \quad Y(s)=\frac{b(s)}{a(s)}U(s)$$

we find

$$\frac{Y(s)}{V(s)}=\frac{\delta(s)b(s)}{a(s)\delta(s)+a(s)p_u(s)+b(s)p_y(s)}$$

Therefore

$$\alpha(s)\delta(s)=a(s)\delta(s)+a(s)p_u(s)+b(s)p_y(s)$$

In our case

$$a(s)=s(s-2), \quad b(s)=s-1, \quad \alpha(s)=(s+1)(s+2), \quad \text{and} \quad \delta(s)=(s+4)^2$$

and polynomials $p_u(s)$ and $p_y(s)$ are assumed to be one degree lower than the polynomial $\delta(s)$. Compare coefficients for the powers of s on both sides. After some algebra, we find:

$$p_u(s)=5s-180 \quad \text{and} \quad p_y(s)=232s-32$$

3.12 Kalman-Bucy filter

When the measurements are noisy, the observer can be designed so that the influence of noise is minimized in the mean-square sense. The optimality requirements for this linear-quadratic-estimator (LQE) reduce to the algebraic Riccati equation. The resulting optimal estimator is called the Kalman-Bucy filter.

Problem 3.12.1 A linear time-invariant system with system and measurement noise inputs is shown in Figure 3.43.

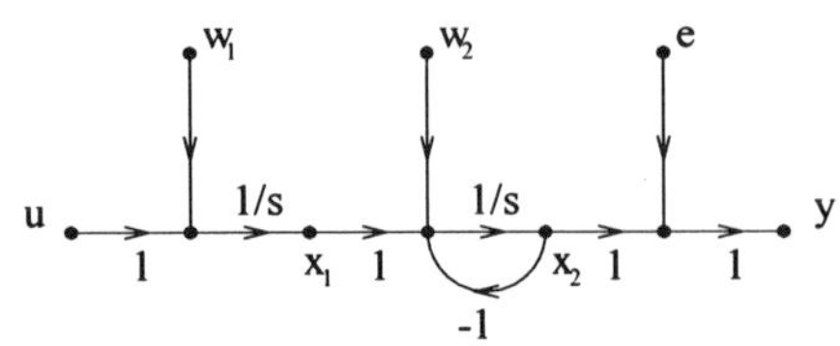

Figure 3.43: LTI with system and measurement noise.

Write the system equation in the form

$$\begin{aligned} \dot{x}(t) &= Ax(t) + Bu(t) + w(t) \\ y(t) &= Cx(t) + e(t) \end{aligned}$$

The noises are zero-mean, white, and Gaussian. The system noise $w(t)$ is uncorrelated with the measurement noise $e(t)$. Let the noise covariances be

$$\begin{aligned} E\left[w(t)w'(\tau)\right] &= Q\,\delta(t-\tau) \\ E\left[e(t)e'(\tau)\right] &= R\,\delta(t-\tau) \end{aligned}$$

where

$$Q = \begin{bmatrix} 16 & 0 \\ 0 & 0 \end{bmatrix} \quad \text{and} \quad R = 1$$

The steady-state Kalman-Bucy estimator is given by (see Figure 3.44)

$$\dot{r}(t) = Ar(t) + Bu(t) + L(y(t) - Cr(t))$$

where $L = PC'R^{-1}$, and P is a solution of the algebraic Riccati equation

$$AP + PA' - PC'R^{-1}CP + Q = 0$$

Find the open-loop eigenvalues, and check controllability, observability, and stability. Solve the Riccati equation, and check that P is positive definite and symmetric. Calculate L. Find the eigenvalues of the Kalman-Bucy estimator. Draw a signal flow graph of the combined system.

Solution: Obviously

$$\begin{bmatrix}\dot{x}_1\\ \dot{x}_2\end{bmatrix} = \begin{bmatrix} 0 & 0 \\ 1 & -1 \end{bmatrix}\begin{bmatrix}x_1\\ x_2\end{bmatrix} + \begin{bmatrix}1\\ 0\end{bmatrix}u + \begin{bmatrix} 1 & 0 \\ 0 & 1 \end{bmatrix}\begin{bmatrix}w_1\\ w_2\end{bmatrix}$$

$$y = \begin{bmatrix} 0 & 1 \end{bmatrix}\begin{bmatrix}x_1\\ x_2\end{bmatrix} + e$$

and the system is controllable, observable, and marginally stable.

With $P = \begin{bmatrix} a & b \\ b & c \end{bmatrix}$ the Riccati equation becomes a system of equations

$$b^2 = 16 \qquad a - b - bc = 0 \qquad 2b - 2c - c^2 = 0$$

whose only positive definite solution is

$$P = \begin{bmatrix} 12 & 4 \\ 4 & 2 \end{bmatrix}$$

Then

$$L = PC'R^{-1} = \begin{bmatrix}4\\ 2\end{bmatrix}$$

The eigenvalues of the Kalman estimator are the eigenvalues of $A - LC$, i.e., $-1.5 \pm 1.32j$.

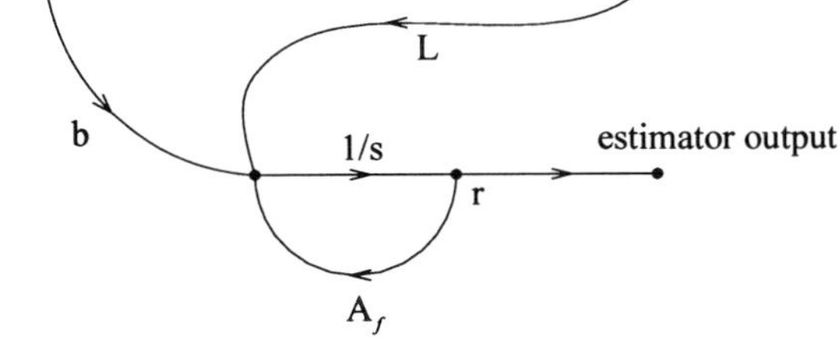

Figure 3.44: Typical configuration of the Kalman-Bucy filter. Note that $A_f = A - LC$.

Matlab note: *We can do the above calculations by a few simple* MATLAB *commands:*

```
A = [0 0; 1 -1];        % Matlab assumes model of the form
B = [1; 0];             % .
G = eye(2);             % x = Ax + Bu + Gw
C = [0 1];              % y = Cx + Du + e
Q = [16 0; 0 0];
R = 1;
[L,P,E] = lqe(A,G,C,Q,R)
   L =                  % Kalman-Bucy gain
    4.0000
    2.0000
   P =                  % Solution of the Riccati equation
   12.0000    4.0000
    4.0000    2.0000
   E =                  % Kalman-Bucy observer eigenvalues
  -1.5000 + 1.3229i
  -1.5000 - 1.3229i
```

Problem 3.12.2 Repeat MATLAB simulations for the inverted pendulum on a cart as in Problem 3.11.4 assuming small process and measurement noises. To account for the noise effects, use a stationary Kalman-Bucy filter. Assume that both components of the output vector are affected by independent zero-mean white Gaussian noises with $\sigma_z = 0.1\,m$ and $\sigma_\theta = 0.1\,rad$, while the process noise has four components, all with $\sigma = 0.2$ (of respective units). Assume initial conditions for the complete system to be $[0 \;\; 0.4 \;\; 0 \;\; 0 \;\; 0 \;\; 0 \;\; 0 \;\; 0]'$.

Solution: The MATLAB program and the plot are shown below (see Figure 3.45).

```
% file simul4.m
%

A = [0     0 1 0;
     0     0 0 1;
     0    -1 0 0;
     0 21.6 0 0];
B = [0;0;1;-2];
C = [1 0 0 0;
     0 1 0 0];
D = [0;0];

K = place(A,B,[-1, -2, -1+j, -1-j]);

G = ones(size(B));
Q = 0.2^2;
R = [0.1^2       0;
     0       0.1^2];

L = lqe(A,G,C,Q,R);       % calculate Kalman gain

% augment B and D to accept noises in simulations and to
% separate feedback control input from the external forcing function

n = size(A,1);
m = size(C,1);
AA = A;
BB = [B,B,G,zeros(n,m)];
CC = C;
DD = [D,D,zeros(size(D)),eye(m)];

sensors = [1 2];
known = [2];            % info on control u1 is internally fed back
control = [1];
```

```
[Ac,Bc,Cc,Dc] = reg(AA,BB,CC,DD,K,L,sensors,known,control);
[Abig,Bbig,Cbig,Dbig] = parallel(AA,BB,CC,DD,Ac,Bc,Cc,Dc,[2],[1],[],[]);
[Abig,Bbig,Cbig,Dbig] = cloop(Abig,Bbig,Cbig,Dbig,[1 2 3],[6 7 -1]);

t0 = 0;
tf = 16;
dt = 0.0001;
t=(t0:dt:tf)';
x0 = [0;0.4;0;0;0;0;0;0];
u1 = zeros(size(t));
u2 = 0.1*ones(size(t));
u3 = 0.2*randn(size(t));
u4 = 0.1*randn(size(t));
u5 = 0.1*randn(size(t));
u6 = zeros(size(t));
u7 = zeros(size(t));
u = [u1 u2 u3 u4 u5 u6 u7];

[y,x] = lsim(Abig,Bbig,Cbig,Dbig,u,t,x0);  % both initial and forced response
y = y(1:1000:size(t),:);
x = x(1:1000:size(t),:);
t = t(1:1000:size(t),:);

subplot(2,1,1)
plot(t,y(:,1),'wo',t,x(:,1),'y-')
title('Inverted Pendulum Simulation 4')
ylabel('z(t) [m]'), grid
subplot(2,1,2)
plot(t,y(:,2),'wo',t,x(:,2),'y-')
ylabel('theta(t) [rad]'), grid
```

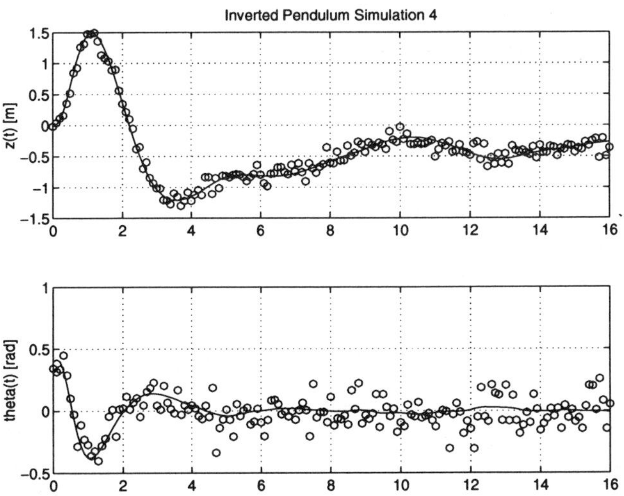

Figure 3.45: The results of the MATLAB Simulation 4, which uses Kalman-Bucy filter to estimate the system states from the noisy measurements. Plots represent the actual values of the system states (continuous lines) and only the samples of the measured values affected by noise (circles). Kalman-Bucy filter helps stabilize the system even when the only available measurements are very noisy.

Problem 3.12.3 In general, for the time-variant (non-stationary) system described by

$$\begin{aligned} \dot{x}(t) &= A(t)x(t) + B(t)u(t) + w(t) \\ y(t) &= C(t)x(t) + e(t) \end{aligned}$$

where

$$E\left[x(0)\right] = x_0, \qquad E\left[(x(0) - x_0)(x(0) - x_0)'\right] = P_0$$

while the noise signals are Gaussian with

$$E\left[w(t)\right] = 0, \qquad E\left[w(t)w'(\tau)\right] = Q(t)\,\delta(t - \tau) \qquad \text{(zero-mean, white)}$$

$$E\left[e(t)\right] = 0, \qquad E\left[e(t)e'(\tau)\right] = R(t)\,\delta(t - \tau) \qquad \text{(zero-mean, white)}$$

and the system noise $w(t)$ is uncorrelated with the measurement noise $e(t)$, i.e.,

$$E\left[w(t)e'(\tau)\right] = 0$$

the Kalman-Bucy filter, which produces $r(t)$, the optimal estimate[10] of $x(t)$ based on the measurements of $y(t)$, and on the available information about the system, is given by

$$\dot{r}(t) = A(t)r(t) + B(t)u(t) + L(t)(y(t) - C(t)r(t))$$

where $L(t) = P(t)C'(t)R^{-1}(t)$, and $P(t)$ is a solution of

$$\dot{P}(t) = A(t)P(t) + P(t)A'(t) + Q(t) - P(t)C'(t)R^{-1}(t)C(t)P(t)$$

It can be proved that if the system is time-invariant, i.e., if $A(t) = A$, $B(t) = B$, $C(t) = C$, and the noises are wide-sense-stationary (WSS), i.e., $Q(t) = Q$ and $R(t) = R$, and if the system is controllable and observable, then

$$\lim_{t \to \infty} P(t) = P$$

and therefore $\lim_{t\to\infty} L(t) = L$.

Derive the formulas for this case.

Solution: If $P(t) \to P$ then $\dot{P}(t) \to 0$. Therefore, P is found from the algebraic Riccati equation $AP + PA' - PC'R^{-1}CP + Q = 0$, while $L = PC'R^{-1}$.

Note: *The derivation of the above formulas is given in Section* 4.12. *Historically, the discrete-time case was considered first by Rudolf Kalman in 1959* [26], *while the continuous-time case was described soon afterwards by Kalman and R. Bucy in their joint paper* [28]. *Kalman's seminal paper* [26] *also dealt with the "surprising" duality of the problem his technique solves, the so-called Wiener problem, i.e., that of optimal estimation of noisy signals, and the problem of the noise-free optimal regulator problem (Section* 3.10*), which was solved a few years earlier, also by Kalman. This duality is important because investigation of properties of Kalman-Bucy filters, e.g., stability, can be conducted by methods used for optimal regulators.*

[10]The index of performance which is minimized by the Kalman-Bucy filter is the mean-squared-error (MSE): $J(t) = E\left[(x(t) - r(t))'(x(t) - r(t))\right] = \mathrm{tr}(E\left[(x(t) - r(t))(x(t) - r(t))'\right])$.

3.13 Reduced-order observers

In this Section the reduced-order observers are introduced. They are used when some of the states are available or measurable and there is no need to estimate all of them.

Problem 3.13.1 When estimating states $x(t)$, we don't need to design a simulator of full order n, because by some appropriate linear transformation of the original states $x(t)$ into $q(t)$ we can make $y(t)$ (in general an $m \times 1$ vector) exactly equal to some m of those n states.

In particular, let the system be given by

$$\begin{aligned}\dot{x}(t) &= Ax(t) + Bu(t) \\ y(t) &= Cx(t)\end{aligned}$$

Let us pick any nonsingular matrix S such that $CS = [O_{m\times(n-m)}\ I_{m\times m}]$. Then

$$\begin{aligned}\dot{q}(t) &= S^{-1}ASq(t) + S^{-1}Bu(t) \\ y(t) &= [q_{n-m+1}(t)\ \dots\ q_n(t)]'\end{aligned}$$

Now it suffices to design a reduced order observer to estimate $q_1(t), \dots, q_{n-m}(t)$, and finally to recombine the q-states back to the x-states by $x(t) = Sq(t)$.

For the sake of simplicity assume that $y(t)$ is a scalar, and that $C = c'$ is already in the desirable form, i.e., $c' = [0\ \dots\ 0\ 1]$. Design the reduced order observer.

Solution: Since $c' = [0\ \dots\ 0\ 1]$ we can write the system equations in the following form

$$\begin{bmatrix}\dot{x}_r(t) \\ \dot{x}_n(t)\end{bmatrix} = \begin{bmatrix} A_r & b_r \\ c_r' & a_{nn}\end{bmatrix}\begin{bmatrix}x_r(t) \\ x_n(t)\end{bmatrix} + \begin{bmatrix} g_r \\ g_n\end{bmatrix} u(t)$$

or equivalently

$$\begin{aligned}\dot{x}_r(t) &= A_r x_r(t) + b_r y(t) + g_r u(t) \\ y_r(t) &= c_r' x_r(t)\end{aligned}$$

where $y_r(t) = \dot{y}(t) - a_{nn}y(t) - g_n u(t)$ is a measurable quantity[11].

Since it can be proved that if the pair $\{c', A\}$ is observable, so is the pair $\{c_r', A_r\}$, now we can set up the observer for the states $x_r(t)$, whose states we shall denote by $r_r(t)$:

$$\dot{r}_r(t) = A_r r_r(t) + b_r y(t) + g_r u(t) - l_r(c_r' r_r(t) - y_r(t))$$

where again $y_r(t) = \dot{y}(t) - a_{nn}y(t) - g_n u(t)$. The gain vector l_r is picked so that the observer, i.e., the eigenvalues of the matrix $A_r - l_r c_r'$, are some designated numbers.

[11] Certainly, it is not desirable to have a differentiation anywhere in the process of estimation, because of the noise effects. We shall get rid of this operation later.

The signal flow graph of this system is shown in Figure 3.46. Along with it shown is its modification which eliminates the differentiation.

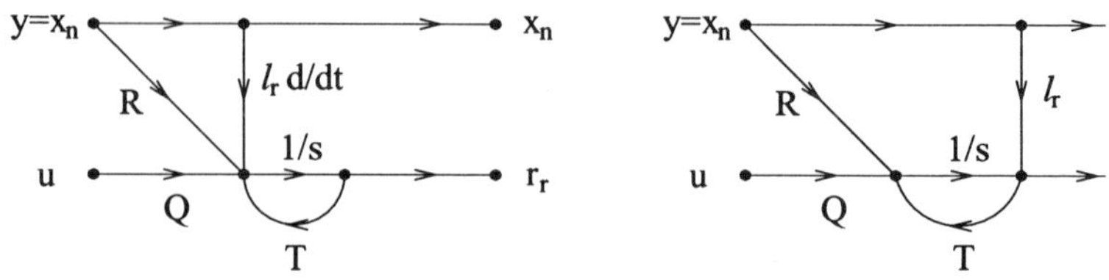

Figure 3.46: Signal flow graphs for two realizations of the reduced observer, one with, and the other without differentiation. $R = b_r - l_r a_{nn}$, $Q = g_r - l_r g_n$, and $T = A_r - l_r c_r'$.

This modification in the signal flow graph is equivalent to the following change of variables:

$$p_r(t) = r_r(t) - l_r y(t)$$

when the observer equation becomes (Figure 3.47)

$$\dot{p}_r(t) = \underbrace{(A_r - l_r c_r')}_{T} p_r(t) + (\underbrace{b_r - l_r a_{nn}}_{R} + \underbrace{(A_r - l_r c_r')}_{T} l_r) y(t) + \underbrace{(g_r - l_r g_n)}_{Q} u(t)$$

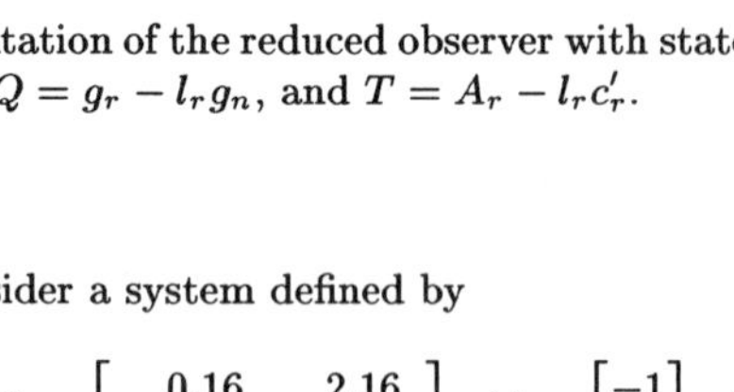

Figure 3.47: Implementation of the reduced observer with states $p_r(t) = r_r(t) - l_r y(t)$. Again $R = b_r - l_r a_{nn}$, $Q = g_r - l_r g_n$, and $T = A_r - l_r c_r'$.

Problem 3.13.2 Consider a system defined by

$$\dot{x}(t) = \begin{bmatrix} 0.16 & 2.16 \\ -0.16 & -1.16 \end{bmatrix} x(t) + \begin{bmatrix} -1 \\ 1 \end{bmatrix} u(t)$$

$$y(t) = \begin{bmatrix} 1 & 1 \end{bmatrix} x(t)$$

Find its open-loop eigenvectors, check the stability, controllability, and observability. Determine a state feedback vector k' such that both closed-loop eigenvalues of the system are at -1. Design a minimum-order observer for this system, with the gain l_r such that the observer has an eigenvalue at -3. What would be the observer eigenvalue if $l_r = 0$?

Solution: The open-loop eigenvalues are $\lambda_1 = -0.2$ and $\lambda_2 = -0.8$, therefore the system is stable. Since

$$\mathcal{C} = \begin{bmatrix} -1 & 2 \\ 1 & -1 \end{bmatrix} \quad \text{and} \quad \mathcal{O} = \begin{bmatrix} 1 & 1 \\ 0 & 1 \end{bmatrix}$$

the system is also controllable and observable.

The characteristic polynomials of the open-loop and the closed-loop system are

$$a(s) = \lambda^2 + \lambda + 0.16 \quad \text{and} \quad \alpha(s) = \lambda^2 + 2\lambda + 1$$

The system is simple enough not to require the use of the Bass-Gura formula. With or without it we find

$$k' = [0.84 \quad 1.84]$$

In order to design a reduced order observer, we need to introduce a nonsingular transformation S such that $c'S = [0 \quad 1]$. This can be accomplished with e.g.,

$$S = \begin{bmatrix} 1 & 0 \\ -1 & 1 \end{bmatrix}$$

when

$$\begin{bmatrix} a_r & b_r \\ c_r & a_{nn} \end{bmatrix} = S^{-1}AS = \begin{bmatrix} -2 & 2.16 \\ -1 & 1 \end{bmatrix} \quad \text{and} \quad \begin{bmatrix} g_r \\ g_n \end{bmatrix} = S^{-1}b = \begin{bmatrix} -1 \\ 0 \end{bmatrix}$$

In this case the reduced observer gain l_r is a scalar, and we find it from

$$\lambda - (a_r - l_r c_r) = \lambda + 3 \quad \Rightarrow \quad l_r = -1$$

Note: *Now it is easy to calculate the remaining parameters of the reduced order observer, and don't forget that the result of the observer operation is a state which needs to be combined with $y(t)$ using the matrix S:*

$$\begin{bmatrix} x_1(t) \\ x_2(t) \end{bmatrix} = S \begin{bmatrix} r_r(t) \\ y(t) \end{bmatrix}$$

If the observer is designed without the feedback ($l_r = 0$), the error would still tend to zero, but slightly slower, with eigenvalue -2, i.e., as e^{-2t}.

Problem 3.13.3 The design of a state feedback for the inverted pendulum on a cart requires the complete state vector

$$x = \begin{bmatrix} z \\ \theta \\ \dot{z} \\ \dot{\theta} \end{bmatrix}$$

In Problems 3.9.9 and 3.11.4 we assumed that all states were available, while in Problem 3.12.2 we estimated them. If we go back to the original assumption that $z(t)$ and $\theta(t)$ are directly available:

$$y = \begin{bmatrix} 1 & 0 & 0 & 0 \\ 0 & 1 & 0 & 0 \end{bmatrix} x$$

we see that we need to estimate only the remaining two states.

Design a reduced-order observer to estimate the two unavailable states. Why a simple differentiation of the two available states is not a satisfactory solution?

Solution: A differentiation of the two available states is not a good idea because even a slight noise can drastically change our estimates of the other two states.

The inverted pendulum on a cart we considered earlier (Problems 3.5.11, 3.7.12, 3.9.9, 3.11.4, and 3.12.2) was described by

$$\dot{x} = \begin{bmatrix} 0 & 0 & 1 & 0 \\ 0 & 0 & 0 & 1 \\ 0 & -1 & 0 & 0 \\ 0 & 21.6 & 0 & 0 \end{bmatrix} x + \begin{bmatrix} 0 \\ 0 \\ 1 \\ -2 \end{bmatrix} u$$

$$y = \begin{bmatrix} 1 & 0 & 0 & 0 \\ 0 & 1 & 0 & 0 \end{bmatrix} x$$

The nonsingular transformation S can be picked as

$$S = \begin{bmatrix} 0 & 0 & 1 & 0 \\ 0 & 0 & 0 & 1 \\ 0 & 1 & 0 & 0 \\ 1 & 0 & 0 & 0 \end{bmatrix} \quad \text{when} \quad S^{-1} = \begin{bmatrix} 0 & 0 & 0 & 1 \\ 0 & 0 & 1 & 0 \\ 1 & 0 & 0 & 0 \\ 0 & 1 & 0 & 0 \end{bmatrix}$$

Indeed,

$$CS = \begin{bmatrix} 0 & 0 & 1 & 0 \\ 0 & 0 & 0 & 1 \end{bmatrix}$$

By calculating $S^{-1}AS$ and $S^{-1}b$ we find

$$A_r = \begin{bmatrix} 0 & 0 \\ 0 & 0 \end{bmatrix}, \quad B_r = \begin{bmatrix} 0 & 21.6 \\ 0 & -1 \end{bmatrix}, \quad C_r = \begin{bmatrix} 0 & 1 \\ 1 & 0 \end{bmatrix}, \quad A_{nn} = \begin{bmatrix} 0 & 0 \\ 0 & 0 \end{bmatrix}$$

$$g_r = \begin{bmatrix} -2 \\ 1 \end{bmatrix}, \quad \text{and} \quad g_n = \begin{bmatrix} 0 \\ 0 \end{bmatrix}$$

In order for both eigenvalues of $A_r - L_r C_r$ to be at -4 we can pick

$$L_r = \begin{bmatrix} 0 & 4 \\ 4 & 0 \end{bmatrix}$$

Therefore (cf. Problem 3.13.1)

$$T = A_r - L_r C_r = \begin{bmatrix} -4 & 0 \\ 0 & -4 \end{bmatrix}$$

$$R + TL_r = B_r - L_r A_{nn} + (A_r - L_r C_r) L_r = \begin{bmatrix} 0 & 5.6 \\ -16 & -1 \end{bmatrix}$$

$$Q = g_r - L_r g_n = \begin{bmatrix} -2 \\ 1 \end{bmatrix}$$

```
% file simul5.m
%

A = [0     0 1 0;
     0     0 0 1;
     0    -1 0 0;
     0  21.6 0 0];
B = [0;0;1;-2];
C = [1 0 0 0;
     0 1 0 0];
D = [0;0];

K = [-0.2041 -15.902 -0.5102 -2.7551];  % picked so that eig(A-B*K)
                                         % are -1, -2, -1+j, -1-j
n = size(A,1);          % order of original system
r = n-size(C,1);        % number of states to estimate

S = [ 0 0 1 0;
      0 0 0 1;
      0 1 0 0;
      1 0 0 0];         % picked so that C*S = [zeros(n-r,r),eye(r)]

Anew = inv(S)*A*S;
Bnew = inv(S)*B;
Cnew = C*S;
Dnew = D;
Knew = K*S;

Ar = Anew(1:r,1:r);
Br = Anew(1:r,r+1:n);
Cr = Anew(r+1:n,1:r);
Ann = Anew(r+1:n,r+1:n);
Gr = Bnew(1:r,1);
Gn = Bnew(r+1:n,1);

Lr = [0 4;
      4 0];     % picked so that eig(Ar-Lr*Cr) are -4, -4

T = Ar-Lr*Cr;
R = Br-Lr*Ann;
Q = Gr-Lr*Gn;

Abig = [Anew, zeros(n,r);
        zeros(r,r), R+T*Lr, T];
Bbig = [Bnew;Q];
Cbig = [Cnew,zeros(n-r,r)];
Dbig = Dnew;

Kbig = [zeros(1,r),Knew(1,r+1:n)+Knew(1,1:r)*Lr,Knew(1,1:r)];

t0 = 0;
tf = 8;
dt = 0.05;
t=(t0:dt:tf)';

% q0 = [0;0;0;0;0;0];            % simulation
% u=0.1*ones(size(t));           % 5a

q0 = [0;0.1;0;0;0;0];        % simulation
u = zeros(size(t));          % 5b

[Yinit,Qinit] = initial(Abig-Bbig*Kbig,Bbig,Cbig,Dbig,q0,t);
[Yinp,Qinp] = lsim(Abig-Bbig*Kbig,Bbig,Cbig,Dbig,u,t);

q = Qinit+Qinp;
y = Yinit+Yinp;

xtrue = (S*q(:,1:n)')';
```

```
xestim = (S*[q(:,n+1:n+r)'+Lr*y'; y'])';

subplot(4,1,1)
plot(t,xtrue(:,1),'y-',t,xestim(:,1),'w.')
title('Inverted Pendulum Simulation 5b')
ylabel('z(t) [m]'), grid
subplot(4,1,2)
plot(t,xtrue(:,2),'y-',t,xestim(:,2),'w.')
ylabel('theta(t) [rad]'), grid
subplot(4,1,3)
plot(t,xtrue(:,3),'y-',t,xestim(:,3),'w.')
ylabel('z_dot(t) [m/s]'), grid
subplot(4,1,4)
plot(t,xtrue(:,4),'y-',t,xestim(:,4),'w.')
xlabel('t [s]'), ylabel('theta_dot(t) [rad/s]'), grid
```

Due to the identical initial conditions of the observer and the system, the two estimated states are identical to the actual states. The results of the MATLAB Simulation 5a are identical to Simulation 2a.

The results of the Simulation 5b slightly differ from Simulation 2b because of discrepancies between initial conditions of the actual and the observer states (see Figure 3.48).

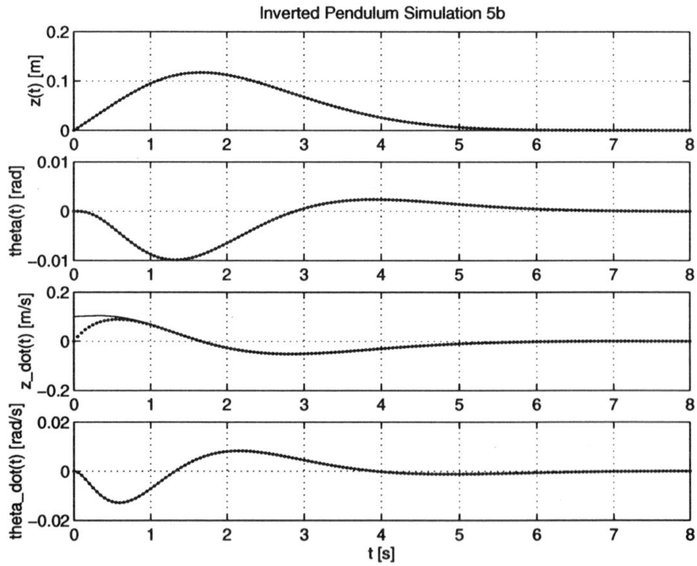

Figure 3.48: The results of the MATLAB Simulation 5b. Since the initial condition for the fourth state was not identical for the observer and the system (the equations describe the p-state, while the estimate of the x-states are the r-states; recall that $p_r(t) = r_r(t) - L_r y(t)$), there is some estimation error, which exponentially dies out. Due to this error, there are some differences with respect to the Simulation 2b.

Problem 3.13.4 For a system given by

$$H(s) = \frac{b(s)}{a(s)} = \frac{s-1}{s(s-2)}$$

(cf. Problem 3.11.6) design a reduced order observer such that the observer error eigenvalue is at -4. The poles of the system should be moved to -1 and -2.

Solution: With $\delta_r(s) = s + 4$ and $\alpha(s) = (s+1)(s+2)$, the condition

$$\alpha(s)\delta_r(s) = a(s)\delta_r(s) + a(s)p_u(s) + b(s)p_y(s)$$

implies

$$p_u(s) = -35 \quad \text{and} \quad p_y(s) = 40s - 8$$

Chapter 4

Discrete linear systems

In this Chapter we present solved problems about discrete-time linear control systems. For the most part it will be a reprise of Chapter 3. It will emphasize both similarities and differences between the discrete-time and the continuous-time systems. In particular, we shall see that many formulas, such as the conditions for controllability and observability, remain the same, while some others, such as the Riccati and Lyapunov equations and stability conditions are changed.

It begins with the background material on linear difference equations and matrices (Sections 4.1, 4.2, and 4.3). It continues with further examples of the advantages of the state-space representation of linear systems over their input-output representation (Sections 4.4 and 4.5). In Sections 4.6 and 4.7 we present three fundamental properties of systems: stability, state controllability, and state observability. We also mention a few peculiarities of discrete-time systems. In Section 4.8 we further illustrate the canonical forms of linear systems. Section 4.9 describes how the poles of the system can be arbitrarily placed using the state feedback. The condition for this so-called modal controllability is, again, the state controllability and observability. Next, in Section 4.10, we investigate the feedback gain which yields the quadratic optimality. In Section 4.11 we explain the design of the state observers. In Section 4.12 we investigate the choice of the observer gain so that the effects of noise are minimized in a mean-square sense. The result is the Kalman filter. Finally, in Section 4.13, we describe the reduced-order observers.

$$
\begin{aligned}
x[k+1] &= Ax[k] + Bu[k] \\
y[k] &= Cx[k] + Du[k]
\end{aligned}
$$

4.1 Simple difference equations

This Section briefly presents the two most commonly used methods for solving linear difference equations with constant coefficients: the time-domain convolution and the z-transform. It also describes the Kronecker's delta sequence.

Problem 4.1.1 Show that the solution of the inhomogeneous difference equation

$$x[k+1] = ax[k] + f[k] \quad (k \geq 0), \qquad \text{with } x[0] = x_0$$

is given by

$$x[k] = x_0 a^k + \sum_{i=0}^{k-1} a^i f[k-1-i]$$

Solution: From the solutions for $k = 0, 1, 2, 3$

$$\begin{aligned} x[0] &= x_0 \\ x[1] &= ax_0 + f[0] \\ x[2] &= a^2 x_0 + af[0] + f[1] \\ x[3] &= a^3 x_0 + a^2 f[0] + af[1] + f[2] \end{aligned}$$

we can easily generalize

$$x[k] = a^k x_0 + a^{n-1} f[0] + \ldots + af[k-2] + f[k-1]$$

i.e.,

$$x[k] = \underbrace{x_0 a^k}_{\text{homogeneous part}} + \underbrace{\sum_{i=0}^{k-1} a^i f[k-1-i]}_{\text{non-homogeneous part}}$$

Note: *The non-homogeneous part is a convolution of two sequences,* $\{a^k\}$ *and* $\{f[k-1]\}$*, a delayed version of* $\{f[k]\}$*. We write*

$$a^k * f[k-1] = \sum_{i=0}^{k-1} a^i f[k-1-i]$$

The next example will throw additional light to the solution of this problem.

Problem 4.1.2 Use the z-transform to solve the difference equation from the previous problem:

$$x[k+1] = ax[k] + f[k] \quad (k \geq 0), \qquad \text{with } x[0] = x_0$$

Solution: The z-transforms of sequences $\{x[k]\}$, $\{x[k+1]\}$, and $\{f[k]\}$ are

$$\begin{aligned}
\mathcal{Z}\{x[k]\} &= x[0]+x[1]z^{-1}+x[2]z^{-2}+x[3]z^{-3}+\ldots &&= X(z)\\
\mathcal{Z}\{x[k+1]\} &= x[1]+x[2]z^{-1}+x[3]z^{-2}+x[4]z^{-3}+\ldots &&= z(X(z)-x[0])\\
\mathcal{Z}\{f[k]\} &= f[0]+f[1]z^{-1}+f[2]z^{-2}+f[3]z^{-3}+\ldots &&= F(z)
\end{aligned}$$

Take a z-transform of both sides of the equation to obtain

$$z(X(z)-x_0) = aX(z)+F(z)$$

Therefore

$$X(z) = \frac{zx_0}{z-a}+\frac{F(z)}{z-a}$$

This form of $X(z)$, although eye-pleasing, is not convenient for expansion into power series of z^{-1}, the unit-delay elements. Thus we write

$$X(z) = \frac{x_0}{1-az^{-1}}+\frac{z^{-1}F(z)}{1-az^{-1}} \tag{4.1}$$

Next, expand the right-hand side into power series:

$$\frac{x_0}{1-az^{-1}} = x_0(1+az^{-1}+a^2z^{-2}+a^3z^{-3}+\ldots)$$

$$\begin{aligned}
\frac{z^{-1}F(z)}{1-z^{-1}a} &= (f[0]z^{-1}+f[1]z^{-2}+f[2]z^{-3}+\ldots)(1+az^{-1}+a^2z^{-2}+\ldots)\\
&= f[0]z^{-1}+(af[0]+f[1])z^{-2}+(a^2f[0]+af[1]+f[2])z^{-3}+\ldots
\end{aligned}$$

Finally, by comparing coefficients next to z^{-k} we find:

$$x[k] = a^kx_0+a^{k-1}f[0]+\ldots+af[k-2]+f[k-1]$$

Note: *We could write the same result directly from* (4.1) *because:*

- *the inverse z-transform of $\frac{x_0}{1-z^{-1}a}$ is the sequence $\{x_0a^k\}$*
- *the inverse of $\frac{z^{-1}F(z)}{1-az^{-1}}$ is a convolution of inverse z-transforms of $z^{-1}F(z)$ (i.e., the sequence $\{f[k-1]\}$ and of $\frac{1}{1-az^{-1}}$ (i.e., the sequence $\{x_0a^k\}$)*

Problem 4.1.3 The homogeneous part of the solution for the higher order difference equations can be found by looking at the roots of its characteristic equation:

For each multiplicity-m root a of the characteristic equation, the homogeneous part of the solution contains the following term(s)

$$\alpha_0a^k+\alpha_1ka^k+\alpha_2k^2a^k+\ldots+\alpha_{m-1}k^{m-1}a^k$$

where $\alpha_0,\ldots,\alpha_{m-1}$ are constants which depend on the initial conditions.

Note: *While the above form may be easy to remember, it is much easier to determine the coefficients if one of the following forms is used:*

$$\beta_0 a^k + \beta_1 k a^k + \beta_2 k(k-1)a^k + \ldots + \beta_{m-1} k(k-1)\ldots(k-m+2)a^k$$

or

$$\gamma_0 a^k + \gamma_1 \frac{(k+1)}{1!} a^k + \gamma_2 \frac{(k+1)(k+2)}{2!} a^k + \ldots + \gamma_{m-1} \frac{(k+1)\ldots(k+m-1)}{(m-1)!} a^k$$

These two forms are also convenient because they are easily summed during derivations (cf. Appendix B.2*). The latter form corresponds to what the z-transform method gives as a solution (see Problem* 4.1.4*).*

First apply and check the above procedure and then derive it for the following homogeneous equations:

a) $x[k+2] = 5x[k+1] - 6x[k]$, $\quad x[0] = x_0,\ x[1] = x_1$

b) $x[k+3] = 2x[k+2] + 4x[k+1] - 8x[k]$, $\quad x[0] = x_0,\ x[1] = x_1,\ x[2] = x_2$

c) $x[k+3] = 6x[k+2] - 12x[k+1] + 8x[k]$, $\quad x[0] = x_0,\ x[1] = x_1,\ x[2] = x_2$

Solution: a) The characteristic equation for this recursion is

$$r^2 - 5r + 6 = 0$$

and since its roots are

$$r_1 = 2 \quad \text{and} \quad r_2 = 3$$

the solution is of the form

$$x[k] = \alpha 2^k + \beta 3^k$$

where α and β can be determined from the initial conditions:

$$\left.\begin{array}{lll} k=0 & \Rightarrow & \alpha + \beta = x_0 \\ k=1 & \Rightarrow & 2\alpha + 3\beta = x_1 \end{array}\right\} \quad \Rightarrow \quad \alpha = 3x_0 - x_1 \quad \text{and} \quad \beta = x_1 - 2x_0$$

It is easy to verify that $x[k] = (3x_0 - x_1)2^k + (x_1 - 2x_0)3^k$ satisfies both the recursion and the initial conditions.

In order to derive the "usual suspects" ($\alpha 2^k$ and $\beta 3^k$) we shall rewrite the recursion so that it reduces to the trivial form $y[k+1] = ay[k]$. With the characteristic equation in mind

$$(r-2)(r-3) = 0$$

which can be rewritten as

$$r^2 - 3r = 2(r-3)$$

we write

$$x[k+2] = 5x[k+1] - 6x[k] \quad \Leftrightarrow \quad \underbrace{x[k+2] - 3x[k+1]}_{y[k+1]} = 2(\underbrace{x[k+1] - 3x[k]}_{y[k]})$$

With a new variable: $y[k] = x[k+1] - 3x[k]$ the recursion becomes

$$y[k+1] = 2y[k] \quad \text{with} \quad y[0] = x_1 - 3x_0 = C$$

Hence

$$y[k] = C2^k$$

This now yields a non-homogeneous difference equation in $x[k]$:

$$x[k+1] = 3x[k] + C2^k$$

whose solution is (directly from this recursion or from Problem 4.1.1)

$$x[k] = 3^k x_0 + \underbrace{(3^{k-1} \cdot 2^0 + 3^{k-2} \cdot 2^1 + \ldots + 3^0 \cdot 2^{k-1})}_{3^{k-1}\left(1+\frac{2}{3}+\ldots+\left(\frac{2}{3}\right)^{k-1}\right)=3^k-2^k} C$$

i.e.,

$$x[k] = \alpha 2^k + \beta 3^k$$

where $\alpha = -C = 3x_0 - x_1$ and $\beta = x_0 + C = x_1 - 2x_0$, as before.

b) The characteristic equation is

$$r^3 - 2r^2 - 4r + 8 = (r+2)(r-2)^2 = 0$$

hence the solution has the following form:

$$x[k] = \alpha(-2)^k + \beta 2^k + \gamma k 2^k$$

where α, β, and γ can be determined from the initial conditions. The details are omitted.

c) The characteristic equation for the recursion in this part is

$$(r-2)^3 = 0$$

hence the solution is of the form

$$x[k] = \alpha 2^k + \beta k 2^k + \gamma k^2 2^k$$

Again, α, β, and γ are constants which can be determined from the initial conditions.

In order to derive this result, we need to simplify the recursion. We first rearrange the characteristic equation to get the idea on what to do with the recursion:

$$r(r-2)^2 = 2(r-2)^2 \quad \text{i.e.,} \quad r^3 - 4r^2 + 4r = 2(r^2 - 4r + 4)$$

We see that with $y[k] = x[k+2] - 4x[k+1] + 4x[k]$ the recursion becomes $y[k+1] = 2y[k]$, with $y[0] = x_2 - 4x_1 + 4x_0$. For the sake of simplicity denote $y[0] = C$. Now we can write $y[k] = C2^k$, therefore the initial recursion can be written in the following form:

$$x[k+2] = 4x[k+1] - 4x[k] + C2^k$$

We see that the order of the recursion has been reduced by one. Its order can be reduced further if we rewrite this as

$$\underbrace{x[k+2] - 2x[k+1]}_{u[k+1]} = 2(\underbrace{x[k+1] - 2x[k]}_{u[k]}) + C2^k$$

when we find that

$$u[k] = B2^k + \frac{C}{2}k2^k$$

i.e.,

$$x[k+1] = 2x[k] + B2^k + \frac{C}{2}k2^k$$

Finally,

$$\begin{aligned} x[k] &= x_0 2^k + \left(\frac{B}{2} - \frac{C}{8}\right) k2^k + \frac{C}{8}k^2 2^k \\ &= \alpha 2^k + \beta k 2^k + \gamma k^2 2^k \end{aligned}$$

Note: *The reader is encouraged to fill in the missing steps and as a check to compare the values of α, β, and γ obtained through this derivation to the values obtained directly from the initial conditions.*

Problem 4.1.4 Repeat the part b) of the previous problem using the z-transform:

$$x[k+3] = 2x[k+2] + 4x[k+1] - 8x[k], \quad x[0] = x_0, x[1] = x_1, x[2] = x_2$$

Solution: Take the z-transform of the equation to obtain

$$z^3X(z) - z^3x_0 - z^2x_1 - zx_2 = 2(z^2X(z) - z^2x_0 - zx_1) + 4(zX(z) - zx_0) - 8X(z)$$

$$X(z)(z^3 - 2z^2 - 4z + 8) = x_0z^3 + (x_1 - 2x_0)z^2 + (x_2 - 2x_1 - 4x_0)z$$

$$\begin{aligned} X(z) &= \frac{x_0z^3 + (x_1 - 2x_0)z^2 + (x_2 - 2x_1 - 4x_0)z}{(z+2)(z-2)^2} \\ &= \frac{(x_2 - 2x_1 - 4x_0)z^{-2} + (x_1 - 2x_0)z^{-1} + x_0}{(1+2z^{-1})(1-2z^{-1})^2} \\ &= \frac{P}{1+2z^{-1}} + \frac{Q}{1-2z^{-1}} + \frac{R}{(1-2z^{-1})^2} \end{aligned}$$

where P, Q, and R depend on the numerator coefficients, i.e., on the initial conditions. They can be determined using MATLAB command **residue**, or, in simple cases such as this one, by equating the last two expressions for $X(z)$. We omit the details here.

Since

$$\begin{aligned}\frac{1}{(1-az^{-1})^2} &= d/dz^{-1}\left(\frac{1}{a(1-az^{-1})}\right)\\ &= d/dz^{-1}\left(\frac{1}{a}(1+az^{-1}+a^2z^{-2}+\ldots)\right)\\ &= 1+2az^{-1}+3a^2z^{-2}+\ldots\\ &= \mathcal{Z}\{(k+1)a^k\}\end{aligned}$$

and in general

$$\frac{1}{(1-az^{-1})^m} = \mathcal{Z}\left\{\frac{(k+1)(k+2)\ldots(k+m-1)}{(m-1)!}a^k\right\}$$

we find

$$\begin{aligned}x[k] &= P(-2)^k + Q2^k + R(k+1)2^k\\ &= P(-2)^k + (Q+R)2^k + Rk2^k\\ &= \alpha(-2)^k + \beta 2^k + \gamma k 2^k\end{aligned}$$

Problem 4.1.5 Solve the following difference equation

$$x[k] - 2x[k-1] - 2x[k-2] = 0$$

with $x[0] = 0$ and $x[1] = 1$.

Solution: The characteristic equation is $r^2 - 2r - 2 = 0$ hence $r_{1,2} = 1 \pm j$. Therefore, the solution is

$$x[k] = \alpha(1+j)^k + \beta(1-j)^k$$

where α and β are complex constants determined from the initial conditions. Calculation of α and β can be simplified if we keep in mind that for $x[k]$ to be real, β must be equal to the complex conjugate of α.

Another way of writing this solution is derived from the polar representation of the characteristic roots: $1 \pm j = \sqrt{2}e^{\pm j\pi/4}$:

$$x[k] = (\alpha+\beta)2^{n/2}\cos\frac{n\pi}{4} + (\alpha-\beta)2^{n/2}j\sin\frac{n\pi}{4}$$

In this particular case $\alpha = -j/2$ and $\beta = j/2$, hence

$$x[k] = 2^{n/2}\sin\frac{n\pi}{4}$$

Note: *In order to use the z-transform in this problem, one must either find the initial conditions x_{-1} and x_{-2} or rewrite the equation as $x[k+2]-2x[k+1]-2x[k] = 0$.*

Problem 4.1.6 A particle is moving in a horizontal line. The distance it travels in each second is equal to two times the distance it travels in the previous second. Let x_k denote the position of the particle at the k-th second.

a) Find a general relation between x_k, x_{k-1}, and x_{k-2}.
b) If $x_0 = 3$ and $x_3 = 10$, find x_k.

Solution: a) Obviously $x_k - x_{k-1} = 2(x_{k-1} - x_{k-2})$, i.e.,

$$x_k - 3x_{k-1} + 2x_{k-2} = 0$$

b) First solve the above difference equation:

$$\left.\begin{array}{l}\text{the “usual suspect”:}\\ \quad x_k = r^k\\ \text{the recursion:}\\ \quad x_k - 3x_{k-1} + 2x_{k-2} = 0\end{array}\right\} \Rightarrow \; r^2 - 3r + 2 = 0 \;\Rightarrow\; r_1 = 1,\; r_2 = 2$$

therefore

$$x_k = \alpha r_1^k + \beta r_2^k$$

i.e.,

$$x_k = \alpha + \beta 2^k$$

From the conditions $x_0 = 3$ and $x_3 = 10$ we get a system of equations in α and β:

$$\left.\begin{array}{l}\alpha + \beta = 3\\ \alpha + 8\beta = 10\end{array}\right\} \;\Rightarrow\; \alpha = 2,\;\; \beta = 1 \;\Rightarrow\; x_k = 2 + 2^k$$

Problem 4.1.7 Investigate the most important properties of the Kronecker's delta $\delta[k]$ with respect to discrete-time convolution and the z-transform.

Solution: The Kronecker's delta impulse sequence is defined as

$$\delta[k] = \begin{cases} 1, & k = 0 \\ 0, & k \neq 0 \end{cases}$$

It is the unity for discrete convolution

$$f[k] * \delta[k] \;=\; \sum_{i=-\infty}^{\infty} f[k-i]\,\delta[i] \;=\; f[k]$$

It's z-transform is obviously

$$\mathcal{Z}\{\delta[k]\} = 1 + 0 \cdot z^{-1} + 0 \cdot z^{-2} + 0 \cdot z^{-2} + \ldots = 1$$

Problem 4.1.8 Write the following third-order difference equation as a system of three first-order difference equations and write them in a matrix form:

$$x[k+3] + 5x[k+2] + 8x[k+1] + 4x[k] = (-1)^k$$

Solution: Use $v[k] = x[k+1]$ and $w[k] = v[k+1]$, to obtain

$$\begin{aligned} x[k+1] &= v[k] \\ v[k+1] &= w[k] \\ w[k+1] &= -4x[k] - 8v[k] - 5w[k] + (-1)^k \end{aligned}$$

i.e.,

$$\begin{bmatrix} x[k+1] \\ v[k+1] \\ w[k+1] \end{bmatrix} = \begin{bmatrix} 0 & 1 & 0 \\ 0 & 0 & 1 \\ -4 & -8 & -5 \end{bmatrix} \begin{bmatrix} x[k] \\ v[k] \\ w[k] \end{bmatrix} + \begin{bmatrix} 0 \\ 0 \\ (-1)^k \end{bmatrix}$$

Note: *Do you recognize the companion matrix? Compare the eigenvalues (or at least the characteristic equations) of the difference equation and the matrix. What are the initial conditions here?*

4.2 More matrix theory

In this Section we present several methods for raising a matrix to an integer power and prove a very important result, the so-called matrix inversion lemma. Matrices are also covered in Section 3.2 and Appendixes B.4 and C.

Problem 4.2.1 Determine A^k for the following matrix

$$A = \begin{bmatrix} 1 & 1 \\ 1 & 0 \end{bmatrix}$$

Solution: Calculate A^k for $k = 2, 3, 4, 5$ and see the regularity:

$$A^2 = \begin{bmatrix} 2 & 1 \\ 1 & 1 \end{bmatrix} \quad A^3 = \begin{bmatrix} 3 & 2 \\ 2 & 1 \end{bmatrix} \quad A^4 = \begin{bmatrix} 5 & 3 \\ 3 & 2 \end{bmatrix} \quad A^5 = \begin{bmatrix} 8 & 5 \\ 5 & 3 \end{bmatrix}$$

We see that the sequence of Fibonacci numbers f_k appears in this result. They are defined by

$$f_{k+2} = f_{k+1} + f_k \qquad f_0 = 0, \; f_1 = 1$$

and start like this

$$f_0 = 0, \; f_1 = 1, \; f_2 = 1, \; f_3 = 2, \; f_4 = 3, \; f_5 = 5, \; f_6 = 8, \; \ldots$$

Indeed, the following can be proved by mathematical induction:

$$A^k = \begin{bmatrix} f_{k+1} & f_k \\ f_k & f_{k-1} \end{bmatrix}$$

Note: *Unfortunately, very rarely do we get such simple and cute results.*

Problem 4.2.2 If $A = QJQ^{-1}$, where J is in Jordan form, then $A^k = (QJQ^{-1})^k = QJ^kQ^{-1}$. In the special case when A is diagonalizable, $A^k = Q\Lambda^kQ^{-1}$, where

$$\Lambda = \text{diag}(\lambda_1, \lambda_2, \ldots, \lambda_n) \quad \text{and} \quad \Lambda^k = \text{diag}(\lambda_1^k, \lambda_2^k, \ldots, \lambda_n^k)$$

Use the results obtained in Problem 3.2.6 to find A^k for

$$A = \begin{bmatrix} -6 & 2 \\ -6 & 1 \end{bmatrix}$$

Solution: In Problem 3.2.6 we found that $\lambda_1 = -3$, $\lambda_2 = -2$ and

$$Q = \begin{bmatrix} 2 & 1 \\ 3 & 2 \end{bmatrix} \quad \text{while} \quad Q^{-1} = \begin{bmatrix} 2 & -1 \\ -3 & 2 \end{bmatrix}$$

Therefore

$$A^k = Q\Lambda^kQ^{-1} = \begin{bmatrix} (4(-3)^k - 3(-2)^k) & (-2(-3)^k + 2(-2)^k) \\ (6(-3)^k - 6(-2)^k) & (-3(-3)^k + 4(-2)^k) \end{bmatrix}$$

Problem 4.2.3 Use the Sylvester interpolation method to determine A^k for

$$A = \begin{bmatrix} -6 & 2 \\ -6 & 1 \end{bmatrix}$$

Solution: As in Problem 3.2.7, since A is 2×2, we can write

$$A^k = \alpha(k)I + \beta(k)A$$

where α and β are found from

$$\begin{aligned} \alpha(k) + \beta(k)\lambda_1 &= \lambda_1^k \\ \alpha(k) + \beta(k)\lambda_2 &= \lambda_2^k \end{aligned}$$

We easily find

$$\alpha(k) = 3(-2)^k - 2(-3)^k \quad \text{and} \quad \beta(k) = (-2)^k - (-3)^k$$

This agrees with the result of the previous problem. Indeed,

$$A^k = \alpha(k)I + \beta(k)A = \begin{bmatrix} (4(-3)^k - 3(-2)^k) & (-2(-3)^k + 2(-2)^k) \\ (6(-3)^k - 6(-2)^k) & (-3(-3)^k + 4(-2)^k) \end{bmatrix}$$

Problem 4.2.4 Use the z-transform to determine A^k for

$$A = \begin{bmatrix} -6 & 2 \\ -6 & 1 \end{bmatrix}$$

Solution: We shall show in Problem 4.3.3 that A^k and $(I - z^{-1}A)^{-1}$ are a z-transform pair:

$$\mathcal{Z}\left\{A^k\right\} = (I - z^{-1}A)^{-1} \quad \text{i.e.,} \quad \mathcal{Z}^{-1}\left\{(I - z^{-1}A)^{-1}\right\} = A^k$$

We obtain the same solution as in previous problems since

$$(I - z^{-1}A)^{-1} = \begin{bmatrix} 1+6z^{-1} & -2z^{-1} \\ 6z^{-1} & 1-z^{-1} \end{bmatrix}^{-1} = \begin{bmatrix} \left(\frac{-3}{1+2z^{-1}} + \frac{4}{1+3z^{-1}}\right) & \left(\frac{2}{1+2z^{-1}} + \frac{-2}{1+3z^{-1}}\right) \\ \left(\frac{-6}{1+2z^{-1}} + \frac{6}{1+3z^{-1}}\right) & \left(\frac{4}{1+2z^{-1}} + \frac{-3}{1+3z^{-1}}\right) \end{bmatrix}$$

hence

$$A^k = \mathcal{Z}^{-1}\left\{(I - z^{-1}A)^{-1}\right\} = \begin{bmatrix} (4(-3)^k - 3(-2)^k) & (-2(-3)^k + 2(-2)^k) \\ (6(-3)^k - 6(-2)^k) & (-3(-3)^k + 4(-2)^k) \end{bmatrix}$$

Problem 4.2.5 Determine A^k for

$$A = \begin{bmatrix} -1 & 0 & 0 \\ 0 & -1 & 0 \\ 1 & 0 & -1 \end{bmatrix}$$

Solution: The eigenvalues of A are $\lambda_{1,2,3} = -1$. Since A is 3×3 and $\lambda = -1$ is a triple eigenvalue, in order to determine the coefficients in

$$A^k = \alpha(k)I + \beta(k)A + \gamma(k)A^2$$

we form the three equations by writing

$$\lambda^k = \alpha(k) + \beta(k)\lambda + \gamma(k)\lambda^2$$

and the first and the second derivatives over λ:

$$k\lambda^{k-1} = \beta(k) + 2\gamma(k)\lambda$$

and

$$k(k-1)\lambda^{k-2} = 2\gamma(k)$$

When we solve this system with $\lambda = -1$, we finally get

$$A^k = \begin{bmatrix} (-1)^k & 0 & 0 \\ 0 & (-1)^k & 0 \\ (-1)^{k-1}k & 0 & (-1)^k \end{bmatrix}$$

Problem 4.2.6 Prove the following very useful result, the so-called matrix inversion lemma: If A, B, C, and D are $n \times n$, $n \times m$, $m \times n$, $m \times m$, respectively, and all necessary inverses exist, then

$$(A + BDC)^{-1} = A^{-1} - A^{-1}B(D^{-1} + CA^{-1}B)^{-1}CA^{-1}$$

Solution: Premultiply the right-hand side of the equation by $A + BDC$:

$$\begin{aligned}
(A + BDC)&(A^{-1} - A^{-1}B(D^{-1} + CA^{-1}B)^{-1}CA^{-1}) = \\
&= I + BDCA^{-1} - B(D^{-1} + CA^{-1}B)^{-1}CA^{-1} - \\
&\qquad - BDCA^{-1}B(D^{-1} + CA^{-1}B)^{-1}CA^{-1} \\
&= I + BDCA^{-1} - (B + BDCA^{-1}B)(D^{-1} + CA^{-1}B)^{-1}CA^{-1} \\
&= I + BDCA^{-1} - BD(D^{-1} + CA^{-1}B)(D^{-1} + CA^{-1}B)^{-1}CA^{-1} \\
&= I
\end{aligned}$$

Note: *Most often we use this lemma with $A = I_n$ and $D = I_m$:*

$$(I_n + BC)^{-1} = I_n - B(I_m + CB)^{-1}C$$

This result has been known among mathematicians, e.g., Woodbury, at least since 1950. The first to use it in the engineering community was Kailath in 1960 (cf. [19]*).*

4.3 Systems of linear difference equations

In this Section we solve systems of difference equations using the matrix notation.

Problem 4.3.1 Write the following system of equations in a matrix form:

$$\begin{aligned} u[k+1] &= u[k] - 2v[k] - 4w[k] + (-1)^k \\ v[k+1] &= u[k] - 3v[k] + 3w[k] + \sin k \\ w[k+1] &= u[k] + 4v[k] + 5w[k] + \cos k \end{aligned}$$

Solution: If we write

$$x[k] = \begin{bmatrix} u[k] \\ v[k] \\ w[k] \end{bmatrix} \quad \text{and} \quad f[k] = \begin{bmatrix} (-1)^k \\ \sin k \\ \cos k \end{bmatrix}$$

the system can be written as

$$x[k+1] = Ax[k] + f[k]$$

where

$$A = \begin{bmatrix} 1 & -2 & -4 \\ 1 & -3 & 3 \\ 1 & 4 & 5 \end{bmatrix}$$

Problem 4.3.2 Show that the solution of the system of inhomogeneous difference equations

$$x[k+1] = Ax[k] + f[k] \quad (k \geq 0), \qquad \text{with } x[0] = \begin{bmatrix} x_{01} \\ \vdots \\ x_{0n} \end{bmatrix}$$

where A is $n \times n$, while $x[k]$ and $f[k]$ are $n \times 1$, is given by

$$x[k] = A^k x_0 + \sum_{i=0}^{k-1} A^i f[k-1-i]$$

Solution: From the solutions for $k = 0, 1, 2, 3$

$$\begin{aligned} x[0] &= x_0 \\ x[1] &= Ax_0 + f[0] \\ x[2] &= A^2 x_0 + Af[0] + f[1] \\ x[3] &= A^3 x_0 + A^2 f[0] + Af[1] + f[2] \end{aligned}$$

we can easily generalize

$$x[k] = A^k x_0 + A^{k-1} f[0] + \ldots + Af[k-2] + f[k-1]$$

i.e.,

$$x[k] = \underbrace{A^k x_0}_{\text{homogeneous part}} + \underbrace{\sum_{i=0}^{k-1} A^i f[k-1-i]}_{\text{non-homogeneous part}}$$

Note: *The non-homogeneous part is a convolution of two sequences,* $\{A^k\}$ *and* $\{f[k-1]\}$*. We write*

$$A^k * f[k-1] = \sum_{i=0}^{k-1} A^i f[k-1-i]$$

Problem 4.3.3 Apply the z-transform to the vector difference equation from the previous problem:

$$x[k+1] = Ax[k] + f[k] \quad (k \geq 0), \qquad \text{with } x[0] = \begin{bmatrix} x_{01} \\ \vdots \\ x_{0n} \end{bmatrix}$$

Solution: The z-transforms of sequences $\{x[k]\}$, $\{x[k+1]\}$, and $\{f[k]\}$ are

$$\begin{aligned} \mathcal{Z}\{x[k]\} &= x[0] + x[1]z^{-1} + x[2]z^{-2} + x[3]z^{-3} + \ldots &&= X(z) \\ \mathcal{Z}\{x[k+1]\} &= x[1] + x[2]z^{-1} + x[3]z^{-2} + x[4]z^{-3} + \ldots &&= z(X(z) - x[0]) \\ \mathcal{Z}\{f[k]\} &= f[0] + f[1]z^{-1} + f[2]z^{-2} + f[3]z^{-3} + \ldots &&= F(z) \end{aligned}$$

Take a z-transform of both sides of the equation to obtain

$$z(X(z) - x_0) = AX(z) + F(z)$$

Therefore

$$X(z) = (zI - A)^{-1} z x_0 + (zI - A)^{-1} F(z)$$

i.e.,

$$X(z) = (I - z^{-1}A)^{-1} x_0 + (I - z^{-1}A)^{-1} z^{-1} F(z)$$

Since z^{-1} corresponds to a time-delay by one sample and the product in the transform domain corresponds to time-domain convolution, the inverse z-transform yields

$$x[k] = A^k x_0 + \sum_{i=0}^{k-1} A^i f[k-1-i]$$

Note: *If we compare results of Problems* 4.1.2 *and* 4.3.2 *we find that* $\{A^k\}$ *and* $(I - z^{-1}A)^{-1}$ *are a z-transform pair:*

$$\mathcal{Z}\left\{A^k\right\} = (I - z^{-1}A)^{-1} \quad \textit{i.e.,} \quad \mathcal{Z}^{-1}\left\{(I - z^{-1}A)^{-1}\right\} = A^k$$

4.4 Input-output representation

In this Section we review the use of input-output representation and transfer function techniques to analysis of discrete-time systems. We also discuss the discretization process, the sampling theorem in particular, and relation between the Laplace and the z-transform.

Problem 4.4.1 Determine the output of a system described by

$$y[k+1] - \frac{1}{2}y[k] = u[k] \quad (k \geq 0)$$

$$y[0] = 5$$

when

a) $u[k] = \cos\frac{2\pi k}{12}$
b) $u[k] = \left(\frac{1}{3}\right)^k$
c) $u[k] = \left(\frac{1}{2} - \varepsilon\right)^k$ where ε is a small positive number
d) $u[k] = \left(\frac{1}{2}\right)^k$

Solution: The homogeneous part of the solution is the same for all four cases. Since the root of the characteristic equation (the *pole* of the system) is $a = \frac{1}{2}$ and $y[0] = 5$

$$y_h[k] = 5\left(\frac{1}{2}\right)^k$$

a) The non-homogeneous part (the *particular* solution) is as in Problem 4.1.1 (using $\cos\varphi = \frac{e^{j\varphi}+e^{-j\varphi}}{2}$ and $1 + x + x^2 + \ldots + x^n = \frac{x^{n+1}-1}{x-1}$)

$$y_{nh}[k] = \sum_{i=0}^{k-1}\left(\frac{1}{2}\right)^i \cos\frac{2\pi(k-1-i)}{12} = \ldots = 0.95\cos\frac{2\pi k}{12} + 1.30\sin\frac{2\pi k}{12} - 0.95\left(\frac{1}{2}\right)^k$$

Finally, the solution is

$$y[k] = y_h[k] + y_{nh}[k] = 4.05\left(\frac{1}{2}\right)^k + 0.95\cos\frac{2\pi k}{12} + 1.30\sin\frac{2\pi k}{12}$$

Note 1: *The first term in the solution is often called the* transient *part of the solution because it approaches zero fast. The remaining terms are then called the* steady-state *part of the solution. Both the initial conditions and the input contribute to the transient part of $y[k]$, through $y_h[k]$ and $y_{nh}[k]$, respectively. The steady-state part, however, comes from the input only and is often called the* forced *output.*

Note 2: *We can solve this equation in other ways, using the z-transform for example. Another method is attractive too: Knowing the root of the characteristic equation and from the form of the input we can immediately write*

$$y[k] = A\left(\frac{1}{2}\right)^k + B\cos\frac{2\pi k}{12} + C\sin\frac{2\pi k}{12}$$

If we substitute this into the original equation (not only its homogeneous part), the initial condition gives us one of three equations for constants A, B, and C:

$$A + B = 5$$

The other two equations are obtained by equating the coefficients next to $\cos\frac{2\pi k}{12}$ *and* $\sin\frac{2\pi k}{12}$ *terms, respectively:*

$$B(\sqrt{3}-1) + C = 2 \quad \textit{and} \quad B - C(\sqrt{3}-1) = 0$$

See also Problem 4.4.2.

b) Similarly, for this input

$$y[k] = 11\left(\frac{1}{2}\right)^k - 6\left(\frac{1}{3}\right)^k$$

c) For $u[k] = (\frac{1}{2} - \varepsilon)^k$ we find

$$y[k] = \left(5 + \frac{1}{\varepsilon}\right)\left(\frac{1}{2}\right)^k - \frac{1}{\varepsilon}\left(\frac{1}{2} - \varepsilon\right)^k$$

Note: *As $\varepsilon \to 0$, i.e., when the input's complex frequency approaches the system's pole, the forced output grows in magnitude. This is* resonance. *Asymptotically (as $\varepsilon \to 0$), the total output behaves like:*

$$\lim_{\varepsilon\to 0}\left(\left(5 + \frac{1}{\varepsilon}\right)\left(\frac{1}{2}\right)^k - \frac{1}{\varepsilon}\left(\frac{1}{2} - \varepsilon\right)^k\right) = 5\left(\frac{1}{2}\right)^k + k\left(\frac{1}{2}\right)^{k-1}$$

d) When $u[k] = (\frac{1}{2})^k$, the input's complex frequency coincides with the pole of the system. The convolution of two similar terms produces a new form. Thus

$$y[k] = 5\left(\frac{1}{2}\right)^k + k\left(\frac{1}{2}\right)^{k-1}$$

Problem 4.4.2 Find the output of a system described by

a) $y[k+3] + 3y[k+2] + 3y[k+1] + y[k] = (-1)^k$
with $y[2] = 1, y[1] = 2$, and $y[0] = 3$

b) $y[k+3] + 3y[k+2] + 4y[k+1] + 12y[k] = (-3)^k$
with $y[2] = 1, y[1] = 1$, and $y[0] = 1$

Solution: a) This system has a triple pole at -1 and the input's complex frequency coincides with this triple pole hence the solution is a linear combination of $(-1)^k$, $k(-1)^k$, $k^2(-1)^k$, and $k^3(-1)^k$ (see also Problem 4.1.3):

$$y[k] = A(-1)^k + Bk(-1)^k + Ck^2(-1)^k + Dk^3(-1)^k$$

Coefficients A, B, C, and D are found from the initial conditions for the whole equation and by substitution of this expression into the equation.

Note: *Much less effort is needed if we use the following form:*

$$y[k] = \alpha(-1)^k + \beta k(-1)^k + \mu k(k-1)(-1)^k + \nu k(k-1)(k-2)(-1)^k$$

b) In this case poles are at $\pm 2j$ and -3, and the input coincides with a pole at -3, therefore

$$y[k] = A(2j)^k + B(-2j)^k + C(-3)^k + Dk(-3)^k$$

Another way to write this is

$$y[k] = \alpha 2^k \cos\frac{k\pi}{2} + \beta 2^k \sin\frac{k\pi}{2} + C(-3)^k + Dk(-3)^k$$

where $\alpha = A + B$ and $\beta = (A - B)j$.

Problem 4.4.3 What is the output of the system described by

$$y[k+1] + 3y[k] = u[k+1] + 2u[k]$$

with $y[0] = 1$ and $u[k] = (-2)^k + \cos\frac{2\pi k}{7}$.

Solution: If we try $y[k] = A(-3)^k + B(-2)^k + C\cos\frac{2\pi k}{7} + D\sin\frac{2\pi k}{7}$ and substitute it into the equation we immediately find that $B = 0$. The complex frequencies for which this happens (in this case only -2) are called the *zeros* of the system. They are the roots of the characteristic equation of the input part of the equation.

Problem 4.4.4 What is the impulse response $h[k]$ of a system? What is the transfer function $T(z)$ of a system? Show that $T(z) = \mathcal{Z}\{h[k]\}$.

Solution: *Impulse response.* The impulse response $h[k]$ of a discrete-time system is the output of the system caused by the Kronecker's delta impulse $\delta[k]$ at the input:

$$\delta[k] = \begin{cases} 1, & k = 0 \\ 0, & k \neq 0 \end{cases}$$

The system is assumed to be at rest when $\delta[k]$ is applied, i.e., all initial conditions are zero.

The impulse response $h[k]$ is important because it completely characterizes the output when input is known. If the initial conditions are non-zero $y_h[k]$ is found as in Problem 4.1.3, while $y_{nh}[k]$ can be characterized in terms of the impulse response as follows.

From the linearity of the system, and from the following decomposition of an arbitrary input for $k \geq 0$

$$u[k] = \sum_{i=0}^{\infty} u[i]\,\delta[k-i]$$

we find (assuming the system is causal, i.e., $h[k] \equiv 0$ for $k < 0$)

$$y_{nh}[k] = \sum_{i=0}^{k} u[i]\, h[k-i]$$

the so-called discrete-time convolution of sequences $h[k]$ and $u[k]$.

In general, if a system is given by a difference equation, the impulse response is most easily obtained as the inverse z-transform of its transfer function. The derivation is given below.

Transfer function. Transfer function is the ratio of the z-transforms of the output and the input of the system, assuming zero initial conditions:

$$T(z) = \frac{Y(z)}{U(z)}$$

If a system is given by

$$y[k] + a_1 y[k-1] + \ldots + a_n y[k-n] = b_0 u[k] + \ldots + b_m u[k-m] \qquad (k \geq 0)$$

and the initial conditions are zero, i.e., $y_{-1} = \ldots = y_{-n} = 0$, then the z-transform yields

$$T(z) = \frac{Y(z)}{U(z)} = \frac{b(z)}{a(z)} = \frac{b_0 z^n + \ldots + b_m z^{n-m}}{z^n + a_1 z^{n-1} + \ldots + a_{n-1} z + a_n}$$

or equivalently

$$T(z) = \frac{b_0 + b_1 z^{-1} + \ldots + b_m z^{-m}}{1 + a_1 z^{-1} + \ldots + a_n z^{-n}}$$

From these expressions we see that $T(z)$ does not depend on the input $u[k]$, only on the coefficients of the difference equation.

Relation between $h[k]$ and $T(z)$. Since $T(z)$ does not depend on the particular choice of $u[k]$, we can pick $u[k] = \delta[k]$ when $U(z) = 1$, $y[k] = h[k]$, and $Y(z) = H(z)$. Then we find

$$T(z) = \frac{Y(z)}{U(z)} = H(z) = \mathcal{Z}\{h[k]\} = \sum_{k=0}^{\infty} h[k] z^{-k}$$

Note: *This is why we often write $H(z)$ instead of $T(z)$. Another way to see this is to use the convolution property of the z-transform: With zero initial conditions*

$$y[k] = h[k] * u[k] \quad \Rightarrow \quad Y(z) = H(z)U(z)$$

Note also that for causal systems ($h[k] \equiv 0$ for $k < 0$) when $z = e^{j\omega}$ the transfer function $T(z)$ becomes the frequency response *$T(e^{j\omega})$ and we find that $h[k]$ and $T(e^{j\omega})$ are a discrete-time Fourier transform (DTFT) pair:*

$$T(e^{j\omega}) = H(e^{j\omega}) = \sum_{k=0}^{\infty} h[k] e^{-j\omega k}$$

Problem 4.4.5 Determine the impulse response of a system described by

$$y[k] + \frac{5}{6}y[k-1] + \frac{1}{6}y[k-2] = u[k-1] + 3u[k-2] \quad (k \geq 0)$$

Solution: Obviously

$$H(z) = \frac{z+3}{z^2 + \frac{5}{6}z + \frac{1}{6}} = \frac{-15}{z+\frac{1}{2}} + \frac{16}{z+\frac{1}{3}} = \frac{-15z^{-1}}{1+\frac{1}{2}z^{-1}} + \frac{16z^{-1}}{1+\frac{1}{3}z^{-1}}$$

Therefore

$$h[k] = -15\left(\frac{1}{2}\right)^{k-1} + 16\left(\frac{1}{3}\right)^{k-1} \quad (k > 0)$$

Matlab note: *To plot this directly from the coefficients of the differential equation (see Figure* 4.1*) do the following:* `dimpulse([1 3],[1 5/6 1/6])`

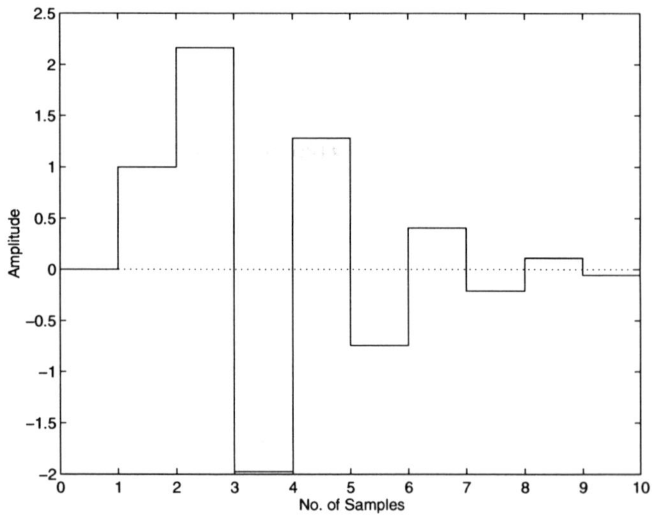

Figure 4.1: The plot produced by the MATLAB command **dimpulse**.

Problem 4.4.6 Determine the impulse response of a system described by

$$y[k] + y[k-1] + \frac{1}{2}y[k-2] = u[k] + 2u[k-1]$$

Solution: Obviously

$$H(z) = \frac{1+2z^{-1}}{1+z^{-1}+\frac{1}{2}z^{-2}} = \frac{1+\frac{1}{2}z^{-1}}{1+z^{-1}+\frac{1}{2}z^{-2}} + 3\frac{\frac{1}{2}z^{-1}}{1+z^{-1}+\frac{1}{2}z^{-2}}$$

From Appendix B.3

$$\{a^k \cos k\omega\} \quad \leftrightarrow \quad \frac{1 - az^{-1}\cos\omega}{1 - 2az^{-1}\cos\omega + a^2 z^{-2}}$$

and

$$\{a^k \sin k\omega\} \quad \leftrightarrow \quad \frac{az^{-1}\sin\omega}{1 - 2az^{-1}\cos\omega + a^2 z^{-2}}$$

hence

$$h[k] = \left(\frac{\sqrt{2}}{2}\right)^k \cos\frac{3k\pi}{4} + 3\left(\frac{\sqrt{2}}{2}\right)^k \sin\frac{3k\pi}{4} \qquad (k \geq 0)$$

Problem 4.4.7 Determine the impulse response of a system described by

$$y[k] - y[k-1] + \frac{1}{4}y[k-2] = u[k] - \frac{1}{3}u[k-1]$$

Solution: Obviously

$$H(z) = \frac{1 - \frac{1}{3}z^{-1}}{(1 - \frac{1}{2}z^{-1})^2} = \frac{2/3}{1 - \frac{1}{2}z^{-1}} + \frac{1/3}{(1 - \frac{1}{2}z^{-1})^2}$$

hence

$$h[k] = \frac{2}{3}\left(\frac{1}{2}\right)^k + \frac{1}{3}(k+1)\left(\frac{1}{2}\right)^k \qquad (k \geq 0)$$

i.e.,

$$h[k] = \left(\frac{1}{2}\right)^k + \frac{k}{3}\left(\frac{1}{2}\right)^k \qquad (k \geq 0)$$

Problem 4.4.8 Discrete-time linear time-invariant systems are often described using linear difference equations with constant coefficients which relate their output $y[k]$ to their input $u[k]$:

$$y[k] + a_1 y[k-1] + \ldots + a_n y[k-n] = b_0 u[k] + \ldots + b_m u[k-m]$$

with initial conditions $y[0], y[1], \ldots, y[n-1]$ given. Discuss the solution of this equation.

Solution: The solution of this equation can be written as

$$y[k] = y_h[k] + y_{nh}[k]$$

where $y_h[k]$ is a homogeneous part of the solution, while $y_{nh}[k]$ is a non-homogeneous (also known as particular) solution:

- $y_h[k]$: For each multiplicity-m root a of the characteristic equation of the difference equation $y_h[k]$ contains the following term(s)

  $$\beta_0 a^k + \beta_1 k a^k + \beta_2 k(k-1)a^k + \ldots + \beta_{m-1} k(k-1)\ldots(k-m+2)a^k$$

  where $\beta_0, \ldots, \beta_{m-1}$ are constants determined from the homogeneous part of the equation

  $$y[k] + a_1 y[k-1] + \ldots + a_n y[k-n] = 0$$

  and the initial conditions. See also the Note after Problem 4.1.3.

- $y_{nh}[k]$: This part of the solution is a convolution of the input $u[k]$ with $h[k]$, the impulse response of the system:

$$y_{nh}[k] = \sum_{i=0}^{k} u[i]\, h[k-i]$$

The impulse response is most easily determined using the inverse z-transform.

Note 1: *If the system is initially at rest, i.e., if all initial conditions are zero, then obviously $y_h[k] \equiv 0$, hence $y[k] = y_{nh}[k]$. On the other hand, if $u[k] \equiv 0$, then $y[k] = y_h[k]$. We say that the non-homogeneous part of the solution is the response to the input, while the homogeneous part of the solution is a response to the initial conditions.*

Note 2: *Show that the convolution formula we derived in Section 4.1 is a special case of this formula.*

Problem 4.4.9 Show that if $u[k] = z_0^k$ is the input and $y[k]$ is the output of a system described by

$$y[k] + a_1 y[k-1] + \ldots + a_n y[k-n] = b_0 u[k] + \ldots + b_m u[k-m]$$

then the output contains a term $T(z_0)z_0^k$, where

$$T(z_0) = \frac{b_0 + b_1 z_0^{-1} + \ldots + b_m z_0^{-m}}{1 + a_1 z_0^{-1} + \ldots + a_n z_0^{-n}} = \frac{b(z_0)}{a(z_0)}$$

It is assumed here that z_0 does not coincide with any of the poles of the system, i.e., roots of $a(z) = 1 + a_1 z^{-1} + \ldots + a_n z^{-n}$.

Solution: With $u[k] = z_0^k$ the above equation becomes

$$y[k] + a_1 y[k-1] + \ldots + a_n y[k-n] = z_0^k (b_0 + b_1 z_0^{-1} + \ldots + b_m z_0^{-m})$$

Its z-transform (assuming all initial conditions are zero) yields

$$Y(z) = \frac{1}{1 - z_0 z^{-1}} \frac{b(z_0)}{a(z)}$$

The partial fraction decomposition of this rational function contains the term $A/(1 - z_0 z^{-1})$ where

$$A = \lim_{z \to z_0} \left((1 - z_0 z^{-1}) Y(z) \right) = \frac{b(z_0)}{a(z_0)} = T(z_0)$$

hence the output $y[k]$ contains the term $T(z_0)z_0^k$.

Note: *Due to linearity, if $u[k] = z_1^k + z_2^k$ then*

$$y[k] = T(z_1)z_1^k + T(z_2)z_2^k + \ldots$$

Problem 4.4.10 Use MATLAB to plot the amplitude of the frequency response and the locations of the poles and the zeros of the following 7th order discrete-time low-pass filters with the cut-off frequency at one third of the sampling frequency: Butterworth, Chebyshev Type I, and Chebyshev Type II.

Solution: Figure 4.2 is easily obtained using the following MATLAB commands: `butter`, `cheby1`, `cheby2`, `freqz`, and `tf2zp`.

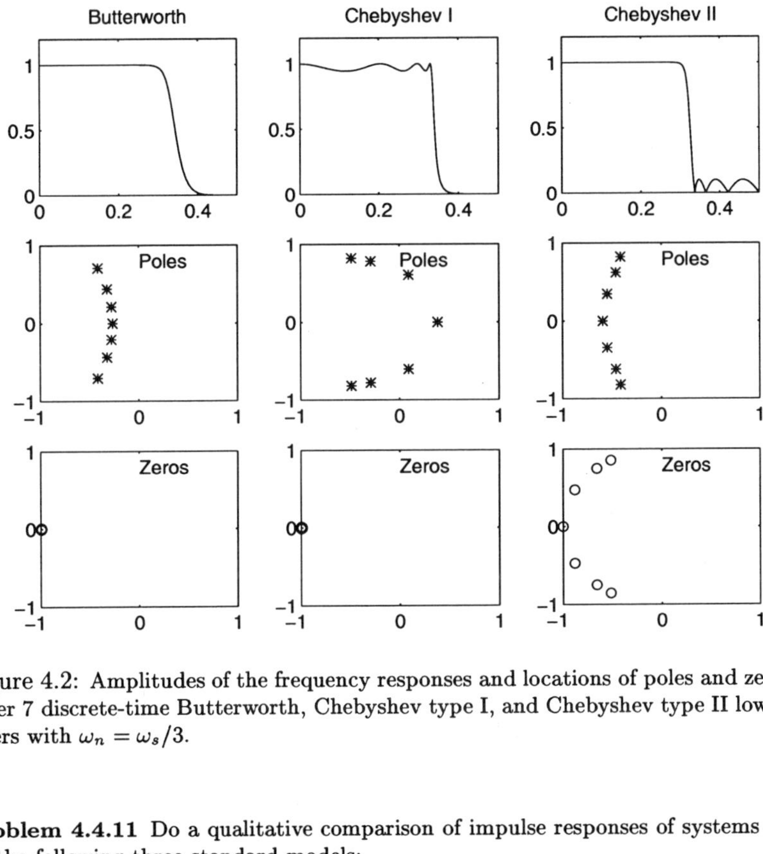

Figure 4.2: Amplitudes of the frequency responses and locations of poles and zeros of order 7 discrete-time Butterworth, Chebyshev type I, and Chebyshev type II low-pass filters with $\omega_n = \omega_s/3$.

Problem 4.4.11 Do a qualitative comparison of impulse responses of systems given by the following three standard models:

Moving Average (MA):

$$y[k] = b_0 u[k] + \ldots + b_m u[k-m]$$

Auto-Regressive (AR):

$$y[k] + a_1 y[k-1] + \ldots + a_n y[k-n] = b_0 u[k]$$

Combined (ARMA):

$$y[k] + a_1 y[k-1] + \ldots + a_n y[k-n] = b_0 u[k] + \ldots + b_m u[k-m]$$

Solution: It is easy to see that for the MA model the impulse response is

$$h_{MA}[k] = b_0\delta[k] + \ldots + b_m\delta[k-m]$$

This is obviously a finite-length sequence, hence such systems are called Finite Impulse Response (FIR) systems. Such systems are inherently stable because their transfer functions have no poles:

$$H_{MA}(z) = b_0 + b_1 z^{-1} + \ldots + b_m z^{-m}$$

They are simple to design but for many applications they don't offer enough flexibility.

For the AR model the impulse response $h_{AR}[k]$ can be found as the inverse z-transform of the transfer function

$$H_{AR}(z) = \frac{b_0}{1 + a_1 z^{-1} + \ldots + a_n z^{-n}}$$

Obviously, $h_{AR}[k]$ is an infinite-length sequence, hence such systems are called Infinite Impulse Response (IIR) systems. Since AR systems have poles, stability is a concern here. Note that this transfer function has poles, but no zeros.

ARMA models have similar properties to AR systems, except that they do have zeros. This provides for additional flexibility in the design, but also adds to the complexity of theoretical analysis. ARMA systems are also IIR. In general

$$h_{ARMA}[k] = \mathcal{Z}^{-1}\{H_{ARMA}(z)\} = \mathcal{Z}^{-1}\left\{\frac{b_0 z^n + \ldots + b_m z^{n-m}}{z^n + a_1 z^{n-1} + \ldots + a_{n-1} z + a_n}\right\}$$

Problem 4.4.12 Derive the Laplace transform of a signal $f(t)$ sampled by a train of Dirac δ-impulses.

Solution: Let the sampled signal be

$$g(t) = f(t)\sum_{k=-\infty}^{\infty}\delta(t-kT) = \sum_{k=-\infty}^{\infty} f(kT)\delta(t-kT)$$

where T is the sampling period. Then

$$\begin{aligned}
G(s) &= \int_{-\infty}^{\infty} g(t)e^{-st}\,dt \\
&= \int_{-\infty}^{\infty}\left(\sum_{k=-\infty}^{\infty} f(kT)\delta(t-kT)\right)e^{-st}\,dt \\
&= \sum_{k=-\infty}^{\infty}\left(\int_{-\infty}^{\infty} f(kT)\delta(t-kT)e^{-st}\,dt\right) \\
&= \sum_{k=-\infty}^{\infty} f(kT)e^{-skT}
\end{aligned}$$

Note 1: *Denote $z = e^{sT}$ and compare this last expression to the z-transform of the sequence $f[k] = f(kT)$. Until now we interpreted the z-transform only as a generating function of sequences. This derivation establishes a close analogy between the two transforms. The relation $z = e^{sT}$ not only explains the equivalence of the $j\omega$ (frequency) axis in the s-plane and the unit circle in the z-plane, but also forms a basis for several design methods in which the results from the continuous-time systems are applied to the discrete-time systems via this or approximate transformations, such as bilinear, in which*

$$z = \frac{1 + \frac{T}{2}s}{1 - \frac{T}{2}s}$$

Note 2: *In a discussion after their 1952 paper* [47] *in which they first introduced the z-transform, J. R. Ragazzini and L. A. Zadeh wrote:*

> In defining z as e^{+sT} rather than e^{-sT}, we have been motivated first by a desire to avoid conflict with the notation used by W. Hurewicz and others, and second by the fact that the alternative choice would make it inconvenient to use the only extensive table of z-transforms now available, namely, the table of so-called generalized Laplace transforms compiled by W. M. Stone. Otherwise, we are in complete agreement with Dr. Salzer's suggestion that it would be preferable to define z as being equal to e^{-sT} rather than e^{+sT}.

This was in response to what J. M. Salzer wrote:

> ...it may be preferable to define z as being equal to e^{-sT} rather than e^{+sT}, when dealing with sampled-data systems ... because the latter corresponds to a time-advance operation, which has no physical meaning in a real-time application. In purely mathematical work one definition is as good as the other, and it is just unfortunate that in previous operational and transform work with difference equations the advance ... operator was given a symbol.

Prior to this paper, W. Hurewicz used generating function methods to analyze sampled-data, i.e., discrete-time systems, while W. K. Linvill applied the Laplace transform to the sampled signals. Ragazzini and Zadeh were the first to unify these two approaches.

Problem 4.4.13 Explain the role of a low-pass filter (LPF) at the input of a system which converts continuous-time signals to discrete-time signals. Why does the sampling frequency ω_s have to be greater than twice the maximum frequency ω_m in the input signal? Derive the sampling theorem.

Solution: The purpose of the low-pass filter is to prevent *aliasing*. Without the filter or if its cut-off frequency is $\omega_m \geq \omega_s/2 = \pi/T$, then more than one input signal $u(t)$ can produce the same sampled signal $y(t)$ (see the diagram in Figure 4.3 and the derivation below).

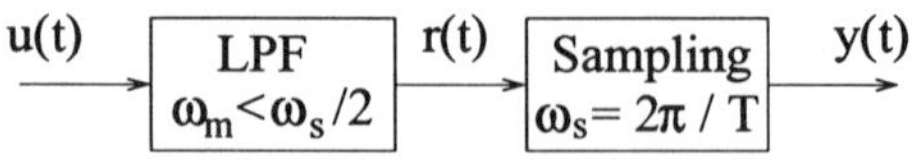

Figure 4.3: Low-pass filtering is necessary before sampling in order to avoid aliasing.

Let us assume that $u(t)$ is filtered so that $r(t)$ has no spectral components above ω_m, i.e., let $R(j\omega) \equiv 0$ for $\omega > \omega_m$. Then, with

$$\delta_T(t) = \sum_{k=-\infty}^{\infty} \delta(t - kT)$$

we have $y(t) = r(t)\delta_T(t)$. Since

$$Y(j\omega) = \frac{1}{2\pi}(R(j\omega) * \Delta_T(j\omega))$$

where

$$\Delta_T(j\omega) = \mathcal{F}\{\delta_T(t)\} = \frac{2\pi}{T} \sum_{k=-\infty}^{\infty} \delta(\omega - k\omega_s) \qquad \text{and} \qquad \omega_s = \frac{2\pi}{T}$$

we finally find

$$Y(j\omega) = \frac{1}{T} \sum_{k=-\infty}^{\infty} R(j(\omega - k\omega_s))$$

Another way to derive this result is to use the Poisson summation formula:

$$\sum_{k=-\infty}^{\infty} \delta(t - kT) = \frac{1}{T} \sum_{k=-\infty}^{\infty} e^{j\omega_s kt}$$

Then

$$Y(j\omega) = \frac{1}{T} \sum_{k=-\infty}^{\infty} \int_{-\infty}^{\infty} r(t)e^{j(\omega - k\omega_s)t}\, dt = \frac{1}{T} \sum_{k=-\infty}^{\infty} R(j(\omega - k\omega_s))$$

We see that $Y(j\omega)$, the spectrum of the sampled signal $y(t)$, is a sum of scaled copies of $R(j\omega)$ shifted by integer multiples of the sampling frequency $\omega_s = 2\pi/T$, as in Figure 4.4.

Unless $\omega_m \leq \omega_s/2$, these shifted copies overlap and the reconstruction of the original signal becomes impossible. If this condition is satisfied, than the sampled signal uniquely corresponds to the input signal. This is the *sampling theorem*. The

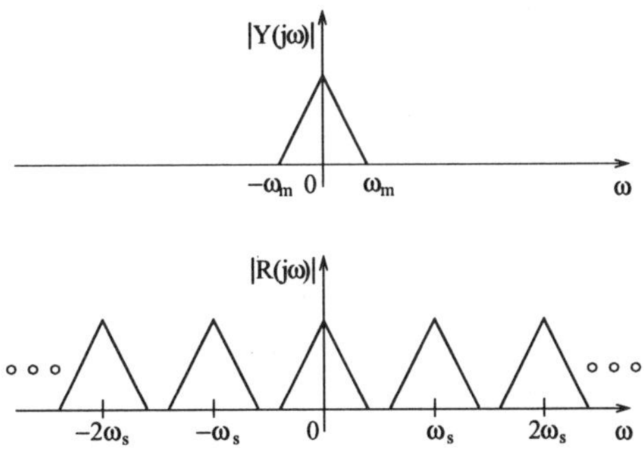

Figure 4.4: In the frequency domain the ideal sampling produces identical copies of the original spectrum centered around $\pm\omega_s$, $\pm 2\omega_s$, $\pm 3\omega_s$, ...

input signal can be reconstructed using its samples by another low-pass filter with a cut-off frequency between ω_m and $\omega_s/2$.

This overlapping is called *aliasing* because whether

$$R_1(t) = \cos\omega_1 t \quad \text{or} \quad R_2(t) = \cos(\omega_s - \omega_1)t$$

the sampled signal $y(t)$ is the same. We say that the higher-frequency signal has taken on the identity (alias) of the lower-frequency signal [43]. This is illustrated in Figure 4.5. We say that the higher-frequency signal $R_2(t)$ has been *undersampled.*

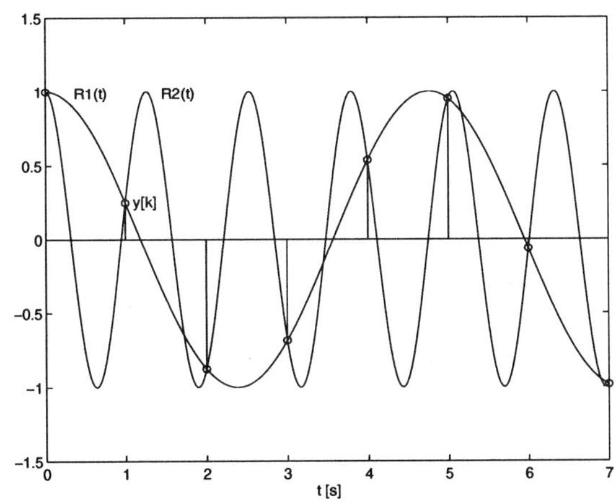

Figure 4.5: Example of two different signals producing the same sequence of samples. Here the sampling period is $T = 1\,\mathrm{s}$ hence the sampling frequency is $\omega_s = 6.28\,\mathrm{rad/s}$, while $\omega_1 = 1.32\,\mathrm{rad/s}$.

Another example of aliasing is the strobe effect in the Western movies, where the stagecoach wheels often seem to rotate slower that they naturally should, sometimes even in the wrong direction.

Note 1: *The minimum sampling frequency which guarantees no aliasing is called* the Nyquist rate*:*

$$\omega_N = 2\omega_m$$

The maximum frequency contained in a signal before sampling, ω_m *is sometimes called* the Nyquist frequency.

Note 2: *The preceding results were known to mathematicians for years. In the communications theory they were first used by H. Nyquist (1928), V. A. Kotelynikov (1933), and D. Gabor (1946). The sampling and reconstruction theorems first explicitly appeared in the communications literature in a seminal paper* [50] *by C. E. Shannon (1949). That is why it is often called* Shannon's sampling theorem.

4.5 State-space representation

In this Section we review the notation and main ideas behind the state-space representation of discrete-time systems. We find many similarities with the continuous-time systems described in Section 3.5.

Problem 4.5.1 A discrete-time system is given by the following state-space equations

$$\begin{aligned} x[k+1] &= Ax[k] + bu[k] \\ y[k] &= c'x[k] + du[k] \end{aligned}$$

where $u[k]$ is the input to the system, $y[k]$ is its output, while $x[k]$ is an $n \times 1$ state vector of the system. A is an $n \times n$ state-transition matrix, while b and c' are $n \times 1$ and $1 \times n$ vectors, respectively. We shall often assume that A has n distinct eigenvalues.

Express $y[k]$ in terms of $u[k]$, A, b, c', d, and the initial conditions $x[0]$. Determine the impulse response in terms of A, b, c', and d.

Solution: From Problem 4.3.2 we know that

$$x[k] = A^k x_0 + \sum_{i=0}^{k-1} A^i \underbrace{bu[k-1-i]}_{f[k-1-i]}$$

Hence

$$y[k] = c'A^k x_0 + (c'A^k b) * (u[k-1]) + du[k]$$

The impulse response is obtained from the above formula by putting $x[0] = 0$ and $u[k] = \delta[k]$:

$$h[k] = y[k]|_{x[0]=0, u[k]=\delta[k]} = c'A^{k-1}b + d\delta[k]$$

Note: *For $k = 1, 2, \ldots$ the impulse response coincides with the Markov parameters of the system (cf. Problem 3.8.4):*

$$h[k] = h_k = c'A^{k-1}b \qquad (k = 1, 2, \ldots)$$

Problem 4.5.2 Solve the state-space equations

$$\begin{aligned} x[k+1] &= Ax[k] + bu[k] \\ y[k] &= c'x[k] + du[k] \end{aligned}$$

in the z-transform domain. Determine the transfer function $H(z)$.

Solution: We know that $X(z) = z(zI - A)^{-1}x_0 + (zI - A)^{-1}bU(z)$ (cf. Problem 4.3.3), therefore

$$Y(z) = c'z(zI - A)^{-1}x_0 + c'(zI - A)^{-1}bU(z) + dU(z)$$

The transfer function is found as

$$H(z) = \left.\frac{Y(z)}{U(z)}\right|_{x[0]=0} = c'(zI - A)^{-1}b + d$$

Problem 4.5.3 Prove that the transfer function of the system given by

$$\begin{aligned} x[k+1] &= Ax[k] + bu[k] \\ y[k] &= c'x[k] \end{aligned}$$

where A has distinct eigenvalues, can be written as

$$H(z) = \sum_{i=1}^{n} \frac{(c'q_i)(p_i'b)z^{-1}}{1-\lambda_i z^{-1}}$$

Solution: This result is a direct consequence of Problem 3.2.15. This representation is very important, because it provides the rational decomposition in the transform domain, thus making the application of the inverse z-transform easy. Since

$$\mathcal{Z}^{-1}\left\{\frac{1}{1-\lambda_i z^{-1}}\right\} = \lambda_i^k$$

for the impulse response of a system with distinct eigenvalues we can write

$$h[k] = \sum_{i=1}^{n} \alpha_i \lambda_i^{k-1} \qquad (k > 0)$$

where $\alpha_i = (c'q_i)(p_i'b) \quad (i = 1, 2, \ldots, n)$.

For systems with multiple eigenvalues the corresponding formula is more complicated. In general, matrix A is not diagonalizable, hence, in notation of Problem 3.2.15, $\sum \lambda_i R_i \neq A$. Then one has to resort to Jordan matrices instead of diagonal matrices, when the impulse response is a linear combination of exponential functions multiplied by polynomials:

$$h[k] = \sum_{i=1}^{n} \alpha_i(k) \lambda_i^{k-1} \qquad (k > 0)$$

The degree of each $\alpha_i(k)$ is equal to the number of generalized eigenvectors corresponding to λ_i, i.e.,

$$\deg(\alpha_i(k)) \;=\; \nu(\lambda_i I - A) - 1 \;=\; n - \rho(\lambda_i I - A) - 1$$

where ρ and ν denote matrix rank and nullity, respectively (cf. Appendix C).

Problem 4.5.4 Derive the state-space equations for a serial, parallel, and a feedback connection of two systems given by triples $\{A_1, B_1, C_1\}$ and $\{A_2, B_2, C_2\}$.

Solution: Let us denote by $x[k]$ the new state vector:

$$x[k] = \begin{bmatrix} x_1[k] \\ x_2[k] \end{bmatrix}$$

In the serial connection the state-space equation is

$$\begin{aligned} x[k+1] &= \begin{bmatrix} A_1 & O \\ B_2C_1 & A_2 \end{bmatrix} x[k] + \begin{bmatrix} B_1 \\ O \end{bmatrix} u[k] \\ y[k] &= \begin{bmatrix} O & C_2 \end{bmatrix} x[k] \end{aligned}$$

In the parallel connection the state-space equation is

$$\begin{aligned} x[k+1] &= \begin{bmatrix} A_1 & O \\ O & A_2 \end{bmatrix} x[k] + \begin{bmatrix} B_1 \\ B_2 \end{bmatrix} u[k] \\ y[k] &= \begin{bmatrix} C_1 & C_2 \end{bmatrix} x[k] \end{aligned}$$

If the system $\{A_1, B_1, C_1\}$ is in the forward loop while $\{A_2, B_2, C_2\}$ is in the feedback loop, the new state-space equation is

$$\begin{aligned} x[k+1] &= \begin{bmatrix} A_1 & -B_1C_2 \\ B_2C_1 & A_2 \end{bmatrix} x[k] + \begin{bmatrix} B_1 \\ O \end{bmatrix} u[k] \\ y[k] &= \begin{bmatrix} O & C_2 \end{bmatrix} x[k] \end{aligned}$$

Note: *State the conditions necessary for matrix size compatibility and verify that in the case of single-input single-output systems the transfer functions are*

$$H_s(z) = H_1(z)H_2(z)$$

$$H_p(z) = H_1(z) + H_2(z)$$

$$H_f(z) = \frac{H_1(z)}{1 + H_1(z)H_2(z)}$$

Problem 4.5.5 Given a continuous-time system

$$\begin{aligned} \dot{x}(t) &= Ax(t) + Bu(t) \\ y(t) &= Cx(t) \end{aligned}$$

derive the equations for the corresponding discrete-time system. The sampling period is T.

Solution: We want to write the discrete-time state equations in the following form

$$\begin{aligned} x[k+1] &= Gx[k] + Hu[k] \\ y[k] &= Cx[k] \end{aligned}$$

From the expression for $x(t)$

$$x(t) = e^{At}x(0) + \int_0^t e^{A(t-\tau)}Bu(\tau)\,d\tau$$

we can directly write

$$x[k+1] = e^{A(k+1)T}x[0] + \int_0^{(k+1)T} e^{A((k+1)T-\tau)}Bu(\tau)\,d\tau$$

and

$$x[k] = e^{AkT}x[0] + \int_0^{kT} e^{A(kT-\tau)}Bu(\tau)\,d\tau$$

Now it is easy to write

$$x[k+1] = e^{AT}x[k] + e^{A(k+1)T}\int_{kT}^{(k+1)T} e^{-A\tau}Bu(\tau)\,d\tau$$

If we assume $u(t) = u[k] \quad (kT \leq t \leq (k+1)T)$, then we see that

$$G = e^{AT} \qquad \text{and} \qquad H = \left(\int_0^T e^{A\tau}\,d\tau\right)B$$

Note: *If A is invertible (i.e., nonsingular), then*

$$H = (e^{AT} - I)A^{-1}B$$

4.6 Stability

This Section presents the stability conditions for discrete-time systems. Although the definitions are practically identical to definitions for continuous-time systems, the conditions are very different: the poles must be inside the unit circle in the transform domain (rather than in the left-hand-side half-plane) and the Lyapunov equation has a different form.

Problem 4.6.1 Define BIBO (bounded-input bounded-output) stability and give the necessary and sufficient condition for a discrete-time system to be BIBO stable.

Solution: A system is BIBO stable if its output to any bounded input remains bounded at all times. Since

$$y[k] = \sum_{i=0}^{\infty} h[i]u[k-i]$$

and

$$|y[k]| = \left| \sum_{i=0}^{\infty} h[i]u[k-i] \right| \leq \sum_{i=0}^{\infty} |h[i]||u[k-i]| \leq C \sum_{i=0}^{\infty} |h[i]|$$

where $C = \max(|u[k]|)$, for a system to be BIBO stable it is sufficient that its impulse response be absolutely summable:

$$\sum_{i=0}^{\infty} |h[i]| < \infty$$

To show that this condition is necessary, suppose $h[i]$ is not absolutely summable. Then for $u[k] = \text{sgn}(h[K-k])$ we have

$$y[K] = \sum_{i=0}^{\infty} h[i]u[K-i] = \sum_{i=0}^{\infty} |h[i]|$$

which is not defined.

Thus, absolute summability is both a necessary and a sufficient condition for BIBO stability of discrete-time linear time-invariant systems.

Problem 4.6.2 Is a system with $h[k] = \frac{(-1)^k}{k+1}$ $(k > 0)$ BIBO stable?

Solution: No, because $h[k]$ is not absolutely summable:

$$\sum_{k=0}^{n} \frac{1}{k+1} \sim \ln n$$

Note: *It is interesting that $h[k]$ is summable but not absolutely summable:*

$$\sum_{k=0}^{\infty} h[k] = \ln 2$$

Problem 4.6.3 If a transfer function $H(z)$ of a discrete-time system is rational, what condition on its poles must be satisfied for BIBO stability?

Solution: If the poles $p_1, p_2, \ldots, p_n$ of $H(z)$ are distinct, then $h[k]$ is given by (cf. Problem 4.5.3)

$$h[k] = \sum_{i=1}^{n} \alpha_i p_i^{k-1} \qquad (k > 0)$$

If $H(z)$ has repeated poles, then (again cf. Problem 4.5.3) the α_i are polynomials in k:

$$h[k] = \sum_{i=1}^{n} \alpha_i(k) p_i^{k-1} \qquad (k > 0)$$

In either case $h[k]$ is absolutely summable if and only if

$$|p_i| < 1 \qquad (i = 1, 2, \ldots, n)$$

i.e., if the poles of the system are inside the unit circle in the z-plane.

See also the note after Problem 3.6.2.

Problem 4.6.4 A system has the impulse response $h[k] = \frac{1}{2^k} - \frac{1}{3^k} \;\; (k > 0)$. Determine the poles of the system. Is this system BIBO stable?

Solution: The poles are obviously $p_1 = 1/2$ and $p_2 = 1/3$. We can already say that the system is stable because $|p_{1,2}| < 1$, but let us verify that the impulse response is absolutely summable:

$$\sum_{k=0}^{\infty} |h[k]| = \sum_{k=1}^{\infty} \left| \frac{1}{2^k} - \frac{1}{3^k} \right| \leq \sum_{k=1}^{\infty} \left(\frac{1}{2^k} + \frac{1}{3^k} \right) = \frac{1/2}{1 - \frac{1}{2}} + \frac{1/3}{1 - \frac{1}{3}} = 1 + \frac{1}{2} = \frac{3}{2}$$

Problem 4.6.5 Use MATLAB to determine whether or not the discrete-time system given by the following recursion is stable:

$$y[k] + \frac{5}{6} y[k-1] + \frac{1}{6} y[k-2] - \frac{1}{8} y[k-3] = u[k-1] + 3u[k-2] \qquad (k \geq 0)$$

Solution: The characteristic equation of this system is

$$z^3 + \frac{5}{6} z^2 + \frac{1}{6} z - \frac{1}{8} = 0$$

Use the following command: `roots([1, 5/6, 1/6, -1/8])` to obtain

$$p_{1,2} = -0.5514 \pm 0.3998j \qquad p_3 = 0.2695$$

It is easy to see that $|p_{1,2,3}| < 1$, hence this system is BIBO stable.

Problem 4.6.6 Although the asymptotic stability in the sense of Lyapunov is defined for discrete-time systems in an analogous way as for continuous-time systems, the Lyapunov equation differs between the two classes of systems. Derive the Lyapunov equation for discrete-time systems.

Solution: Starting from a symmetric positive definite matrix P which determines the Lyapunov function

$$V(x[k]) = x'[k]Px[k]$$

for the asymptotic Lyapunov stability we require that $\Delta V(x[k]) < 0$, or at least $\Delta V(x[k]) \leq 0$ with $\Delta V(x[k]) \not\equiv 0$ along any possible system trajectory. Since

$$\Delta V(x[k]) = V(x[k+1]) - V(x[k]) = x'(A'PA - P)x$$

with $Q = -(A'PA - P)$, we require that Q is positive definite or at least positive semi-definite with the above condition that $\Delta V(x[k]) \not\equiv 0$ along any possible system trajectory.

As in continuous-time systems, we often start with any symmetric positive definite matrix Q, solve the Lyapunov equation for P, and test it for positive definiteness. The system is stable, i.e., A is discrete-time stability matrix if and only if the solution P of the discrete-time Lyapunov equation

$$Q = P - A'PA$$

is symmetric and positive definite.

Problem 4.6.7 The state transition matrix A has all eigenvalues inside the unit circle if and only if for an arbitrary positive definite symmetric matrix Q there exists a positive definite symmetric matrix P such that $A'PA - P = -Q$.

Consider two state transition matrices:

$$A_1 = \begin{bmatrix} 0.25 & 0.5 \\ 0.5 & 0 \end{bmatrix} \qquad A_2 = \begin{bmatrix} 1 & 0.2 \\ 0 & 1 \end{bmatrix}$$

In each case find P for $Q = I$. Check if P is positive definite. Calculate the eigenvalues to verify the results.

Solution: With

$$P = \begin{bmatrix} a & b \\ b & c \end{bmatrix}$$

(note the inherent symmetry) in the first case we obtain

$$A'PA - P = -I \qquad \Rightarrow \qquad P = \begin{bmatrix} 1.5 & 0.25 \\ 0.25 & 1.375 \end{bmatrix}$$

Since $1.5 > 0$ and $\det(P) > 0$, P is positive definite, i.e. $P > 0$. This agrees with the eigenvalues of A having magnitudes

$$|\lambda_{1,2}| = \left| \frac{1 \pm \sqrt{17}}{8} \right| < 1$$

In the second case matrix P does not exist, hence A is not a stability matrix for discrete-time systems. Indeed

$$\lambda_{1,2} = 1$$

Problem 4.6.8 Use MATLAB to generate a random 4×4 matrix and determine its eigenvalues. Is it a stability matrix in discrete-time? Use **dlyap** to solve the discrete Lyapunov equation and if MATLAB doesn't say that the solution is not unique, investigate the positive definiteness of the solution by looking at its eigenvalues (note that the solution is symmetric).

Solution: Do the following in MATLAB:

```
A = rand(4,4)

    A = 0.2190    0.9347    0.0346    0.0077
        0.0470    0.3835    0.0535    0.3834
        0.6789    0.5194    0.5297    0.0668
        0.6793    0.8310    0.6711    0.4175

eig(A)

    ans =

       1.4095
       0.1082 + 0.4681i
       0.1082 - 0.4681i
      -0.0763
```

Since $1.4095 > 1$ we already know that A is not a discrete-time stability matrix. Therefore, the solution of the Lyapunov equation will either be non-unique or will not be symmetric and positive definite.

```
P = dlyap(A',eye(4))

    P =

        1.0265   -0.7360    0.0069   -0.6269
       -0.7360   -0.5062   -0.7541   -1.2460
        0.0069   -0.7541    1.0462   -0.4689
       -0.6269   -1.2460   -0.4689    0.5550
```

Since P is symmetric, we can use the Rayleigh-Ritz criterion and look at its eigenvalues to see if it is positive definite.

```
eig(P)

   ans =

       -1.9091
        1.6373
        1.0037
        1.3896
```

Since $-1.9091 < 0$ matrix P is not positive definite. Actually, since it has both positive and negative eigenvalues, it is indefinite.

Problem 4.6.9 The discrete-time equivalent of a continuous-time system

$$\begin{aligned}\dot{x}(t) &= Ax(t) + Bu(t)\\ y(t) &= Cx(t)\end{aligned}$$

is given by (Problem 4.5.5)

$$\begin{aligned}x[k+1] &= Gx[k] + Hu[k]\\ y[k] &= Cx[k]\end{aligned}$$

where

$$G = e^{AT} \qquad \text{and} \qquad H = \left(\int_0^T e^{A\tau}\, d\tau\right) B$$

Recall that if A is nonsingular then $H = (e^{AT} - I)A^{-1}B$.

Show that if the continuous-time system is asymptotically stable, then the corresponding discrete-time system is also asymptotically stable. In other words if for all eigenvalues of A

$$\text{Re}\{\lambda_i\} < 0 \qquad (i = 1, 2, \ldots, n)$$

then for all eigenvalues of G

$$|\mu_i| < 1 \qquad (i = 1, 2, \ldots, n)$$

Solution: From the note in Problem 3.2.5 we know that $G = e^{AT}$ implies $\mu_i = e^{\lambda_i T}$. Then obviously

$$\text{Re}\{\lambda_i\} < 0 \quad \Rightarrow \quad |\mu_i| = |e^{\lambda_i T}| = e^{\text{Re}\{\lambda_i\} T} < 1 \qquad (i = 1, 2, \ldots, n)$$

4.7 Controllability and observability

Although the conditions for state controllability and observability are identical to the continuous-time case, there are some subtle differences in derivations. These differences lead to definitions of reachability and constructibility.

Problem 4.7.1 In Section 3.7 we derived the condition for state controllability. We said that, since e^{At} is always nonsingular, without loss of generality we could consider a special case when $x(t_f) = 0$. In discrete-time that is not so, because the corresponding matrix A^k is not necessarily nonsingular. Furthermore, if A is a *nilpotent* matrix, A^k could be O, a null matrix, for some k.

Show that the system given by

$$x[k+1] = \begin{bmatrix} 0 & 1 & 0 \\ 0 & 0 & 1 \\ 0 & 0 & 0 \end{bmatrix} x[k] + \begin{bmatrix} 0 \\ 1 \\ 0 \end{bmatrix} u[k]$$

$$y[k] = \begin{bmatrix} 1 & 1 & 0 \end{bmatrix} x[k]$$

can be taken from any initial state $x[0]$ to the origin even though its controllability matrix has rank 2.

Solution: Since, in general,

$$x[k] = A^k x[0] + \sum_{i=0}^{k-1} A^{k-1-i} b u[i]$$

and here A is *nilpotent* (because $A^k = O$ for $k \geq 3$; see also the note below), with $u[k] \equiv 0$ any initial state $x[0]$ of this system goes to the origin after only 3 time units.

The controllability matrix is

$$\mathcal{C} = \begin{bmatrix} 0 & 1 & 0 \\ 1 & 0 & 0 \\ 0 & 0 & 0 \end{bmatrix} \qquad \text{and} \qquad \rho(\mathcal{C}) = 2$$

Note 1: *Matrix A is said to be* nilpotent *if for some $k < \infty$ we have $A^k = O$. We shall prove here that a matrix is nilpotent if and only if all of its eigenvalues are zero:*

- *If $\lambda_i = 0 \quad (i = 1, 2, \ldots, n)$ then the characteristic equation of A is $\lambda^n = 0$ and according to the Cayley-Hamilton theorem we also have $A^n = O$.*
- *If $A^k = O$ for some $k < \infty$ then consider any of its eigenvalues and an eigenvector corresponding to it: λ and p. From $pA^k = 0$ and $pA^k = p\lambda^k$ we conclude that $\lambda = 0$ because, by definition, $p \neq 0$.*

Note 2: *Discrete-time systems with a nilpotent transition matrix A are usually called* deadbeat *systems.*

Problem 4.7.2 Show that the transition matrix does not have to be nilpotent if a system is to be controllable-to-the-origin but not state controllable. Do this by considering a singular but non-nilpotent transition matrix A.

Solution: Consider the following example:

$$x[k+1] = \begin{bmatrix} 0 & 0 \\ 1 & 1 \end{bmatrix} x[k] + \begin{bmatrix} 0 \\ 1 \end{bmatrix} u[k]$$

Any initial condition

$$x[0] = \begin{bmatrix} x_1[0] \\ x_2[0] \end{bmatrix}$$

is taken to the origin by $u[0] = -(x_1[0] + x_2[0])$. On the other hand

$$\det(\mathcal{C}) = \begin{vmatrix} 0 & 0 \\ 1 & 1 \end{vmatrix} = 0$$

Problem 4.7.3 Consider a single-input single-output system given by

$$\begin{aligned} x[k+1] &= Ax[k] + bu[k] \\ y[k] &= c'x[k] + du[k] \end{aligned}$$

where x is $n \times 1$, u, y, and d are scalars, while A is $n \times n$, b is $n \times 1$, and c' is $1 \times n$.

We say that a system is state controllable if application of a proper input $u[k]$ can change its state from any given state to any other given state in a finite amount of time.

Show that this system is state controllable if and only if

$$\rho(\mathcal{C}) = n, \qquad \text{where} \qquad \mathcal{C} = [b \;\; Ab \;\; A^2b \;\; \ldots \;\; A^{n-1}b]$$

Solution: Since

$$x[k] = A^k x[0] + \sum_{i=0}^{k-1} A^{k-1-i} bu[i]$$

we can write

$$A^{k_f} x[0] - x[k_f] = - \sum_{i=0}^{k_f - 1} A^{k_f - 1 - i} bu[i]$$

This vector equation is actually a system of n simultaneous equations in k_f unknowns $u[i]$, hence, in general, it is necessary that $k_f \geq n$. Finally, the solution exists if and only if

$$\rho(\mathcal{C}) = n, \qquad \text{where} \qquad \mathcal{C} = [b \;\; Ab \;\; A^2b \;\; \ldots \;\; A^{n-1}b]$$

Note 1: *In the case of single-input systems the controllability matrix $\mathcal{C}$ is $n \times n$, therefore we could write the above condition as* $\det(\mathcal{C}) \neq 0$. *The reason we didn't is that the validity of the above condition can be extended to the systems with m-dimensional inputs, when $\mathcal{C}$ is $n \times mn$.*

Note 2: *The state controllability was first considered in connection with the* finite settling time *problem in which a control input $u[k]$ is designed to return the perturbed system to the origin. While in continuous-time this so-called* controllability-to-the-origin[1] *is equivalent to the state controllability, in discrete-time $\rho(\mathcal{C}) = n$ is only a sufficient condition, as we saw in Problems* 4.7.1 *and* 4.7.2. *On the other hand,* controllability-from-the-origin[2] *is always equivalent to the state controllability.*

In discrete-time systems the controllability-to-the-origin is equivalent to the state controllability if and only if A is nonsingular, i.e., if $\det(A) \neq 0$.

Often, the word controllability *is used to mean controllability-to-the-origin, while controllability-from-the-origin is called* reachability. *Since a system is reachable if and only if $\rho(\mathcal{C}) = n$, the matrix $\mathcal{C}$ is sometimes called the* reachability matrix *instead of* controllability matrix.

Note 3: *It can be shown that the condition for observability in discrete-time is $\rho(\mathcal{O}) = n$. The property dual to the controllability-to-the-origin is called* constructibility *and it refers to the ability to determine the state vector from past outputs. For a discrete-time system with singular transition matrix A observability is sufficient but not necessary for constructibility. If A is nonsingular than observability is equivalent to constructibility. In continuous-time systems this distinction does not exist because e^{At} is always nonsingular.*

Problem 4.7.4 A system given by

$$x[k+1] = \begin{bmatrix} -2 & 1 & 0 \\ -3 & 0 & 1 \\ -4 & -1 & 2 \end{bmatrix} x[k] + \begin{bmatrix} 1 \\ 0 \\ 0 \end{bmatrix} u[k]$$

$$y[k] = \begin{bmatrix} 1 & 0 & 0 \end{bmatrix} x[k]$$

is deadbeat, i.e., even with $u[k] = 0$ $(k \geq 0)$ the state goes to the origin in at most $n = 3$ steps, regardless of the initial state $x[0] = x_0$.

a) Determine the initial conditions x_0 which guarantee that with $u[k] = 0$ $(k \geq 0)$ the output $y[k]$ is zero for $k \geq 2$.

b) Repeat part a) if the requirement is $y[k] = 0$ $(k \geq 1)$.

Solution: With $u[k] = 0$ $(k \geq 0)$, the expression for $y[k]$ becomes

$$y[k] = c' A^k x_0$$

a) Since the system is deadbeat, $y[k] = 0$ for $k \geq 3$ for any x_0. With $x_0 = [x \;\; y \;\; z]'$ the condition for $k = 2$ is

$$c' A^2 x_0 = 0 \qquad \text{i.e.} \qquad x - 2y + z = 0$$

a plane in the 3-D space of the initial state x_0.

[1]It is also called the *controllability p.s.t.o. (pointwise state to the origin).*

[2]Also called the *controllability p.s.f.o. (pointwise state from the origin).*

b) If we require $y[k] = 0$ $(k \geq 1)$, then in addition to the condition obtained in part a) we also need $y[1] = 0$, i.e.,

$$c' A x_0 = 0 \qquad \text{i.e.} \qquad -2x + y = 0$$

Combined with the previous condition $x - 2y + z = 0$, we find that x_0 must lie on the line given by

$$x = -\frac{1}{3}z \quad y = -\frac{2}{3}z \quad z \text{ any real number}$$

Note: *Are the conditions any different if we required $x[1] = 0$ and $x[2] = 0$? Would the solution change if we allowed $u[k] \neq 0$?*

Problem 4.7.5 A system given by

$$x[k+1] = \begin{bmatrix} 1 & 2 & 2 \\ 0 & 2 & 2 \\ 0 & 0 & 3 \end{bmatrix} x[k] + \begin{bmatrix} 1 \\ 0 \\ 1 \end{bmatrix} u[k]$$

$$y[k] = \begin{bmatrix} 1 & 0 & 0 \end{bmatrix} x[k]$$

is controllable, hence using the appropriate input $u[k]$, any initial condition x_0 can be taken to the origin in at most $n = 3$ steps.

a) Given the initial state x_0 determine the input $u[k]$ to take the state $x[k]$ to the origin in $n = 3$ steps.

b) There are infinitely many input sequences $u[k]$ which can drive the state from x_0 to the origin in $n + 1 = 4$ steps. Determine the one such sequence with minimum energy $\sum u^2[k]$.

c) Find x_0 which can be taken to the origin in just one step using any $u[k]$.

d) Repeat part c) if the input is restricted by $|u[k]| \leq 1$.

Solution: The expression for $x[k]$ is

$$x[k] = A^k x_0 + A^{k-1} b u[0] + \ldots + A b u[k-2] + b u[k-1]$$

a) The condition $x[3] = 0$ becomes

$$A^2 b u[0] + A b u[1] + b u[2] = -A^3 x_0$$

i.e.,

$$\mathcal{C} \begin{bmatrix} u[2] & u[1] & u[0] \end{bmatrix}' = -A^3 x_0$$

Thus

$$\begin{bmatrix} u[2] & u[1] & u[0] \end{bmatrix}' = -\mathcal{C}^{-1} A^3 x_0 = \begin{bmatrix} -1.50 & -9.00 & -4.50 \\ 1.25 & 13.50 & 9.75 \\ -0.25 & -3.50 & -5.75 \end{bmatrix} x_0$$

b) With $\mathcal{C}_4 = [b \;\; bA \;\; bA^2 \;\; bA^3]$ we can write the condition $x[4] = 0$ as

$$\mathcal{C}_4 \begin{bmatrix} u[3] & u[2] & u[1] & u[0] \end{bmatrix}' = -A^4 x_0$$

Since $\rho(\mathcal{C}_4) < 4$ there are infinitely many solutions $[u[3] \;\; u[2] \;\; u[1] \;\; u[0]]'$. To obtain the minimum energy solution (the *minimum norm solution* in the language of matrix algebra), we use the right pseudoinverse (see Section C.7):

$$\left[\begin{array}{cccc} u^*[3] & u^*[2] & u^*[1] & u^*[0] \end{array}\right]' = -\mathcal{C}_4' \left(\mathcal{C}_4\mathcal{C}_4'\right)^{-1} A^4 x_0$$

c) From $0 = x[1] = Ax[0] + bu[0]$ we find

$$x_0 = -A^{-1}bu[0] = \left[\begin{array}{ccc} -1 & 1/3 & -1/3 \end{array}\right]' u[0]$$

hence only the initial states $x_0 = [x \;\; y \;\; z]$ on the line given by

$$x = -u[0] \qquad y = u[0]/3 \qquad z = -u[0]/3$$

where $u[0]$ can be any real number, can be brought to the origin in just one sampling period.

d) Similarly, we obtain that the initial state must be on the line segment of the same line as in part c), between points $(-1, 1/3, -1/3)$ and $(1, -1/3, 1/3)$.

Problem 4.7.6 The discrete-time equivalent of a continuous-time system

$$\begin{aligned} \dot{x}(t) &= Ax(t) + Bu(t) \\ y(t) &= Cx(t) \end{aligned}$$

is given by (Problem 4.5.5)

$$\begin{aligned} x[k+1] &= Gx[k] + Hu[k] \\ y[k] &= Cx[k] \end{aligned}$$

where

$$G = e^{AT} \qquad \text{and} \qquad H = \left(\int_0^T e^{A\tau}\, d\tau\right) B$$

Recall that if A is nonsingular then

$$H = (e^{AT} - I)A^{-1}B$$

If the continuous-time system is controllable (observable), then the corresponding discrete-time system is also controllable (observable) if and only if the eigenvalues of A satisfy the following condition:

$$\text{Re}\{\lambda_i\} = \text{Re}\{\lambda_j\} \qquad \Rightarrow \qquad \text{Im}\{\lambda_i - \lambda_j\} \neq \frac{2m\pi}{T} \quad (m = 0, \pm 1, \pm 2, \ldots)$$

For a system given by

$$\dot{x}(t) = \left[\begin{array}{cc} 0 & 1 \\ -4 & 0 \end{array}\right] x(t) + \left[\begin{array}{c} 1 \\ 0 \end{array}\right] u(t)$$

$$y(t) = \left[\begin{array}{cc} 1 & 1 \end{array}\right] x(t)$$

determine the corresponding discrete-time system and determine the values of T which should be avoided in order to preserve controllability and observability.

Solution: Using the methods from Section 3.2 we find that

$$G = e^{AT} = \begin{bmatrix} \cos 2T & \frac{1}{2}\sin 2T \\ -2\sin 2T & \cos 2T \end{bmatrix}$$

while (since A is invertible)

$$H = (e^{AT} - I)A^{-1}B = \begin{bmatrix} \frac{1}{2}\sin 2T \\ -2\sin^2 T \end{bmatrix}$$

Since the eigenvalues of A are $\lambda_{1,2} = \pm 2j$, with an unfortunate choice of the sampling period T we may lose controllability or observability. According to the result stated above, we should make sure that

$$\mathrm{Im}\{2j - (-2j)\} \neq \frac{2m\pi}{T} \quad (m = 0, \pm 1, \pm 2, \dots)$$

In other words,

$$T \neq \frac{m\pi}{2} \quad (m = 0, \pm 1, \pm 2, \dots)$$

Note: *To establish this result directly from the system equations the reader should form the controllability and observability matrices and examine their ranks with respect to T.*

Problem 4.7.7 Let the discrete-time versions of systems given by $\{A_1, B_1, C_1\}$ and $\{A_2, B_2, C_2\}$ be $\{G_1, H_1, C_1\}$ and $\{G_2, H_2, C_2\}$, respectively. Show that if systems $\{A_1, B_1, C_1\}$ and $\{A_2, B_2, C_2\}$ are related through a nonsingular similarity transformation S, then the same is true for the corresponding discrete-time systems.

Use this result to show that for matrices with distinct non-zero eigenvalues, controllability and observability are preserved under sampling if and only if the eigenvalues of A satisfy the following condition:

$$\mathrm{Re}\{\lambda_i\} = \mathrm{Re}\{\lambda_j\} \quad \Rightarrow \quad \mathrm{Im}\{\lambda_i - \lambda_j\} \neq \frac{2m\pi}{T} \quad (m = 0, \pm 1, \pm 2, \dots)$$

Solution: If the sampling period is T, then

$$G_2 = e^{A_2 T} = e^{SA_1S^{-1}T} = Se^{A_1 T}S^{-1} = SG_1S^{-1}$$

Similarly

$$H_2 = \left(\int_0^T e^{SA_1S^{-1}\tau}\, d\tau\right) SB_1 = S\left(\int_0^T e^{A_1\tau}\, d\tau\right) S^{-1}SB_1 = SH_1$$

Now consider a controllable continuous-time system with distinct non-zero eigenvalues given by $\{A, b, c'\}$. A can be diagonalized using the matrix of its left eigenvectors P. Due to the above result, we can simplify the derivation by assuming that A is already in the diagonal form. Then, due to the assumed controllability, vector b has no zeros. To further simplify the notation, assume $b = [1 \ \ \dots \ \ 1]'$. Then

$$G = e^{AT} = e^{\text{diag}(\lambda_1,\ldots,\lambda_n)T} = \text{diag}(e^{\lambda_1 T},\ldots,e^{\lambda_n T})$$

while (recall the assumption that all eigenvalues are non-zero, hence A is invertible)

$$\begin{aligned} H &= (e^{AT} - I)A^{-1}B \\ &= \text{diag}((e^{\lambda_1 T} - 1),\ldots,(e^{\lambda_n T} - 1))\ \text{diag}(1/\lambda_1,\ldots,1/\lambda_n)\ [1\ \ \ldots\ \ 1]' \\ &= [(e^{\lambda_1 T} - 1)/\lambda_1\ \ \ldots\ \ (e^{\lambda_n T} - 1)/\lambda_n]' \end{aligned}$$

The determinant of the controllability matrix is

$$\det(\mathcal{C}) = |H\ \ GH\ \ \ldots\ \ G^{n-1}H| = \prod_{i=1}^{n} \frac{(e^{\lambda_i T} - 1)}{\lambda_i}\, V(e^{\lambda_1 T},\ldots,e^{\lambda_n T})$$

where V denotes the Vandermonde determinant. Since

$$V(e^{\lambda_1 T},\ldots,e^{\lambda_n T}) = \prod_{l>k}(e^{\lambda_l T} - e^{\lambda_k T})$$

we see that $\det(\mathcal{C}) \neq 0$ if and only if

$$\text{Re}\{\lambda_k\} = \text{Re}\{\lambda_l\} \quad \Rightarrow \quad \text{Im}\{\lambda_l - \lambda_k\} \neq \frac{2m\pi}{T} \quad (m = 0, \pm 1, \pm 2, \ldots)$$

The proof for observability is very similar.

Note: *Although this result is true in general, this proof covers only the case when A has distinct non-zero eigenvalues. For the proof of the general case PBH tests for controllability and observability can be used.*

4.8 Canonical realizations

All properties of canonical forms described in Section 3.8 hold for discrete-time systems as well. In this Section we describe some further properties of canonical forms and illustrate them through numerical examples.

Problem 4.8.1 A system is given by

$$\begin{aligned} y[k] - \frac{154}{120}y[k-1] + \frac{71}{120}y[k-2] - \frac{14}{120}y[k-3] + \frac{1}{120}y[k-4] &= \\ = u[k-1] - \frac{63}{90}u[k-2] + \frac{14}{90}u[k-3] - \frac{1}{90}u[k-4] \end{aligned}$$

Determine its poles and zeros. Determine its transfer function $H(z)$. Write the system in the order-4 canonical forms: controller, observer, controllability, observability, and modal. Show that in this case the controllable forms are not observable and vice versa, the observable forms are not controllable. Can this system be written in a form which is neither controllable nor observable?

Solution: To simplify the notation we will use the following:

$$a_0 = 1 \quad a_1 = -\frac{154}{120} \quad a_2 = \frac{71}{120} \quad a_3 = -\frac{14}{120} \quad a_4 = \frac{1}{120}$$

and

$$b_1 = 1 \quad b_2 = -\frac{63}{90} \quad b_3 = \frac{14}{90} \quad b_4 = -\frac{1}{90}$$

We can use MATLAB to determine the poles and zeros of this system:

`roots([a0 a1 a2 a3 a4])` and `roots([b1 b2 b3 b4])`

tell us that the poles of the system are at

$$p_1 = \frac{1}{2}, \quad p_2 = \frac{1}{3}, \quad p_3 = \frac{1}{4}, \quad \text{and} \quad p_4 = \frac{1}{5}$$

while the zeros are at

$$z_1 = \frac{1}{3}, \quad z_2 = \frac{1}{5}, \quad \text{and} \quad z_3 = \frac{1}{6}$$

The system is stable, but due to pole-zero cancellations it is not minimal, therefore some of its state-space representations will not be controllable, while others will not be observable.

The transfer function can be written directly from the equation:

$$H(z) = \frac{b_1 z^3 + b_2 z^2 + b_3 z + b_4}{z^4 + a_1 z^3 + a_2 z^2 + a_3 z + a_4}$$

The following are several canonical realizations of this system:

- Controller (in MATLAB: `[Ac,Bc,Cc,Dc] = tf2ss(num,den)`):

$$
\begin{aligned}
A_c &= \begin{bmatrix} -a_1 & -a_2 & -a_3 & -a_4 \\ 1 & 0 & 0 & 0 \\ 0 & 1 & 0 & 0 \\ 0 & 0 & 1 & 0 \end{bmatrix} \qquad b_c = \begin{bmatrix} 1 \\ 0 \\ 0 \\ 0 \end{bmatrix} \\
c_c' &= \begin{bmatrix} b_1 & b_2 & b_3 & b_4 \end{bmatrix}
\end{aligned}
$$

This realization is always controllable. In this case it is not observable. The modal realization similar to this realization is obtained using $S = Q^{-1}$, where (as we discussed in Problem 3.2.1) Q is the matrix of right eigenvectors of A_c (in MATLAB: `[Q,D] = eig(Ac)`):

$$
\begin{aligned}
A_{d_c} &= \begin{bmatrix} 1/2 & 0 & 0 & 0 \\ 0 & 1/3 & 0 & 0 \\ 0 & 0 & 1/4 & 0 \\ 0 & 0 & 0 & 1/5 \end{bmatrix} \qquad b_{d_c} = \begin{bmatrix} -92.2 \\ -572.7 \\ 991.5 \\ -510.3 \end{bmatrix} \\
c_{d_c}' &= \begin{bmatrix} -0.0145 & 0 & -0.000336 & 0 \end{bmatrix}
\end{aligned}
$$

The lack of observability is obvious in this realization because of the zeros in the output vector c_{d_c}'.

- Observer:

$$
\begin{aligned}
A_o &= \begin{bmatrix} -a_1 & 1 & 0 & 0 \\ -a_2 & 0 & 1 & 0 \\ -a_3 & 0 & 0 & 1 \\ -a_4 & 0 & 0 & 0 \end{bmatrix} \qquad b_o = \begin{bmatrix} b_1 \\ b_2 \\ b_3 \\ b_4 \end{bmatrix} \\
c_o' &= \begin{bmatrix} 1 & 0 & 0 & 0 \end{bmatrix}
\end{aligned}
$$

This realization is always observable. In this case it is not controllable. The modal realization similar to this realization is given by

$$
\begin{aligned}
A_{d_o} &= \begin{bmatrix} 1/2 & 0 & 0 & 0 \\ 0 & 1/3 & 0 & 0 \\ 0 & 0 & 1/4 & 0 \\ 0 & 0 & 0 & 1/5 \end{bmatrix} \qquad b_{d_o} = \begin{bmatrix} 1.7147 \\ 0 \\ -0.492 \\ 0 \end{bmatrix} \\
c_{d_o}' &= \begin{bmatrix} 0.7776 & -0.7109 & 0.6773 & -0.6571 \end{bmatrix}
\end{aligned}
$$

The lack of controllability is obvious here because of the zeros in the input vector b_{d_o}. These realizations are not equivalent to the controllable realizations.

- Let us construct a realization which is neither controllable nor observable. The impulse response of the system in modal form is

$$
h[k] = h_k = c_d' A_d^{k-1} b_d = \sum_{i=1}^{n} b_i c_i \lambda_i^{k-1} \qquad (k > 0)
$$

If in the first of the two modal representations above we set zeros in b_{d_c} to match the zeros in c_{d_c}', the products $b_i c_i$ remain unchanged, but the system also becomes uncontrollable. In addition we do some balancing between the input and the output vector, keeping the products $b_i c_i$ same as before:

$$A_d = \begin{bmatrix} 1/2 & 0 & 0 & 0 \\ 0 & 1/3 & 0 & 0 \\ 0 & 0 & 1/4 & 0 \\ 0 & 0 & 0 & 1/5 \end{bmatrix} \quad b_d = \begin{bmatrix} 4/3 \\ 0 \\ -1/3 \\ 0 \end{bmatrix}$$

$$c_d' = \begin{bmatrix} 1 & 0 & 1 & 0 \end{bmatrix}$$

This realization of $H(s)$ is not similar to any of the other realizations above.

Problem 4.8.2 Show that if a system given by $\{A, b, c'\}$ is observable, it can be transformed into the observability form using the following transformation matrix:

$$S_{ob} = \mathcal{O}$$

where

$$\mathcal{O} = \begin{bmatrix} c' \\ c'A \\ \vdots \\ c'A^{n-1} \end{bmatrix}$$

is the observability matrix of the original system.

Solution: We need to show that $S_{ob}AS_{ob}^{-1} = A_{ob}$, $S_{ob}b = b_{ob}$, and $c'S_{ob}^{-1} = c_{ob}'$. To show that $S_{ob}AS_{ob}^{-1} = A_{ob}$ we will prove that $\mathcal{O}A = A_{ob}\mathcal{O}$. Indeed

$$\mathcal{O}A = \begin{bmatrix} c' \\ c'A \\ \vdots \\ c'A^{n-1} \end{bmatrix} A = \begin{bmatrix} c'A \\ c'A^2 \\ \vdots \\ c'A^{n} \end{bmatrix}$$

while

$$A_{ob}\mathcal{O} = \begin{bmatrix} c'A \\ c'A^2 \\ \vdots \\ c'(-a_nI - a_{n-1}A - \ldots - a_1A^{n-1}) \end{bmatrix}$$

From the Cayley-Hamilton theorem $-a_nI - a_{n-1}A - \ldots - a_1A^{n-1} = A^n$ and therefore $\mathcal{O}A = A_{ob}\mathcal{O}$.

To show $S_{ob}b = [h_1 \ \ h_2 \ \ \ldots \ \ h_n]'$ observe that

$$S_{ob}b = [c'b \ \ c'Ab \ \ \ldots \ \ c'A^{n-1}b]' = [h_1 \ \ h_2 \ \ \ldots \ \ h_n]'$$

where h_i's are Markov parameters.

Problem 4.8.3 Show that an observable system given by $\{A, b, c'\}$ can be transformed into the observer form if we use

$$S_o = \mathcal{O}_o^{-1}\mathcal{O} \qquad (\mathcal{O}_o = \mathsf{a}_-^{-1})$$

Solution: Using the result of Problem 4.8.2 we know that the transformation from observer into the observability form is given by $S = \mathcal{O}_o = \mathsf{a}_-^{-1}$. Thus, to go from any observable form into the observer form we can go via the observability form, when we find

$$S_o = \mathsf{a}_-\mathcal{O}$$

4.9 State feedback

This Section illustrates the pole placement using the state feedback for discrete-time systems. The results of Section 3.9 apply here without any changes.

Problem 4.9.1 Use MATLAB to discretize the equations for the inverted pendulum on a cart given in Problem 3.7.12. Choose the sampling period T so that it is 5 times smaller than the smallest time constant of the continuous-time system. Check the stability, controllability, and observability of the discrete-time system. Determine the feedback gain vector f so that all eigenvalues of the system are halved in magnitude. Simulate the system under conditions set in Problem 3.9.9.

Solution: The continuous-time parameters are as follows:

$$A = \begin{bmatrix} 0 & 0 & 1 & 0 \\ 0 & 0 & 0 & 1 \\ 0 & -1 & 0 & 0 \\ 0 & 21.6 & 0 & 0 \end{bmatrix} \qquad B = \begin{bmatrix} 0 \\ 0 \\ 1 \\ -2 \end{bmatrix} \qquad C = \begin{bmatrix} 1 & 0 & 0 & 0 \\ 0 & 1 & 0 & 0 \end{bmatrix}$$

The eigenvalues are $\lambda_1 = 0$, $\lambda_2 = 0$, $\lambda_3 = 4.65$, $\lambda_4 = -4.65$, hence we pick $T = \tau/5 = 1/(5|\lambda_4|) = 40\,ms$. The parameters of the discrete time system can be obtained using c2d, the MATLAB command for conversion from continuous-time to discrete-time: `[G, H] = c2d(A,B,T)`. Using this command we obtain

$$G = \begin{bmatrix} 1 & -0.0008 & 0.0400 & 0 \\ 0 & 1.0173 & 0 & 0.0402 \\ 0 & -0.0402 & 1 & -0.0008 \\ 0 & 0.8690 & 0 & 1.0173 \end{bmatrix} \qquad H = \begin{bmatrix} 0.0008 \\ -0.0016 \\ 0.0400 \\ -0.0805 \end{bmatrix}$$

Using MATLAB commands `abs(eig(G))`, `rank(ctrb(G,H))`, and `rank(obsv(G,C))` we find that, just like the continuous-time system, the discretized system is controllable, observable, and unstable, with eigenvalues at

$$p_1 = 1, \quad p_2 = 1, \quad p_3 = 1.2043, \quad p_4 = 0.8304$$

To determine the feedback gain f so that the eigenvalues are halved in magnitude, note that g, the vector of coefficients of the characteristic polynomial $g(z)$, can be obtained using `g = poly(eig(G))`, while γ, the vector of coefficients of the desired characteristic polynomial $\gamma(z)$ is obtained from `gamma = poly(eig(G)/2)`. Be careful, however, because the Bass-Gura formula doesn't use the first coefficient of these polynomials.

```
A = [0 0 1 0; 0 0 0 1; 0 -1 0 0; 0 21.6 0 0];
B = [0; 0; 1; -2];
C = [1 0 0 0; 0 1 0 0];
T = 0.04;
[G, H] = c2d(A,B,T);
g = poly(eig(G));
gamma = poly(eig(G)/2);
CCc = inv(toeplitz([1 0 0 0],g(1:4)));
CC = ctrb(G,H);
f = ((gamma(2:5) - g(2:5))*CCc*inv(CC))';
```

Thus

$$f = [-1155.9 \quad -912.9 \quad -310.9 \quad -173.0]'$$

The same result is obtained by MATLAB implementation of the Ackermann formula:

```
f = acker(G,H,eig(G)/2)'
```

Let us simulate the behavior of the discretized system under the conditions given in Problem 3.9.9. There, the desired eigenvalues were set to $\lambda_1 = -1$, $\lambda_2 = -2$, $\lambda_3 = -1 + j$, and $\lambda_4 = -1 - j$. This means that the eigenvalues of the discretized system are $\mu_i = e^{\lambda_i T}$ $(i = 1, 2, 3, 4)$.

The first simulation is for zero initial conditions and the unit step function at the input. The graph obtained by simulation is shown in Figure 4.6. Compare this graph to the graph obtained in Simulation 2a in Problem 3.9.9.

```
A = [0 0 1 0; 0 0 0 1; 0 -1 0 0; 0 21.6 0 0];
B = [0; 0; 1; -2];
C = [1 0 0 0; 0 1 0 0];
T = 0.04;
[G, H] = c2d(A,B,T);
f = acker(G,H,exp([-1,-2,-1+j,-1-j]*T))';
t = 0:T:8;
u = (ones(size(t)))';
x0= [0 0 0 0]';
sys = ss(G-H*f',H,C,0,T);
lsim(sys,u,t,x0)
```

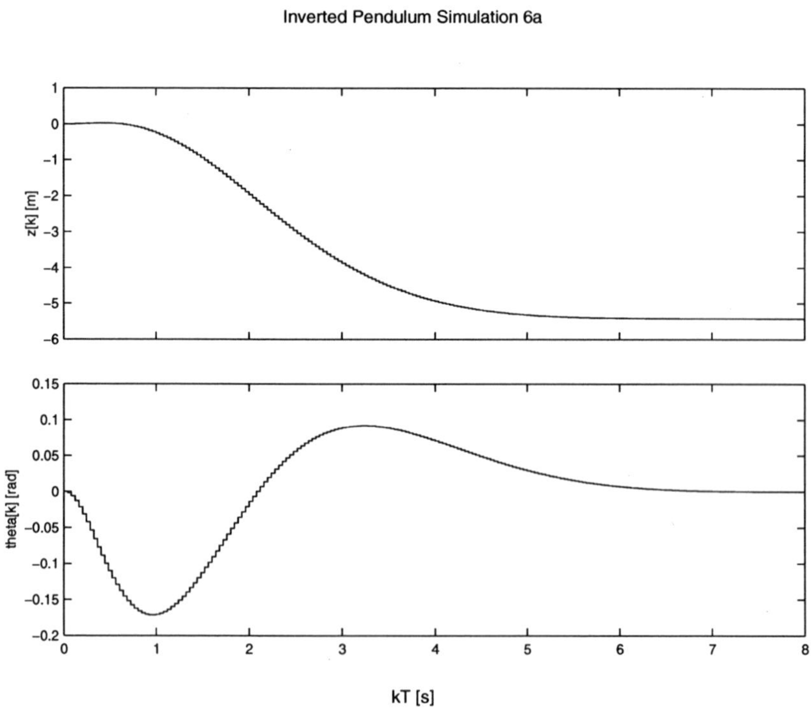

Figure 4.6: The results of the MATLAB Simulation 6a.

The second simulation is for zero input and non-zero initial conditions. The graph obtained by simulation is shown in Figure 4.7. Compare this graph to the graph obtained in Simulation 2b in Problem 3.9.9.

```
A = [0 0 1 0; 0 0 0 1; 0 -1 0 0; 0 21.6 0 0];
B = [0; 0; 1; -2];
C = [1 0 0 0; 0 1 0 0];
T = 0.04;
[G, H] = c2d(A,B,T);
f = acker(G,H,exp([-1,-2,-1+j,-1-j]*T))';
t = 0:T:8;
u = (zeros(size(t)))';
x0= [0 0.1 0 0]';
sys = ss(G-H*f',H,C,0,T);
lsim(sys,u,t,x0)
```

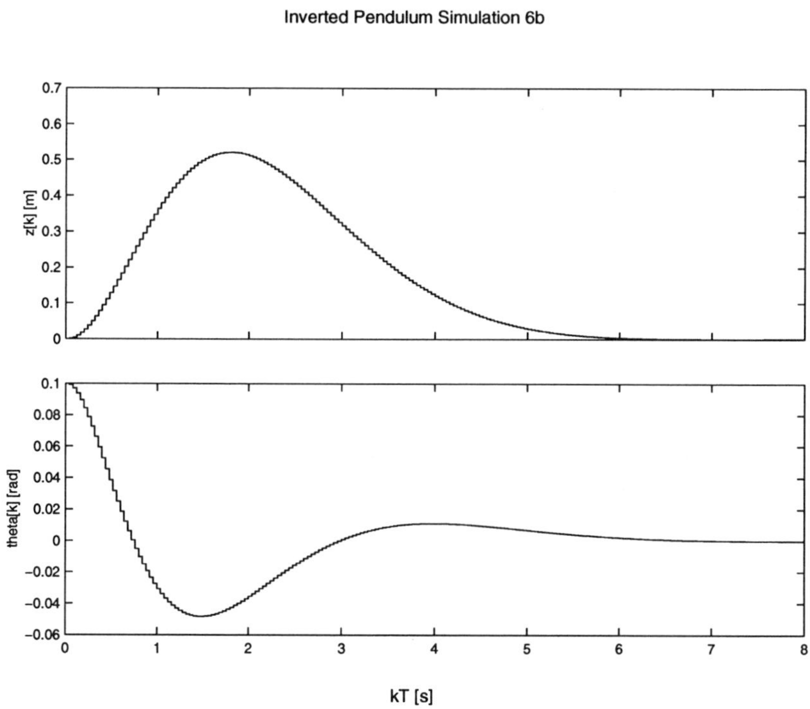

Figure 4.7: The results of the MATLAB Simulation 6b.

Problem 4.9.2 For a system given by

$$x[k+1] = \begin{bmatrix} 1 & 2 & 2 \\ 0 & 2 & 2 \\ 0 & 0 & 3 \end{bmatrix} x[k] + \begin{bmatrix} 1 \\ 0 \\ 1 \end{bmatrix} u[k]$$

$$y[k] = \begin{bmatrix} 1 & 0 & 0 \end{bmatrix} x[k]$$

check for controllability and observability and then, assuming all states are available (we will design observers in Section 4.10, and there observability will play a critical role, just like controllability does here), determine the feedback gain vector f so that the closed-loop system with

$$u[k] = -f'x[k]$$

is deadbeat, i.e., its eigenvalues are $\mu_{1,2,3} = 0$.

Do this by writing the desired characteristic equation in terms of the feedback gain vector $f = [f_1 \ f_2 \ f_3]'$. Compare the result to the output of the MATLAB command

```
f = acker(A,b,[0 0 0])
```

Solution: It is easy to verify that the system is controllable and observable. The desired characteristic polynomial is $\alpha(z) = z^3$. In terms of the feedback gain $f = [f_1 \ f_2 \ f_3]'$ it can be written as

$$\alpha(z) = \det(zI - (A - bf'))$$

i.e.,

$$\alpha(z) = z^3 + (f_1 + f_3 - 6)z^2 + (-3f_1 + 2f_2 - 3f_3 + 11)z + 6f_1 - 2f_2 + 2f_3 - 6$$

When we equate all coefficients in this polynomial to zero, we obtain a system of three equations in f_1, f_2, and f_3 whose solution is

$$f = [0.25 \quad 3.50 \quad 5.75]'$$

Note: *The same result is obtained using the Bass-Gura formula*

$$f' = (\alpha' - a')\mathcal{C}_c\mathcal{C}^{-1}$$

or the Ackermann formula

$$f' = [0 \ \ldots \ 0 \ 1]\mathcal{C}^{-1}\alpha(A)$$

Recall that $\mathcal{C}_c = \mathsf{a}_-^{-T}$, *where* a_- *is as defined in Problem* 3.8.4.

4.10 Optimal control

Although discrete-time systems can achieve the deadbeat response using the feedback design techniques described in Section 4.9, the resulting system may not be acceptable because it may require large values of input signal. The solution is to use the quadratic cost function which weighs both the settling time and the magnitude of the input. Thus we determine the linear feedback gain to achieve optimal control (discrete-linear-quadratic-regulator — DLQR).

Problem 4.10.1 In Problem 4.9.2 we verified that the system given by

$$x[k+1] = \begin{bmatrix} 1 & 2 & 2 \\ 0 & 2 & 2 \\ 0 & 0 & 3 \end{bmatrix} x[k] + \begin{bmatrix} 1 \\ 0 \\ 1 \end{bmatrix} u[k]$$

$$y[k] = \begin{bmatrix} 1 & 0 & 0 \end{bmatrix} x[k]$$

is both controllable and observable. Assuming all states of the system are available design the stationary state feedback

$$u[k] = -Fx[k]$$

which minimizes the cost function given by

$$J = \sum_{k=0}^{\infty} \left(x'[k]Qx[k] + u'[k]Ru[k] \right)$$

where $Q = I_{3\times 3}$ and $R = 1$.

The optimal feedback gain is found from the following:

$$F = (B'PB + R)^{-1}B'PA$$

where P is the real symmetric positive definite solution of the matrix algebraic Riccati equation

$$P = A'P(I - B(B'PB + R)^{-1}B'P)A + Q$$

Note that the solution is guaranteed to exist if Q and R are symmetric and Q is positive semi-definite, while R is positive definite.

The minimum cost is given by $J = x'[0]Px[0]$.

Solution: To solve the Riccati equation, assume P is symmetric

$$P = \begin{bmatrix} p_{11} & p_{12} & p_{13} \\ p_{12} & p_{22} & p_{23} \\ p_{13} & p_{23} & p_{33} \end{bmatrix}$$

and substitute that into the equation. The solution is

$$P = \begin{bmatrix} 4.17 & 16.13 & 6.88 \\ 16.13 & 150.84 & 108.09 \\ 6.88 & 108.09 & 103.22 \end{bmatrix}$$

Then the optimal feedback gain is

$$F = (B'PB + R)^{-1}B'PA = [0.09 \quad 2.21 \quad 4.92]$$

The eigenvalues of the optimized closed-loop system are at

$$\mu_1 = 0.20 \qquad \mu_2 = 0.39 + 0.29j \qquad \mu_3 = 0.39 - 0.29j$$

Matlab note: *The above results were obtained using the following* MATLAB *command:* `[F,P,mu] = dlqr(A,B,Q,R);`

Problem 4.10.2 For a discrete-time system given by

$$x[k+1] = Ax[k] + Bu[k]$$

$$y[k] = Cx[k]$$

the cost of control on the interval $0 \leq k \leq N$ is given by

$$J_N = \sum_{k=0}^{N} \left(x'[k]Qx[k] + u'[k]Ru[k]\right)$$

where Q and R are symmetric and Q is positive semi-definite, while R is positive definite. Derive the optimal control law from the *Optimality Principle*[3] which can be paraphrased as follows:

> If a system state at some time instance is on the optimal trajectory, its motion from that point to the final state along this trajectory will be optimal.

Solution: In the following we shall use the following identities from the matrix calculus (Appendix C):

For M symmetric[4]

$$\frac{\partial}{\partial x}(x'Mx) = 2Mx \qquad \frac{\partial}{\partial x}(x'My) = My \qquad \frac{\partial}{\partial y}(x'My) = M'x$$

First define $S_j = J_N - J_{N-j}$, the control cost over $j \leq k \leq N$. Then $S_{j+1} = S_j + \Delta J_{N-j}$, i.e.,

$$S_{j+1} = S_j + x'[N-j]Qx[N-j] + u'[N-j]Ru[N-j]$$

Note that S_1 corresponds to the control cost of the last control period. As the index of S_j increases, we go back in time. The convenience of such notation will become clear shortly.

[3]The importance of this principle was independently discovered and used by several mathematicians over the last several centuries: Jakob Bernoulli (1697), Johann Bernoulli (1706), L. Euler (1744), C. Carathéodory (1930s), and R. Bellman (1950s). Applied to our system the Optimality Principle states that if $u^*[k] = f(x[k])$ is optimal over $0 \leq k \leq N$, then it is also optimal over $j \leq k \leq N$, where $0 \leq j \leq N$.

[4]Actually, only the first identity requires $M = M'$.

From the Optimality Principle, S^*_{j+1}, the minimum value of S_{j+1} is obtained by using $S_j = S^*_j$ and from

$$\frac{\partial S_{j+1}}{\partial u[N-j]} = 0$$

Consider first $j = 0$:

$$S_1 = J_N - J_{N-1} = x'[N]Qx[N] + u'[N]Ru[N]$$

Since $x[N]$ depends only on $u[k]$ for $k < N$, obviously S_1 is minimized with $u^*[N] = 0$. Hence

$$S^*_1 = x'[N]Qx[N]$$

For $j = 1$ and $S_1 = S^*_1 = x'[N]Qx[N]$ we have

$$S_2 = S^*_1 + x'[N-1]Qx[N-1] + u'[N-1]Ru[N-1]$$

Now use $x[N] = Ax[N-1] + Bu[N-1]$ in the expression for S^*_1. Then the condition $\partial S_2/\partial u[N-1] = 0$ yields

$$2B'Q(Ax[N-1] + Bu^*[N-1]) + 2Ru^*[N-1] = 0$$

i.e.,

$$u^*[N-1] = -\underbrace{(B'QB + R)^{-1}B'QA}_{\text{call this } F_{N-1}}\,x[N-1]$$

Substitute this back into the expression for S_2 to obtain the expression for its minimum

$$S^*_2 = x'[N-1]\underbrace{(A'Q(A - BF_{N-1}) + Q)}_{\text{call this } P_{N-1}}x[N-1]$$

With this notation we have $F_N = 0$ and $P_N = Q$.

To derive the general recursive procedure for calculating F_{N-j} for $j = 0, 1, 2, \ldots, N$ (note the sequence of calculation: $F_N, F_{N-1}, \ldots, F_0$), consider the following. The minimum value of S_j has the following form:

$$S^*_j = x'[N-j+1]P_{N-j+1}x[N-j+1]$$

i.e.,

$$S^*_j = (Ax[N-j] + Bu[N-j])'P_{N-j+1}(Ax[N-j] + Bu[N-j])$$

With $S_{j+1} = S_j + \Delta J_{N-j}$ and $S_j = S^*_j$ the condition $\partial S_{j+1}/\partial u[N-j] = 0$ yields

$$2B'P_{N-j+1}Ax[N-j] + 2(B'P_{N-j+1}B + R)u^*[N-j]) = 0$$

i.e.,

$$u^*[N-j] = -\underbrace{(B'P_{N-j+1}B + R)^{-1}B'P_{N-j+1}A}_{F_{N-j}}\, x[N-j]$$

The expression for S_{j+1}^* now becomes

$$S_{j+1}^* = x'[N-j]P_{N-j}x[N-j]$$

where P_{N-j} is determined recursively from

$$P_{N-j} = A'P_{N-j+1}(A - BF_{N-j}) + Q$$

This calculation is necessary not only to determine S_{j+1}^*, but also in the next step, in the calculation of F_{N-j-1}.

The recursion starts at $j = 0$, when, as we found in the beginning, $F_N = 0$ and $P_N = Q$.

The minimum cost of control is

$$J_N^* \;=\; S_N^* + J_0^* \;=\; S_{N+1}^* \;=\; x'[0]P_0x[0]$$

Note: *It is important to note that the optimal control law is linear:* $u^*[k] = f(x[k]) = -F_kx[k]$. *This is not because we required that condition, but it followed directly from our derivations.*

However, even though the system is stationary (i.e., matrices A, B, and C are assumed to be constant), the optimal control law is not, i.e., the values of the feedback gain change with time.

If the system is controllable than for large values of N the sequence $\{F_{N-j}\}$ *converges (except for* $j = 0$*) to a constant sequence:*

$$\lim_{N\to\infty} F_{N-j} = F \qquad (j = 1, 2, \ldots)$$

This fact can be used to significantly simplify the design of the optimal controller, because the equations simplify to

$$u^*[k] = -Fx[k]$$

where

$$F = (B'PB + R)^{-1}B'PA$$

and P is a real symmetric solution of the matrix algebraic Riccati equation

$$P = A'P(I - B(B'PB + R)^{-1}B'P)A + Q$$

The minimum control cost is then

$$J^* = x'[0]Px[0]$$

4.11 State observers

In this Section we design state observers. The basic results are identical to those derived for the continuous-time systems in Section 3.11. Unlike in feedback design, large signals are an acceptable side-effect of the deadbeat response, because observers are usually implemented using software and computers. The only drawback of this design is its sensitivity to noise. The solution to this is presented in Section 4.12.

Problem 4.11.1 Consider a discrete-time system

$$\begin{aligned} x[k+1] &= \begin{bmatrix} 0 & 3 & 0 \\ 0 & 2 & 1 \\ 1 & 1 & 0 \end{bmatrix} x[k] + \begin{bmatrix} 2 \\ 1 \\ -1 \end{bmatrix} u[k] \\ y[k] &= \begin{bmatrix} 0 & 2 & 1 \end{bmatrix} x[k] \end{aligned}$$

Is this system observable? Rewrite system equations in observer form. Design a state observer in observer form of the system such that all three eigenvalues of the observer are at $\lambda = 0$. Obtain the final observer equations to observe the original state vector $x[k]$.

Solution: Since

$$\mathcal{O} = \begin{bmatrix} c' \\ c'A \\ c'A^2 \end{bmatrix} = \begin{bmatrix} 0 & 2 & 1 \\ 1 & 5 & 2 \\ 2 & 15 & 5 \end{bmatrix}$$

we have $\det(\mathcal{O}) \neq 0$, hence this system is observable.

The characteristic equation of the system is

$$a(z) = \det(zI - A) = z^3 - 2z^2 - z - 3$$

hence

$$A_o = \begin{bmatrix} 2 & 1 & 0 \\ 1 & 0 & 1 \\ 3 & 0 & 0 \end{bmatrix}$$

The transformation matrix is

$$S_o = \mathcal{O}_o^{-1}\mathcal{O} = \mathsf{a}_-\mathcal{O} = \begin{bmatrix} 0 & 2 & 1 \\ 1 & 1 & 0 \\ 0 & 3 & 0 \end{bmatrix}$$

hence

$$b_o = S_o b = \begin{bmatrix} 1 \\ 3 \\ 3 \end{bmatrix} \qquad c_o' = [1 \;\; 0 \;\; 0]$$

The system in observer form is

$$\begin{aligned} x_o[k+1] &= A_o x_o[k] + b_o u[k] \\ y[k] &= c_o' x_o[k] \end{aligned}$$

The original characteristic equation is $a(z) = z^3 - 2z^2 - z - 3$, while the desired characteristic equation is $\alpha(z) = (z-0)^3 = z^3$. With

$$\alpha = [0 \;\; 0 \;\; 0]' \qquad \text{and} \qquad a = [-2 \;\; 1 \;\; -3]'$$

the observer gain is

$$l = \mathcal{O}^{-1}\mathcal{O}_o(\alpha - a) = -S_o^{-1} a = [0 \;\; 1 \;\; 0]'$$

The observer equation is

$$\hat{x}[k+1] = (A - lc')\hat{x}[k] + bu[k] + ly[k]$$

where $\hat{x}[0] = \hat{x}_0$ and

$$A - lc' = \begin{bmatrix} 0 & 3 & 0 \\ 0 & 0 & 0 \\ 1 & 1 & 0 \end{bmatrix}$$

Figure 4.8: Rudolf E. Kalman at the Kyoto Prize ceremony in 1985. Photo courtesy of the Inamori Foundation (`www.inamori-f.or.jp`).

4.12 Kalman filter

The discrete-time observer (the more appropriate term here is the *estimator*) that minimizes the mean-squared error due to noisy measurements is called the discrete-linear-quadratic-estimator – DLQE) or, more commonly, the Kalman filter. In this Section we derive the Kalman filter equations. Careful reader will notice many similarities between the derivation of Kalman filter and the derivation of optimal controller in Section 4.10. This is due to the duality first described by Kalman [26]. It is very similar to the duality between the concepts of controllability and observability and to the duality of pole placement techniques for controllers and observers. We also show that in the limit the Kalman-Bucy filter (cf. Section 3.12) is obtained. It is interesting that unlike most of the other results presented in this book, here the discrete-time case (Kalman filter) preceded the continuous-time case (Kalman-Bucy filter).

Many extensions of basic Kalman filtering are available. We cannot describe them in the present book, but let us just mention that they deal with many possible variations on the basic theme described here: colored and/or cross-correlated noises, partially known system models, etc. Another reason why this Section does not give full justice to these important techniques is that applications of Kalman filtering rely heavily on the incredible computing power of modern computers, and that is impossible to illustrate in a textbook. Let us name just a few applications of Kalman filtering: satellite and rocket navigation, automated landing of jumbo jets, Global Positioning System (GPS).

Finally, a few words about the man himself: Rudolf Emil Kalman was born in 1930 in Budapest, Hungary. He received the bachelor's and the master's degrees in electrical engineering from MIT in 1953 and 1954, respectively, and the DSc degree from Columbia in 1957. He held research positions at IBM and at the Research Institute for Advanced Studies in Baltimore. From 1962 to 1971, he was at Stanford. After that he worked at the University of Florida, Gainesville, and the ETH in Zurich, Switzerland. R. E. Kalman is a member of the U.S. National Academy of Sciences, the U.S. National Academy of Engineering, and the American Academy of Arts and Sciences. He is a foreign member of the French, Hungarian, and Russian Academies of Sciences and a recipient of numerous honorary doctorates. His work has been recognized by highest engineering and scientific awards, including the IEEE Medal of Honor (1974), the IEEE Centennial Medal (1984), the Steele Prize of the American Mathematical Society (1987), and the Bellman Prize of the American Automatic Control Council (1997).

In 1985 he was awarded the Kyoto Prize in Advanced Technology[5] for his fundamental contributions to modern control theory, which include the concepts of controllability and observability and the solution to Wiener's problem of system dynamics estimation in a noisy environment – the Kalman filter.

[5]The Kyoto Prize is awarded annually since 1985 by the Inamori Foundation to honor lifetime achievements in the fields of Advanced Technology, Basic Sciences, and Creative Arts and Moral Sciences. It is sometimes called the *Japanese Nobel Prize*. It is funded from a grant given by Dr. Kazuo Inamori, the founder of Kyocera Corporation.

Problem 4.12.1 Consider a discrete-time system in noisy environment:

$$\begin{aligned} x[k+1] &= Ax[k] + Bu[k] + w[k] \\ y[k] &= Cx[k] + e[k] \end{aligned}$$

where $w[k]$ is the system noise and $e[k]$ is the output measurement noise.

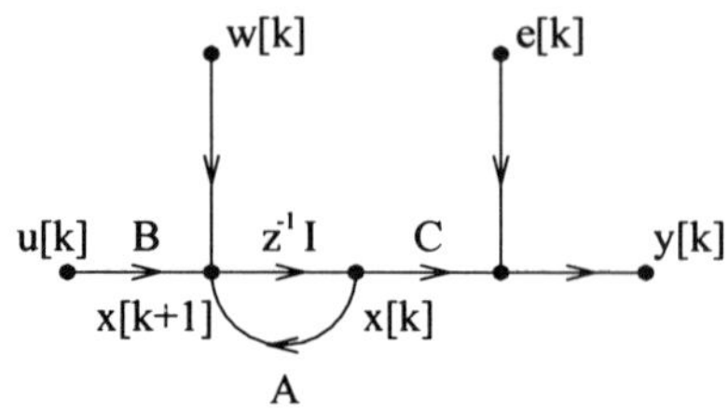

Figure 4.9: Discrete-time system in noisy environment.

Assume that the noises are independent of $x[k]$ and have the following properties:

- Both $w[k]$ and $e[k]$ are Gaussian random signals and

$$\begin{aligned} E\{w[k]\} = 0, \quad & E\{w[k]w'[l]\} = Q\delta[k-l] \quad \text{(zero-mean, white)} \\ E\{e[k]\} = 0, \quad & E\{e[k]e'[l]\} = R\delta[k-l] \quad \text{(zero-mean, white)} \end{aligned}$$

- The system noise $w[k]$ is uncorrelated with the measurement noise $e[k]$, i.e.,

$$E\{w[k]e'[l]\} = 0$$

Additionally, assume the following for the initial value $x[0]$:

$$E\{x[0]\} = x_0 \quad \text{and} \quad E\{(x[0]-x_0)(x[0]-x_0)'\} = P_0$$

Demonstrate that in the state estimator

$$\hat{x}[k] = z[k] + L_k(y[k] - Cz[k])$$

where

$$z[k] = A\hat{x}[k-1] + Bu[k-1] \qquad (z[0] = x_0)$$

the gain L_k which minimizes the mean-squared error is given by

$$L_k = N_k C'(R + CN_kC')^{-1},$$

where $N_0 = P_0$ and N_k is calculated recursively from the following recursion

$$N_k = Q + AN_{k-1}A' - A\underbrace{N_{k-1}C'(R + CN_{k-1}C')^{-1}}_{L_{k-1}}CN_{k-1}A$$

Note: *The estimator with this choice of L_k is called the Kalman filter.*

Solution: Our goal is to estimate the state of the system $\hat{x}[k]$ so that the mean-squared estimation error

$$\text{MSE}\{\hat{x}[k]\} = E\left\{\sum_{i=1}^{n}(x_i[k] - \hat{x}_i[k])^2\right\}$$

is minimized. Since

$$\text{MSE}\{\hat{x}[k]\} = \text{tr}(P_k) \qquad \text{(a scalar)}$$

where

$$P_k = E\{\tilde{x}[k]\tilde{x}'[k]\} \qquad \text{(an } n \times n \text{ matrix)}$$

and

$$\tilde{x}[k] = x[k] - \hat{x}[k] \qquad \text{(estimation error)}$$

and the fact that when P_k is minimized so is $\text{tr}(P_k)$, we will minimize P_k.

Note: *To find the minimum matrix in a set of matrices means to find the matrix P for which the quadratic form $\alpha' P \alpha$ is minimum for any choice of vector α. To show that when P is minimum in some set of matrices then $\text{tr}(P)$ is also minimum in that set, consider the n unit vectors e_i $(i = 1, 2, \ldots, n)$:*

$$\begin{array}{lll} e_1'Pe_1 \text{ is minimum} & \Rightarrow & p_{11} \text{ is minimum} \\ e_2'Pe_2 \text{ is minimum} & \Rightarrow & p_{22} \text{ is minimum} \\ & \cdots & \\ e_n'Pe_n \text{ is minimum} & \Rightarrow & p_{nn} \text{ is minimum} \end{array}$$

hence also $\text{tr}(P) = \sum_1^n p_{ii}$ is minimum over that set of matrices.

It can be shown that the optimum form of the estimator is

$$\hat{x}[k] = z[k] + L_k(y[k] - Cz[k])$$

where $z[k] = A\hat{x}[k-1] + Bu[k-1]$ is the estimate of $x[k]$ based only on the previous measurements and the system model, while $\hat{x}[k]$ takes also into account the correction based on the latest measurement.

For $k = 0$

$$\hat{x}[0] = z[0] + L_0(y[0] - Cz[0])$$

If we choose $z[0] = x_0$ then

$$\hat{x}[0] = x_0 + L_0(y[0] - Cx_0)$$

Some algebraic manipulations yields the state equation for the estimation error (defined above as $\tilde{x}[k] = x[k] - \hat{x}[k]$):

$$\tilde{x}[k+1] = (I - L_{k+1}C)A\tilde{x}[k] + v[k]$$

where $v[k] = (I - L_{k+1}C)w[k] - L_{k+1}e[k+1]$ is independent of $\tilde{x}[k]$. Hence

$$E\{\tilde{x}[k+1]\} = (I - L_{k+1}C)AE\{\tilde{x}[k]\}$$

Since also

$$E\{\tilde{x}[0]\} = (L_0C - I)(z[0] - x_0) = 0$$

we have

$$E\{\tilde{x}[k]\} = 0, \qquad \text{i.e., this estimator is unbiased.}$$

To determine P_k in terms of L_k we start from

$$P_{k+1} = E\{\tilde{x}[k+1]\tilde{x}'[k+1]\}$$

Using the state equation for $\tilde{x}[k]$ and since

$$E\{v[k]v'[k]\} = (I - L_{k+1}C)Q(I - L_{k+1}C)' + L_{k+1}RL'_{k+1}$$

we have

$$P_{k+1} = (I - L_{k+1}C)(AP_kA' + Q)(I - L_{k+1}C)' + L_{k+1}RL'_{k+1}$$

With $N_{k+1} = AP_kA' + Q$ (the covariance matrix of $\hat{x}[k+1]$), some further algebraic manipulation, and with a change of the time-variable $(k+1) \mapsto k$ we can write

$$P_k = N_k + T_k(R + CN_kC')T'_k - N_kC'(R + CN_kC')^{-1}CN_k$$

where $T_k = L_k - N_kC'(R + CN_kC')^{-1}$. The quadratic form $\alpha' P \alpha$ is minimized when $T_k = 0$, i.e.,

$$L_k = N_kC'(R + CN_kC')^{-1}$$

Then

$$P_k = N_k - N_kC'(R + CN_kC')^{-1}CN_k$$

Since $N_{k+1} = AP_kA' + Q$, the last recursion can be rewritten as

$$N_{k+1} = Q + AN_kA' - AN_kC'(R + CN_kC')^{-1}CN_kA$$

where $N_0 = P_0 = E\{(x[0] - x_0)(x[0] - x_0)'\}$.

Note: *When the system is time-invariant and the noises are wide-sense-stationary (i.e., Q and R do not change over time), then*

$$\lim_{k \to \infty} N_k = N$$

Actually, this convergence is so fast that we can often decide to simplify the design by using the stationary Kalman filter which is derived from the algebraic Riccati equation:

$$N = Q + ANA' - ANC'(R + CNC')^{-1}CNA$$

and

$$L = NC'(R + CNC')^{-1}$$

Problem 4.12.2 Given that the relation between the noise covariance matrices in the continuous-time formulation ($Q_{(c)}$ and $R_{(c)}$) and their counterparts in the discrete-time case ($Q_{(d)}$ and $R_{(d)}$) is

$$Q_{(d)} = Q_{(c)}T \qquad \text{and} \qquad R_{(d)} = \frac{R_{(c)}}{T}$$

where T is the sampling period, derive the Kalman-Bucy (continuous-time) equations from the Kalman filter (discrete-time) equations by considering their behavior when $T \to 0$.

Solution: When $T \to 0$

$$N_{k+1} = A_{(d)}P_k A'_{(d)} + Q_{(d)} \;\to\; P_k$$

hence there is no need to distinguish between the *a priori* and the *a posteriori* error covariance matrices. Then from

$$L_k = N_k C'(R_{(d)} + CN_kC')^{-1} \qquad \text{and} \qquad R_{(c)}/T \gg CN_kC'$$

we have

$$L_k = \underbrace{N_k C' R_{(c)}^{-1}}_{\text{call this } L} T = LT$$

We shall see shortly that L is the Kalman-Bucy filter gain because $N_k \to P$.

When $T \to 0$, then $A_{(d)} = I + A_{(c)}T$ so the error covariance equation becomes (recall that $L_k \sim T$)

$$\begin{aligned} N_{k+1} &= A_{(d)}P_kA'_{(d)} + Q_{(d)} \\ &= A_{(d)}(I - L_kC)N_kA'_{(d)} + Q_{(d)} \\ &= A_{(d)}N_kA'_{(d)} - A_{(d)}L_kCN_kA'_{(d)} + Q_{(d)} \\ &= N_k + A_{(c)}N_kT + N_kA'_{(c)}T - L_kCN_k + Q_{(d)} \end{aligned}$$

Finally,

$$\lim_{T\to 0} \frac{N_{k+1} - N_k}{T} = A_{(c)}N_k + N_kA'_{(c)} - N_kC'R_{(c)}^{-1}CN_k + Q_{(c)}$$

If we denote $P = N_k$, then this becomes the familiar covariance error equation for the Kalman-Bucy filter:

$$\dot{P}(t) = A_{(c)}P(t) + P(t)A'_{(c)} - P(t)C'R_{(c)}^{-1}CP(t) + Q_{(c)} \qquad (P(0) = P_0)$$

Finally, the state estimation equation

$$\hat{x}[k] = z[k] + L_k(y[k] - Cz[k])$$

where

$$z[k] = A_{(d)}\hat{x}[k-1] + B_{(d)}u[k-1] \qquad (z[0] = x_0)$$

becomes (with $A_{(d)} = I + A_{(c)}T$, $B_{(d)} = B_{(c)}T$, and $L_k = LT$)

$$\hat{x}[k] - \hat{x}[k-1] = A_{(c)}\hat{x}[k-1]T + B_{(c)}Tu[k-1] + LT(y[k] - Cx[k-1] - B_{(c)}Tu[k-1])$$

Dividing by T, taking the limit $T \to 0$, and denoting $r(t) = \hat{x}$, we obtain the familiar Kalman-Bucy estimation equation

$$\dot{r}(t) = Ar(t) + Bu(t) + L(t)(y(t) - Cr(t))$$

where $L(t) = P(t)C'R^{-1}$, and $P(t)$ is a solution of

$$\dot{P}(t) = AP(t) + P(t)A' + Q - P(t)C'R^{-1}CP(t)$$

Note: *To derive the relations between $Q_{(c)}$ and $R_{(c)}$ on one side and $Q_{(d)}$ and $R_{(d)}$ on the other, consider the following:*

When $T \to 0$ we can write $e^{A_{(c)}T} = I$, hence

$$Q_{(d)} = \iint_T E[w(\xi)w(\eta)]\, d\xi\, d\eta = Q_{(c)}T$$

The derivation for $R_{(d)} = R_{(c)}/T$ requires us to model the measurement process as averaging, i.e., write the output sample $y[k]$ as

$$y[k] = \frac{1}{T}\int_{(k-1)T}^{kT} y(t)\, dt = \frac{1}{T}\int_{(k-1)T}^{kT} (Cx(t) + e(t))\, dt = Cx[k] + \frac{1}{T}\int_{(k-1)T}^{kT} e(t)\, dt$$

Therefore we can write

$$e[k] = \frac{1}{T}\int_{(k-1)T}^{kT} e(t)\, dt$$

hence

$$R_{(d)} = \frac{1}{T^2}\iint_T E[e(\xi)e(\eta)]\, d\xi\, d\eta = \frac{R_{(c)}}{T}$$

Problem 4.12.3 The following is an example of Kalman filtering for identifying the parameters of a communication channel.

If the channel is time-invariant (stationary) and linear, then it is modelled as

$$y[k] = \sum_{i=0}^{p-1} h_i u[k-i] + e[k]$$

Assume that the sequences $u[k]$ and $y[k]$ are known and the channel parameters h_i $(i = 0, 1, 2, \ldots, p-1)$ are to be estimated.

The matrix form of the above model is given by

$$\underbrace{\begin{bmatrix} y[0] \\ y[1] \\ y[2] \\ \vdots \\ y[N-1] \end{bmatrix}}_{y\ (N\times 1)} = \underbrace{\begin{bmatrix} u[0] & 0 & 0 & \dots & 0 \\ u[1] & u[0] & 0 & \dots & 0 \\ u[2] & u[1] & u[0] & \dots & 0 \\ \vdots & \vdots & \vdots & \ddots & \vdots \\ u[N-1] & u[N-2] & u[N-3] & \dots & u[0] \end{bmatrix}}_{H\ (N\times p)} \underbrace{\begin{bmatrix} h_0 \\ h_1 \\ h_2 \\ \vdots \\ h_{p-1} \end{bmatrix}}_{h\ (p\times 1)} + \underbrace{\begin{bmatrix} e[0] \\ e[1] \\ e[2] \\ \vdots \\ e[N-1] \end{bmatrix}}_{e\ (N\times 1)}$$

If $e[k]$ is a white Gaussian noise, then the minimum variance unbiased (MVU) estimate of the channel parameters coincides with their least squares (LS) estimate, and is given by

$$\hat{h} = (H'H)^{-1}H'y$$

Recall that $(H'H)^{-1}H'$ is the left pseudoinverse of H and that it minimizes the magnitude of the squared error $|Hh - y|^2$ (see Appendix C.7).

Note: *Obviously, H depends on the choice of the input sequence $u[k]$ and if we can design it, the best choice would be a pseudo-random sequence, because it has the widest and the flattest possible spectrum.*

If the channel is time-variant, i.e.,

$$y[k] = \sum_{i=0}^{p-1} h_i[k]u[k-i] + e[k]$$

the previous approach quickly produces more unknowns than equations. If the channel variations are slow, we can model them as

$$h[n+1] = Ah[n] + w[n]$$

Write the Kalman filter equations to adaptively identify the channel parameters. Use $p = 3$ with

$$A = \begin{bmatrix} 0.99 & 0 & 0 \\ 0 & 0.999 & 0 \\ 0 & 0 & 0.997 \end{bmatrix}$$

and with

$$Q = E\{w[k]w'[k]\} = \begin{bmatrix} 10^{-3} & 0 & 0 \\ 0 & 10^{-4} & 0 \\ 0 & 0 & 10^{-4} \end{bmatrix} \quad \text{and} \quad R = E\{e[k]e'[k]\} = 10^{-2}$$

Solution: The solution and MATLAB simulations are left to the reader. A very useful MATLAB command for this is **dlqe**.

4.13 Reduced-order observers

In this Section we illustrate the concept of reduced-order observers applied to discrete-time systems.

Problem 4.13.1 Consider a discrete-time system defined by the equations

$$\begin{aligned} x[k+1] &= \begin{bmatrix} 0.16 & 2.16 \\ -0.16 & -1.16 \end{bmatrix} x[k] + \begin{bmatrix} -1 \\ 1 \end{bmatrix} u[k] \\ y[k] &= \begin{bmatrix} 1 & 1 \end{bmatrix} x[k] \end{aligned}$$

Find the open-loop eigenvalues and check the controllability and observability of this system. Determine a state feedback vector f' such that the closed-loop system with $u[k] = -f'x[k] + r[k]$ has eigenvalues at $0.6 \pm 0.4j$. Finally, design a minimum-order observer for this system. Let the desired eigenvalue for the observer be equal to zero, i.e., the observer is a deadbeat system. Note that the concept of lower order observer works exactly the same way for discrete-time systems as it applies to continuous-time systems. Rather than derivatives, we have terms at time $k+1$.

Solution: It is easy to see that $\lambda_1 = -0.2$ and $\lambda_1 = -0.2$. Since $|\lambda_{1,2}| < 1$, this system is stable. It is also controllable and observable because $\det(\mathcal{C}) \neq 0$ and $\det(\mathcal{O}) \neq 0$.

The desired eigenvalues are $\mu_{1,2} = 0.6 \pm 0.4j$, therefore the desired characteristic equation is $\mu^2 - 1.2\mu + 0.52$. Let us write the desired characteristic equation in terms of the state feedback vector $f' = [f_1 \quad f_2]$:

$$\mu^2 - 1.2\mu + 0.52 = \det(\mu I - A_f) \quad \text{where} \quad A_f = A - bf'$$

This reduces to

$$\mu^2 - 1.2\mu + 0.52 = \mu^2 + (f_2 - f_1 + 1)\mu + f_1 + 0.16$$

and finally

$$f' = [0.36 \quad -1.84]$$

Now introduce a nonsingular transformation S such that $c'S = [0 \quad 1]$, e.g.,

$$S = \begin{bmatrix} 1 & 0 \\ -1 & 1 \end{bmatrix}$$

when (cf. Section 3.13)

$$\begin{bmatrix} a_r & b_r \\ c_r & a_{nn} \end{bmatrix} = S^{-1}AS = \begin{bmatrix} -2 & 2.16 \\ -1 & 1 \end{bmatrix} \quad \text{and} \quad \begin{bmatrix} g_r \\ g_n \end{bmatrix} = S^{-1}b = \begin{bmatrix} -1 \\ 0 \end{bmatrix}$$

Finally, the reduced observer gain l_r is found from

$$\lambda - (a_r - l_r c_r) = \lambda - 0 \quad \Rightarrow \quad l_r = 2$$

Note: *Calculate the remaining parameters of the reduced order observer. Don't forget that the output of the observer is a state which needs to be combined with $y(t)$ using the matrix S:*

$$\begin{bmatrix} x_1(t) \\ x_2(t) \end{bmatrix} = S \begin{bmatrix} r_r(t) \\ y(t) \end{bmatrix}$$

Chapter 5

Exercise problems

This Chapter contains exercise problems. They are given without solutions, in order to challenge the reader to go through the solution process alone. If necessary, the reader may look at derivations in Chapter 2 or similar problems in Chapters 3 and 4.

$$\mathcal{OC} = \mathcal{M}$$

5.1 Miscellaneous problems

Problem 5.1.1 Given matrix A

$$A = \begin{bmatrix} 3 & 4 \\ 2 & 1 \end{bmatrix}$$

and

$$x[k+1] = Ax[k], \quad x[0] = \begin{bmatrix} 1 \\ 1 \end{bmatrix}$$

we have

$$x[k] = A^k x[0]$$

Calculate $x[k]$ for $k = 1, 2, 3, 4, 5$. Find eigenvalues λ_1 and λ_2 and eigenvectors p_1 and p_2. Express $x[5]$ in the form $x[5] = \alpha_1 \lambda_1^5 p_1 + \alpha_2 \lambda_2^5 p_2$. Determine a similar expression for $x[k]$.

Problem 5.1.2 For the transfer function given by

$$H(s) = \frac{s+3}{s^3 + 9s^2 + 24s + 18}$$

find a controller realization. Determine its controllability matrix. Is the system controllable? Determine its observability matrix. Is the system observable? Repeat this problem with the observer form.

Problem 5.1.3 Consider the following state equations

$$\dot{x} = \begin{bmatrix} 0 & 1 & 0 \\ 0 & 0 & 1 \\ -6 & 11 & -6 \end{bmatrix} x + \begin{bmatrix} 0 \\ 0 \\ 1 \end{bmatrix} u$$

$$y = \begin{bmatrix} -1 & 0 & 1 \end{bmatrix} x$$

Calculate eigenvalues and right eigenvectors of A. Form matrix P with eigenvectors as columns. Calculate $Q = P^{-1}$ and verify that the rows of Q are the left eigenvectors of A. Calculate $(sI - A)^{-1}$ directly using matrix inversion. Compare your result to what is obtained using

$$(sI - A)^{-1} = \sum_{i=1}^{3} \frac{p_i q_i}{s - \lambda_i}$$

Finally, determine the transfer function from

$$H(s) = \sum_{i=1}^{3} \frac{(c' p_i)(q_i b)}{s - \lambda_i}$$

Problem 5.1.4 Given a system matrix

$$A = \begin{bmatrix} 0 & 1 & 0 \\ 0 & 0 & 1 \\ -6 & 11 & -6 \end{bmatrix}$$

determine its eigenvalues and eigenvectors. Calculate e^{At} as

$$e^{At} = \mathcal{L}^{-1}\{(sI - A)^{-1}\}$$

Compare this result to the result of the following procedure: diagonalize A using P (the matrix of its right eigenvectors)

$$\Lambda = \begin{bmatrix} \lambda_1 & 0 & 0 \\ 0 & \lambda_2 & 0 \\ 0 & 0 & \lambda_3 \end{bmatrix} = P^{-1}AP$$

Then

$$e^{At} = Pe^{\Lambda t}P^{-1}$$

Finally, use Cayley-Hamilton theorem to calculate e^{At}.

Problem 5.1.5 Given a continuous-time system

$$\dot{x}(t) = Ax(t) + bu(t)$$

$$y(t) = c'x(t)$$

where

$$A = \begin{bmatrix} 1 & 2 & 0 \\ 0 & 1 & 0 \\ 3 & 0 & 1 \end{bmatrix}, \quad b = \begin{bmatrix} 0 \\ 1 \\ 1 \end{bmatrix}, \quad c' = [1 \;\; 2 \;\; 0]$$

determine its eigenvalues. Determine the transfer function of this system. What are the poles and zeros of this transfer function? Is $\{A, b, c'\}$ a minimal realization of that transfer function? With that in mind and if you are told that this system is controllable, is it observable? Verify your answer by direct calculation of $\mathcal{O}$. Calculate the parameters of a discretized system using sampling period $T = 0.01s$. Is the resulting system controllable? Is it controllable for any other value of T? Is it observable?

Problem 5.1.6 Given a continuous-time system

$$\dot{x}(t) = Ax(t) + bu(t)$$

$$y(t) = c'x(t)$$

where

$$A = \begin{bmatrix} 2 & 1 & 0 \\ 0 & 2 & 1 \\ 1 & 0 & 2 \end{bmatrix}, \quad b = \begin{bmatrix} 1 \\ 1 \\ 1 \end{bmatrix}, \quad c' = [1 \;\; 0 \;\; 2]$$

determine its eigenvalues λ_1, λ_2, λ_3. Design a feedback vector k such that the eigenvalues of the closed-loop system with $u(t) = -k'x(t) + v(t)$ are the mirror images of the open-loop eigenvalues, i.e., $\mu_i = -\lambda_i$ $(i = 1, 2, 3)$.

If a cost function is given by

$$J = \int_0^\infty (x'(\tau)Qx(\tau) + Ru^2(\tau))\, d\tau$$

where $Q = I_{3\times3}$, while $R = 10$, determine the optimal feedback k_{opt} to minimize the cost of control. Compare the results of these two designs. Try to show that in general, for large values of R, the optimal feedback moves the unstable eigenvalues of the open-loop system to their mirror images in the left half-plane. What is the corresponding result for discrete-time systems?

Problem 5.1.7 Discretize a continuous-time system given by

$$\dot{x}(t) = Ax(t) + bu(t)$$
$$y(t) = c'x(t)$$

where

$$A = \begin{bmatrix} -1 & 1 & 0 \\ 0 & -1 & 1 \\ 1 & 0 & -1 \end{bmatrix}, \quad b = \begin{bmatrix} 1 \\ 0 \\ 1 \end{bmatrix}, \quad c' = [1 \;\; 1 \;\; 2]$$

using the sampling period $T = 1ms$. Discuss the stability of both the continuous-time and the discrete-time systems using Lyapunov's stability theory.

Problem 5.1.8 Discuss controllability and observability of a continuous-time system given by

$$\dot{x}(t) = Ax(t) + bu(t)$$
$$y(t) = c'x(t)$$

where

$$A = \begin{bmatrix} -1 & 1 & 0 \\ 0 & -1 & 1 \\ 1 & 0 & -1 \end{bmatrix}, \quad b = \begin{bmatrix} 1 \\ 0 \\ 1 \end{bmatrix}, \quad c' = [1 \;\; 1 \;\; 2]$$

Transform it into the controller form and calculate the feedback vector k_c which moves the eigenvalues of the controller form to -1, -2, and -3. Transform this feedback gain back to the original state space. Compare the result to the result of the Bass-Gura formula. Design the state observer with eigenvalues at -6. What are the eigenvalues of the combined controller-observer system? Design a reduced-order observer with eigenvalues at -6.

Problem 5.1.9 Solution P of the Lyapunov matrix equation

$$A'P + PA = -Q$$

for any given positive definite symmetric matrix Q is unique and symmetric positive definite itself if and only if A is Hurwitz. Use the Lyapunov theory to discuss the stability of the system given by

$$\dot{x} = \begin{bmatrix} 1 & 2 \\ 3 & 1 \end{bmatrix} x$$

Use $Q = I$.

Problem 5.1.10 Consider a discrete-time system

$$x[k+1] = Gx[k] + Hu[k]$$

with

$$G = \begin{bmatrix} 1 & 1 \\ 1 & 0 \end{bmatrix}, \quad H = \begin{bmatrix} 1 \\ 0 \end{bmatrix}, \quad x[0] = \begin{bmatrix} 1 \\ 0 \end{bmatrix}$$

Define the performance index

$$V = \sum_{k=0}^{\infty} (x'[k]Qx[k] + u'[k]Ru[k])$$

Let $Q = I$ and $R = 1$. The optimal control law is given by

$$u[k] = -F'x[k]$$

where $F' = (R + H'PH)^{-1}H'PGx[k]$ and the matrix P is given by the steady-state discrete Riccati equation

$$P = Q + G'P(I + HR^{-1}H'P)^{-1}G$$

Since G is nonsingular in this problem, the Riccati equation may be rewritten as

$$(P - Q)G^{-1}(I + HR^{-1}H'P) = G'P$$

This is convenient because the unknown of the equation, P, is not being inverted.

Find the positive definite symmetric solution P. Calculate the feedback gain. Determine the open-loop and the closed-loop eigenvalues. Calculate the performance index using

$$V = \frac{1}{2}x'[0]Px[0]$$

Problem 5.1.11 Consider a continuous-time system given by

$$\dot{x} = Ax + Bu \qquad y = Cx + Du$$

where

$$A = \begin{bmatrix} -1 & -2 & -2 \\ 0 & -1 & 1 \\ 1 & 0 & -1 \end{bmatrix}, \quad B = \begin{bmatrix} 2 \\ 0 \\ 1 \end{bmatrix}, \quad C = [1 \ \ 1 \ \ 0], \quad D = 0$$

Use MATLAB to calculate the eigenvalues and ranks of controllability and observability matrices. Plot the unit step response of this system. Design the state feedback gain vector to move all eigenvalues to -2. Plot the unit step response of the closed-loop system. Calculate the full-order state observer gain vector L assuming the desired eigenvalues of the observer are at -4. Write the matrix equation for the total system, determine its eigenvalues (there should be three of them at -2 and three of them at -4) and plot its unit step response.

Problem 5.1.12 Consider a discrete-time system in noisy environment:

$$\begin{aligned} x[k+1] &= Ax[k] + bu[k] + w[k] \\ y[k] &= c'x[k] + e[k] \end{aligned}$$

where

$$A = \begin{bmatrix} 1 & 0.1 & 0 \\ 0 & 1 & 0.1 \\ 0.1 & 0 & 1 \end{bmatrix}, \quad b = \begin{bmatrix} 1 \\ 2 \\ 1 \end{bmatrix}, \quad c' = [1 \ \ 1 \ \ 1]$$

while $w[k]$ is the system noise and $e[k]$ is the output measurement noise. Let the noises be independent of $x[k]$ and mutually uncorrelated. In addition, assume they are both zero-mean, white, Gaussian random signals with covariance matrices given by

$$Q = E\{w[k]w'[k]\} = \begin{bmatrix} 10^{-3} & 0 & 0 \\ 0 & 10^{-3} & 0 \\ 0 & 0 & 10^{-3} \end{bmatrix} \quad \text{and} \quad R = E\{e[k]e'[k]\} = 10^{-2}$$

Additionally, assume the following for the initial value $x[0]$:

$$x_0 = E\{x[0]\} = \begin{bmatrix} 1 \\ 1 \\ 1 \end{bmatrix} \quad \text{and} \quad P_0 = E\{(x[0] - x_0)(x[0] - x_0)'\} = \begin{bmatrix} 10^{-2} \\ 10^{-2} \\ 10^{-2} \end{bmatrix}$$

First design a state feedback vector f such that the closed-loop eigenvalues are all stable and real, and then design the Kalman filter to estimate the states needed for feedback control.

Problem 5.1.13 Matrix A is Hurwitz if and only if for any given positive definite symmetric matrix Q there exists a positive definite symmetric matrix P such that

$$A'P + PA = -Q$$

As an extension to Lyapunov's equation show that all eigenvalues of the matrix A have real parts less than $-\mu < 0$ if and only if for any given positive definite symmetric matrix Q there exists a positive definite symmetric matrix P that satisfies

$$A'P + PA + 2\mu P = -Q$$

Hint: *If the eigenvalue of A is λ, find the eigenvalue of $A + \mu I$.*

Problem 5.1.14 The Lyapunov stability theory has been applied to study the long term behavior of artificial neural nets [20]. In Hopfield nets the N neurons are connected to each other. The dynamic equations describing the net are given by

$$C_i \frac{du_i}{dt} = \sum_{j=1}^{N} t_{ij} V_j - \frac{u_i}{R_i} + I_i \quad (i = 1, 2, \ldots, N)$$

where u_i represent the state variables of individual neurons, while V_i are the neuron outputs. This system of equations models neurons as leaky capacitances: the equation for neuron i expresses the total charging current of that neuron ($C_i du_i/dt$) as a sum of the following components: the current induced in neuron i by the output of neuron j, summed over all j, then the leakage current due to finite input resistance R_i of neuron i, and the input current I_i from other external sources. We will assume that $t_{ij} = t_{ji}$ and $t_{ii} = 0$, hence the matrix T of elements t_{ij} is symmetric with elements on the main diagonal equal to zero.

In a Hopfield network, the output of a neuron V_i is a characteristic of a nonlinear amplifier associated with neuron i. It is a monotonically increasing and bounded function of the state u_i. Hopfield used $-1 \leq V_i \leq 1$. Additionally, V_i is such that

$$V_i = g_i(u_i), \quad g_i(0) = 0 \quad (i = 1, 2, \dots, N)$$

and

$$u_i = g_i^{-1}(V_i) = f_i(V_i) \quad (i = 1, 2, \dots, N)$$

Since g_i are assumed to be monotonically increasing and $g_i(0) = 0$, f_i are also monotonically increasing, $f_i(0) = 0$, and all g_i and f_i lie in the first and the third quadrants only. Hence

$$\int_0^{V_i} f(v)\, dv \geq 0$$

To prove the asymptotic stability of this system, Hopfield used the neuron outputs V_i as state variables and considered the following candidate Lyapunov function:

$$E = -\frac{1}{2} \sum_i \sum_j t_{ij} V_i V_j + \sum_i \frac{1}{R_i} \int_0^{V_i} f_i(v)\, dv - \sum_i V_i I_i$$

with R and C parameters positive.

To complete the proof first show that E is positive definite (or at least bounded from below). Then calculate dE/dt and show that this quantity is negative semi-definite. Finally, show that $dE/dt \equiv 0$ happens only at the equilibrium point.

Problem 5.1.15 Consider a cart of mass M with two inverted pendulums on it. Let their lengths be l_1 and l_2, respectively, both with bobs of mass m. If angles θ_1 and θ_2 describe the deviation of pendulums from the vertical, then for small values of $|\theta_1|$ and $|\theta_2|$ the linearized equations are (cf. [22], p. 103)

$$\dot{x}(t) = Ax(t) + bu(t)$$

where

$$x = \begin{bmatrix} \theta_1 \\ \theta_2 \\ \dot{\theta}_1 \\ \dot{\theta}_2 \end{bmatrix}, \quad A = \begin{bmatrix} 0 & 0 & 1 & 0 \\ 0 & 0 & 0 & 1 \\ a_1 & a_2 & 0 & 1 \\ a_3 & a_4 & 0 & 0 \end{bmatrix}, \quad b = \begin{bmatrix} 0 \\ 0 \\ -1/(Ml_1) \\ -1/(Ml_2) \end{bmatrix}$$

while

$$a_1 = \frac{(M+m)g}{Ml_1}, \quad a_2 = \frac{mg}{Ml_1}, \quad a_3 = \frac{mg}{Ml_2}, \quad a_4 = \frac{(M+m)g}{Ml_2}$$

Show that this system is controllable if and only if $l_1 \neq l_2$. What does it mean for someone trying to vertically balance two sticks on the same finger?

Hint: *First show that*

$$\det(\mathcal{C}) = \frac{2\alpha\beta}{M^2 l_1 l_2} - \frac{\beta^2}{M^2 l_1^2} - \frac{\alpha^2}{M^2 l_2^2}$$

where

$$\alpha = \frac{a_1}{Ml_1} + \frac{a_2}{Ml_2} \qquad \textit{and} \qquad \beta = \frac{a_3}{Ml_1} + \frac{a_4}{Ml_2}$$

Problem 5.1.16 Consider a continuous-time system given by $\{A, b, c'\}$. Write the parameters of the corresponding discrete-time system obtained using the sampling period T. Determine the impulse response of a discrete-time system in terms of the parameters of the original continuous-time system.

Problem 5.1.17 Let p be a right eigenvector of an $n \times n$ matrix A and λ the corresponding eigenvalue, i.e.,

$$Ap = \lambda p$$

Show that λ^n is an eigenvalue of A^n with eigenvector p, i.e.,

$$A^n p = \lambda^n p$$

Problem 5.1.18 Show that

$$\mathrm{tr}(A) = \sum_{i=1}^{n} \lambda_i$$

and

$$\mathrm{tr}(A^n) = \sum_{i=1}^{n} \lambda_i^n$$

Problem 5.1.19 Show that polynomials

$$a(z) = a_0 z^n + a_1 z^{n-1} + \ldots + a_{n-1} z + a_n$$

and

$$b(z) = b_0 z^m + b_1 z^{m-1} + \ldots + b_{m-1} z + b_m$$

have no common factor if and only if there exist two unique polynomials $f(z)$ and $g(z)$ such that

$$a(z)f(z) + b(z)g(z) = 1$$

and $\deg(f(z)) < m$ and $\deg(g(z)) < n$.

Hint: *Use the Euclidean algorithm for polynomials.*

Problem 5.1.20 Consider

$$a(z) = a_0 z^n + a_1 z^{n-1} + \ldots + a_{n-1} z + a_n$$

and

$$b(z) = b_0 z^m + b_1 z^{m-1} + \ldots + b_{m-1} z + b_m$$

such that $a_0, b_0 \neq 0$ and $m, n \geq 1$. Let α_i $(i = 1, \ldots, n)$ and β_j $(j = 1, \ldots, m)$ be the roots of $a(z)$ and $b(z)$, respectively. The *resultant* of polynomials $a(z)$ and $b(z)$ is defined as

$$R(a, b) = a_0^m b(\alpha_1) \ldots b(\alpha_n)$$

Obviously, $a(z)$ and $b(z)$ have no common factor if and only if $R(a, b) \neq 0$. Show that

$$R(a, b) = a_0^m b_0^n \prod_{i=1}^{n} \prod_{j=1}^{m} (\alpha_i - \beta_j)$$

Problem 5.1.21 Consider the system determinant $S(a, b)$ of the following system of $m + n$ linear equations in variables $z^{n+m-1}, \ldots, z, 1$

$$z^{m-1} a(z) = 0 \quad \ldots \quad z a(z) = 0 \quad a(z) = 0$$

$$z^{n-1} b(z) = 0 \quad \ldots \quad z b(z) = 0 \quad b(z) = 0$$

For example, if $n = 3$ and $m = 2$

$$S(a, b) = \begin{vmatrix} a_0 & a_1 & a_2 & a_3 & 0 \\ 0 & a_0 & a_1 & a_2 & a_3 \\ b_0 & b_1 & b_2 & 0 & 0 \\ 0 & b_0 & b_1 & b_2 & 0 \\ 0 & 0 & b_0 & b_1 & b_2 \end{vmatrix}$$

Show that

$$S(a, b) = R(a, b)$$

and conclude that $a(z)$ and $b(z)$ have no common factor if and only if $S(a, b) \neq 0$. This determinant is called *Sylvester's resultant* and the corresponding matrix is *Sylvester's matrix.*

Hint: *Prove that both $S(a, b)$ and $R(a, b)$ have the following recursive properties:*

$$S(a, 0) = 0 \quad \textit{and} \quad R(a, 0) = 0$$

$$S(a, b) = (-1)^{mn} S(b, a) \quad \textit{and} \quad R(a, b) = (-1)^{mn} R(b, a)$$

if $n \geq m$ and $a(z) = q(z) b(z) + r(z)$ then

$$S(a, b) = b_0^{n-m} S(r, b) \quad \textit{and} \quad R(a, b) = b_0^{n-m} R(r, b)$$

Problem 5.1.22 Throughout the book we considered polynomials $a(z)$ and $b(z)$ as above but with $m = n$, $a_0 = 1$, and $b_0 = 0$. Without any loss of generality we use these conventions again. Hence, consider

$$a(z) = z^n + a_1 z^{n-1} + \ldots + a_{n-1} z + a_n$$

and

$$b(z) = b_1 z^{n-1} + \ldots + b_{n-1} z + b_n$$

Then, using the notation from Problem 3.8.4, the *Bezoutian matrix* is defined as

$$\mathcal{B} = \tilde{I}(\mathsf{a}_+ \mathsf{b}_- - \mathsf{b}_+ \mathsf{a}_-)$$

Prove that

$$S(a,b) \neq 0 \quad \Leftrightarrow \quad \det(\mathcal{B}) \neq 0$$

hence the *Bezoutian resultant*, $\det(\mathcal{B})$, can be used as another test for common factors of $a(z)$ and $b(z)$.

Hint: *Show that*

$$S(a,b) = \det \begin{bmatrix} \mathsf{a}_- & \mathsf{b}_- \\ \mathsf{a}_+ & \mathsf{b}_+ \end{bmatrix}$$

Finally,

$$\begin{bmatrix} \mathsf{a}_- & \mathsf{b}_- \\ \mathsf{a}_+ & \mathsf{b}_+ \end{bmatrix} \begin{bmatrix} I & \mathsf{b}_- \\ O & -\mathsf{a}_- \end{bmatrix} = \begin{bmatrix} \mathsf{a}_- & O \\ \mathsf{a}_+ & \tilde{I}\mathcal{B} \end{bmatrix}$$

Problem 5.1.23 Show that the observability matrix of the controller form can be written as

$$\mathcal{O}_c = (\mathsf{b}_+ - \mathsf{a}_-^{-1} \mathsf{b}_- \mathsf{a}_+)\tilde{I}$$

Hint: *Write $H(z)a(z)z^{-k} = b(z)z^{-k}$ for $k = 0, 1, \ldots, n-1$ as a single matrix equation, for example for $n = 3$*

$$\begin{bmatrix} 0 & 0 & 0 & 0 & 0 & 0 \\ h_1 & 0 & 0 & 0 & 0 & 0 \\ h_2 & h_1 & 0 & 0 & 0 & 0 \\ h_3 & h_2 & h_1 & 0 & 0 & 0 \\ h_4 & h_3 & h_2 & h_1 & 0 & 0 \\ h_5 & h_4 & h_3 & h_2 & h_1 & 0 \end{bmatrix} \begin{bmatrix} 1 & 0 & 0 \\ a_1 & 1 & 0 \\ a_2 & a_1 & 1 \\ a_3 & a_2 & a_1 \\ 0 & a_3 & a_2 \\ 0 & 0 & a_3 \end{bmatrix} \doteq \begin{bmatrix} 0 & 0 & 0 \\ b_1 & 0 & 0 \\ b_2 & b_1 & 0 \\ b_3 & b_2 & b_1 \\ 0 & b_3 & b_2 \\ 0 & 0 & b_3 \end{bmatrix}$$

and deduce that

$$\mathcal{M} = (\mathsf{b}_+ - \mathsf{b}_- \mathsf{a}_-^{-1} \mathsf{a}_+) \mathsf{a}_-^{-1} \tilde{I}$$

Finally, use $\mathcal{M} = \mathcal{O}_c \mathcal{C}_c$ and $\mathcal{C}_c = \mathsf{a}_-^{-T}$.

Problem 5.1.24 Derive the following expression for the similarity transformation from the controller to the observer form of a minimal system (cf. Problem 3.8.4):

$$S = -\tilde{I}\mathcal{B}\tilde{I}$$

Hint: *First show that*

$$a_-b_+ + a_+b_- = b_+a_- + b_-a_+$$

Problem 5.1.25 Show that $\det(\mathcal{M})$ can be called *Markov's resultant.*

Hint: *Recall that* $\mathcal{O}_{co} = \mathcal{M}$.

Problem 5.1.26 Show that $a(z)$ and $b(z)$ are coprime if and only if

$$\det(b(A_c)) \neq 0$$

where A_c is the top companion matrix of $a(z)$. This resultant was discovered independently by Barnett, Kalman, and Macdaffee (not necessarily in that order).

Hint: *Recall that the top companion matrix of $a(z)$ is the system matrix of the controller realization.*

Problem 5.1.27 Let

$$a(z) = \det(zI - A) = z^n + a_1 z^{n-1} + \ldots + a_{n-1}z + a_n$$

Verify that

$$\mathrm{adj}(zI - A) = a(z)(zI - A)^{-1} = R_1 z^{n-1} + \ldots + R_{n-1}z + R_n$$

where

$$\begin{aligned}
R_1 &= I \\
R_2 &= AR_1 + a_1 I = A + a_1 I \\
R_3 &= AR_2 + a_2 I = A^2 + a_1 A + a_2 I \\
&\vdots \\
R_n &= AR_{n-1} + a_{n-1} I = A^{n-1} + a_1 A^{n-2} + \ldots + a_{n-2}A + a_{n-1}I
\end{aligned}$$

Also show that

$$AR_n + a_n I = 0$$

and that

$$a_i = -\frac{1}{i}\mathrm{tr}(AR_i)$$

Show how this last formula can be included in the recursion for R_i to eliminate the need to calculate the coefficients of $a(z)$ beforehand. This procedure for determining adj$(zI - A)$ by recursively calculating matrices R_i is called Leverrier-Souriau-Faddeeva-Frame algorithm.

Hint: *Compare coefficients on both sides of*

$$a(z)I = (zI - A)(R_1 z^{n-1} + \ldots + R_{n-1}z + R_n)$$

Problem 5.1.28 Let

$$a(z) = \det(zI - A) = z^n + a_1 z^{n-1} + \ldots + a_{n-1}z + a_n$$

Define

$$\delta(\Phi, \Psi) = (a(\Phi) - a(\Psi))(\Phi - \Psi)^{-1}$$

and show that

$$\delta(\Phi, \Psi) = \Phi^{n-1} + (\Psi + a_1 I)\Phi^{n-2} + (\Psi^2 + a_1\Psi + a_2 I)\Phi^{n-3} + \ldots$$

Use the following substitutions

$$\Phi = zI \quad \text{and} \quad \Psi = A$$

and then

$$\Phi = A \quad \text{and} \quad \Psi = zI$$

to derive the following *resolvent* identities

$$\begin{aligned} \mathrm{adj}(zI - A) &= z^{n-1}I + (A + a_1 I)z^{n-2} + \ldots + (A^{n-1} + a_1 A^{n-2} + \ldots + a_{n-1}I) \\ &= A^{n-1} + (z + a_1)A^{n-2} + \ldots + (z^{n-1} + a_1 z^{n-2} + \ldots + z_{n-1})I \end{aligned}$$

Compare them to the identity in Problem 5.1.27.

Part III

Appendixes

Appendix A

A quick introduction to MATLAB

A.1 Introduction

MATLAB is a computer programming language whose only data types are matrices[1] of various sizes. Many engineering problems are most concisely phrased using the matrix notation, therefore the popularity of MATLAB among students, engineers, and scientists should not be surprising. MATLAB is available for all major platforms, including Unix, Mac, and Windows.

MATLAB supports all basic control structures (`for` loops, `if - then` constructs, etc.), but majority of its commands are calls to the state-of-the-art routines for matrix operations[2]. In addition to program control commands and mathematical commands, MATLAB has very easy-to-use commands for plotting graphs, and many toolboxes aimed for use in various branches of science and engineering. To mention just a few, available are Signal Processing, Controls, System Identification, Image Processing, Neural Networks, Symbolic Math, Statistics, Wavelets, and many other toolboxes.

The most important commands in MATLAB are certainly `help` and `quit`. Commands can be issued directly in MATLAB's command prompt, but if we wish to run a sequence of commands, frequently repeating them with possible slight changes, it is much more convenient to create a file and name it, for example, `progr01.m`. Then the commands from that file, i.e., the program stored in it, can be executed by typing `progr01` in MATLAB's command prompt. Note that the variables need not be declared or dimensioned, this job is done automatically by MATLAB.

[1] Hence its name: MATRIX LABORATORY.

[2] MATLAB was first written as an outgrowth of LINPACK and EISPACK, the public domain, state-of-the-art software packages for numerical analysis, written in FORTRAN. The first version of MATLAB was written in the late 1970's at the University of New Mexico and Stanford, by Cleve Moler and Jack Little. In 1984 they founded The MathWorks, Inc., and since then successfully commercialized and developed their product.

A.2 Basic matrix operations

Quite informally, matrices are tables of numbers. A matrix A given by

$$A = \begin{bmatrix} 2 & 4 & 5 & -1 \\ 7 & 4 & 1 & 2 \\ 4 & 2 & 6 & 0 \end{bmatrix}$$

is said to have three rows and four columns, i.e., to be 3×4. We also write $A = [a_{ij}]_{3\times4}$, with $a_{11} = 2$, $a_{21} = 7$, etc.

In MATLAB A can be defined as

```
A = [2 4 5 -1; 7 4 1 2; 4 2 6 0]
  A =
     2     4     5    -1
     7     4     1     2
     4     2     6     0
```

If we add a semicolon at the end, i.e.,

```
A = [2 4 5 -1; 7 4 1 2; 4 2 6 0];
```

the result of this command will not appear on the screen, but will be kept in the memory. A ";" at the end of a command suppresses printing to the screen. Elsewhere it has a different meaning.

To transpose A, we write $B = A'$, or in MATLAB

```
B = A';
```

In general, for complex matrices, the prime denotes the Hermitian operator, i.e., the conjugate transpose.

Matrices include vectors and sequences

Special cases of matrices are vectors and scalars. The element a_{21} can be extracted from A as follows:

```
a21 = A(2,1);
```

The first row of A can be written as

```
r1 = [2 4 5 -1];
```

It can also be extracted from A directly by writing

```
r1 = A(1,:);
```

Similarly, the second column of A can be written as

```
c2 = A(:,2);
```

If `r1`, `r2`, `r3`, and `c1`, `c2`, `c3`, `c4` are the rows and columns of `A`, then `A` can also be defined as `A = [r1; r2; r3];` or as `A = [c1, c2, c3, c4];` or just `A = [c1 c2 c3 c4];` (without the commas).

Addition

To add two matrices, they must have the same size. For example, redefine B as

```
B = 3*A;
```

This multiplication produces a matrix whose each element is three times the corresponding element of A. Since A and B have the same sizes, now we can add them

```
C = A + B
  C =
     8    16    20    -4
    28    16     4     8
    16     8    24     0
```

Often we need to increase all elements of A by the same amount, for example by 2. Although mathematically this is incorrect

$$D = 2 + A \quad \text{(mathematically incorrect)}$$

the MATLAB syntax allows us to write

```
D = 2 + A;
```

Products

There are several types of multiplications available in MATLAB. The simplest is the scalar product, which we already used:

```
B = 3*A
  B =
     6    12    15    -3
    21    12     3     6
    12     6    18     0
```

Its result is a matrix of the same size as A.

The scalar product of two vectors

$$a = [a_1 \quad a_2 \quad a_3 \quad a_4] \quad \text{and} \quad b = [b_1 \quad b_2 \quad b_3 \quad b_4]$$

is a scalar calculated as

$$a_1b_1 + a_2b_2 + a_3b_3 + a_4b_4$$

In MATLAB it can be evaluated using the matrix product, which is to be presented next.

If we multiply A and B, the matrices of sizes $m \times n$ and $p \times q$, respectively, then for this product to be well defined, the inner dimensions of the two matrices must be equal, i.e., $n = p$. The result of multiplication AB is a matrix C, which is $m \times q$, whose element c_{ij} is equal to the scalar product of the ith row of A and the jth column of B.

For example,

```
A = [1 2 3 4;
     5 6 7 8];
B = [1  2  3;
     4  5  6;
     7  8  9;
    10 11 12];
C = A*B
  C =
     70     80     90
    158    184    210
```

To calculate the scalar product of two vectors, we can use the matrix product operator, but we have to make sure that the left vector is in a row form, while the right vector is in the column form:

```
a = [1 2 3 4]
  a =
     1      2      3      4

b = [7 8 9 10]'
  b =
      7
      8
      9
     10

c = a*b
  c =
     90
```

If we multiply these two vectors so that the left vector is column, while the right is row, the result will be a matrix, calculated according to the rules of the matrix product. For example

```
d = b*a
  d =
      7     14     21     28
      8     16     24     32
      9     18     27     36
     10     20     30     40
```

Sometimes we need to multiply the corresponding elements of two equally sized matrices. This *Hadamard* product is denoted by ".*" in MATLAB.

```
A = [1 2 3;
     4 5 6];
B = [0 1 2;
     2 1 0]
C = A.*B        % note:    .*
  C =
      0      2      6
      8      5      0
```

If for some reason we want to create a matrix which contains all possible products of the elements of the two given matrices A and B, we use the *Kronecker* product $C = A \otimes B$:

```
A = [1 2 3 4;
     5 6 7 8];
B = [1 2;
     3 4;
     5 6]
C = kron(A,B)
  C =
     1     2     2     4     3     6     4     8
     3     4     6     8     9    12    12    16
     5     6    10    12    15    18    20    24
     5    10     6    12     7    14     8    16
    15    20    18    24    21    28    24    32
    25    30    30    36    35    42    40    48
```

To square a matrix means to multiply it by itself, hence due to the constraints on the dimensions of matrices in a matrix product, a matrix must be square in order for the squaring operation to be well defined.

```
A = [1 2;
     3 4];
B = A^2
  B =
     7    10
    15    22
```

We may also need to square each element of a matrix. In this case, of course, a matrix does not have to be square.

```
A = [1 2 3;
     4 5 6];
B = A.^2
  B =
     1     4     9
    16    25    36
```

Functions

Similarly, we can do many other operations on the elements of matrices:

```
A = [1 2 3 4 5;
     6 7 8 9 10];
X = sin(A)          % argument assumed to be in radians
  X =
    0.8415    0.9093    0.1411   -0.7568   -0.9589
   -0.2794    0.6570    0.9894    0.4121   -0.5440

Y = log(A)          % natural base (e)
  Y =
         0    0.6931    1.0986    1.3863    1.6094
    1.7918    1.9459    2.0794    2.1972    2.3026

Z = log10(A)        % base 10
  Z =
         0    0.3010    0.4771    0.6021    0.6990
    0.7782    0.8451    0.9031    0.9542    1.0000

W = 1./A
  W =
    1.0000    0.5000    0.3333    0.2500    0.2000
    0.1667    0.1429    0.1250    0.1111    0.1000
```

Note that (Problem 3.2.4) `exp` and `expm` are different functions: `B=exp(A)` is a matrix such that $b_{ij} = e^{a_{ij}}$, while `C=expm(A)` is a matrix exponential of A. For example

```
A = [1 1; 0 1];
B = exp(A)
  B =
    2.7183    2.7183
    1.0000    2.7183
C = expm(A)
  C =
    2.7183    2.7183
         0    2.7183
```

Matrix inversion, eigenvectors, etc.

Important feature of MATLAB is its ability to seamlessly invert large matrices, find matrix rank, determinant, eigenvalues, eigenvectors, and singular values, and all that using the state-of-the-art algorithms. The appropriate commands are `inv`, `rank`, `det`, `eig`, `svd`. Use `help` for more details and various options for using these commands.

A.3 Plotting graphs

Often we want to analyze the spectral components of a measured signal. We can put the measured values into a vector, use the `fft` command, and plot the magnitude of the components of the resulting vector. Here we create the input vector as a sum of few sinusoidal signals and some additive noise (the result is shown in Figure A.1):

```
Ts = 0.01;              % sampling interval
N = 300;                % number of samples
Tf = N*Ts;              % final time
t = 0:Ts:Tf;            % time vector
A1 = 2.5; f1 = 12;      % amplitudes and frequencies
A2 = 1.0; f2 = 23;
A3 = 4.0; f3 = 17;
x = A1*sin(2*pi*f1*t) + A2*sin(2*pi*f2*t) ...       % note: ... means the command will be
    + A3*sin(2*pi*f3*t) + 0.5*randn(size(t));       % continued in the next line
X = fft(x);
subplot(2,1,1)
plot(t,x)
xlabel('time [s]')
ylabel('signal')
subplot(2,1,2)
plot((0:1/N:1)/Ts,abs(X)/N)
xlabel('frequency [Hz]')
ylabel('spectrum')
```

Here are a few more useful commands when making nice plots: `figure`, `stem`, `axis`, `xlabel`, `ylabel`, `title`, `text`, `grid`. Again, use `help` for more details.

For example, we may not be satisfied by the MATLAB's choice of the ranges for the axis of the graph. Then we use the `axis` command. Commands like `xlabel`, `ylabel`, `title`, and `text` allow us to put some words of explanation at

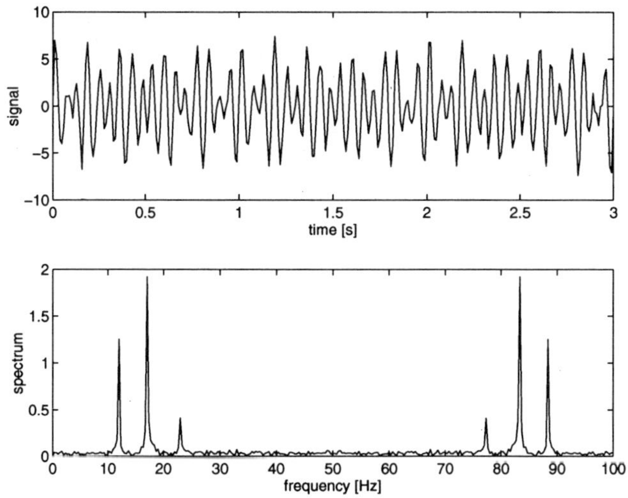

Figure A.1: Plot of the signal and its spectrum.

different places on the graph, **grid** often makes the graph more readable, while **stem** is used to draw discrete sequences.

A.4 Data analysis

To create a sequence of increasing integers, the appropriate structure is a vector. For example

```
n = 4;
seq = 1:n
    seq =
    1     2     3     4
```

or

```
n = 4;
seq = (1:n)'
    seq =
    1
    2
    3
    4
```

In data analysis often used commands are **mean**, **std**, **min**, **max**. For example

```
x = rand(1,1000);      % vector of 1000 U(0,1) random numbers
mean(x)
  ans =
    0.4966             % theoretically    1/2
```

```
std(x)                  %
  ans =                 %                  ------
    0.2802              % theoretically  V 1/12  = 0.2887

m=3;
v=4;
x = m + sqrt(v)*(randn(1,1000));  % vector of 1000 N(m,v) random numbers
mean(x)
  ans = 2.9884               % theoretically   m   (=3)
std(x)                       %                ---
  ans = 2.0137               % theoretically V v   (=2)
```

Note that if the argument of `mean`, `std`, `min`, or `max` is a matrix, the results are row vectors of means, std's, min's, and max's of each column. For example,

```
X = [1 11 56;
     5 12 53;
     3 10 54;
     4 11 53];
mean(X)
  ans =
    3.2500   11.0000   54.0000
max(max(X))
  ans =
    56
```

A.5 Data management and I/O operations

Command `who` lists all currently used variables. Command `whos` does the same, and, in addition, gives a few more details about each variable (size, dimensions, etc.).

To remove some of the variables, type e.g., `clear A B`. To remove all variables, just type `clear`.

To save all variables to a file, type `save`. This creates a file called `matlab.mat` on the disk. To load the values saved in it, type `load`. See `help` for further details on how to save (load) to (from) a file with a different name, or how to specify the variables or a format to be used.

To print the current figure to a file, type e.g., `print -deps fig01.eps`. See `help` for more details, other formats, and options.

A.6 Exercises

1. Create a sequence of even numbers from 0 to 20.

2. Create a sequence of first 10 squares: $1, 4, 9, \ldots, 100$.

3. Create the following length-1024 sequence: $0, 1, 0, -1, \ldots, 0, 1, 0, -1$. (Use vector concatenation.)

4. Check that for $n = 1, 2, \ldots, 1000$ the following equality holds $1 + 2 + \ldots + n = \frac{n(n+1)}{2}$. (Use `sum`.)

5. Check that for $n = 1, 2, \ldots, 1000$ the following equality holds $(1 + 2 + \ldots + n)^2 = 1^3 + 2^3 + \ldots + n^3$. (Use `sum`.)

6. Euler found that $\frac{1}{1^2} + \frac{1}{2^2} + \frac{1}{3^2} + \ldots = \frac{\pi^2}{6}$. Calculate the sum on the left for the first 100 terms and compare it to the value on the right. Calculate the sum for the first n terms, where n runs from 1 to 1000, and plot the calculated values. (Use `for` loop or `cumsum`.)

7. Construct a 7×7 matrix with a Pascal triangle below its main diagonal. (Use `for` loop and the rule for creating the triangle: $P_{i,j} = P_{i-1,j-1} + P_{i-1,j}$, with $P_{i,1} = 1$ for any $i = 1, 2, \ldots$)

8. Calculate A^n for $n = 1, 2, 3, 4, 5, 6$, where

$$A = \begin{bmatrix} 1 & 1 \\ 1 & 0 \end{bmatrix}$$

9. Form the Fibonacci sequence $f_1, f_2, \ldots, f_{10}$. Use the following recursion: $f_1 = f_2 = 1, \quad f_k = f_{k-2} + f_{k-1}$. (Use `for` loop.)

10. Harmonic numbers are defined as $H_n = \frac{1}{1} + \frac{1}{2} + \ldots + \frac{1}{n}$. Find the first harmonic number greater than 3. (Use `cumsum` and `find`.)

11. Given are points $A_1(-1, 7)$, $A_2(2, 3)$, $A_3(4, 7)$, $A_4(2, 4)$, and $A_5(4, 3)$. Form a 5×2 matrix A with their coordinates. Calculate a 5×5 matrix D of distances between these points, i.e., let $d_{i,j} = d(A_i, A_j)$. Check that D is symmetric, and find the two most distant points. If asked to find the two points closest to each other, how would you avoid the presence of zeros in D? (Use `for` loop and `find`.)

12. Write the following system of linear equations in the matrix form, and use the `inv` command to solve it:

$$\begin{aligned} 2x - y + z - w &= -3 \\ x + y + z + w &= 10 \\ x - y + z + w &= 4 \\ 3x - y + 2z - w &= 0. \end{aligned}$$

13. Modify the system above to see how the `pinv` (pseudo-inverse) command can be used to solve underdetermined and overdetermined systems.

14. Use the Sieve of Eratosthenes to generate all primes less than 10000.

15. Implement the Euclid's algorithm.

Solutions

1.

```
s = 0:2:20
  s =
     0     2     4     6     8    10    12    14    16    18    20
```

2.

```
s = (1:10).^2
  s =
     1     4     9    16    25    36    49    64    81   100
```

3. Here are three different solutions:

```
a = [0 1 0 -1];
b = [a a a a];
c = [b b b b];
d = [c c c c];
s1 = [d d d d];              % first solution
size(s1)
  ans =
           1        1024

s2 = sin(pi*(0:1023)/2);     % second solution

s3 = imag(j.^(0:1023));      % third solution
```

4.

```
for n = 1:1000
  x = sum(1:n);              % x = 1 + 2 + ... + n
  y = n*(n+1)/2;
  if x~=y                    % if x <> y
    found = n                % print such n to the screen
  end
end
```

5.

```
for n = 1:1000
  x = (sum(1:n))^2;);        % x = (1 + 2 + ... + n)^2
  y = sum((1:n).^3);         % y = 1^3 + 2^3 + ... + n^3
  if x~=y                    % if x <> y
    found = n                % print such n to the screen
  end
end
```

6.

```
s = 0;
for i = 1:100
  s = s + 1/i^2;
end
```

```
s
  s =
    1.6350

pi^2/6
  ans =
    1.6449
```

Much faster calculation (approximately 5 times faster) is as follows:

```
n = 100;
s = sum(1./(1:n).^2)
  s =
    1.6350
```

To measure the time needed for some operation, the following commands may be used: `tic`, `toc`, `cputime`. For example:

```
t = cputime;
n = 100000;
x = sum(1./(1:n).^2)
cputime-t
  ans =
    0.3333                    % in seconds
```

The plot is obtained as follows (Figure A.2)

```
n = 100;
plot((1:n),cumsum(1./(1:n).^2),'.')         % cumsum is a vector of partial sums
hold on                                     % draw over the current plot
plot((1:n),pi^2/6*ones(1,n))                % draw a horizontal line y = pi^2/6
hold off
```

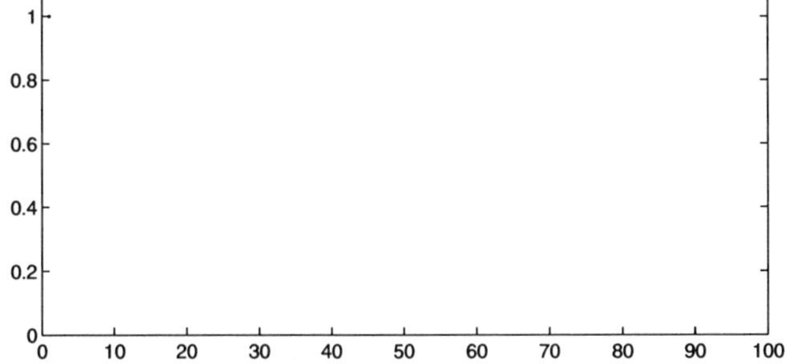

Figure A.2: Convergence of the Euler's sum.

7.

```
n = 7;
P = zeros(n,n);                           % zero matrix nxn
P(:,1) = ones(n,1);                       % the left-most column is all ones
for i = 2:n
  for j = 2:i                             % do nothing above the main diagonal
    P(i,j) = P(i-1,j-1) + P(i-1,j);      % recursion for binomial coefficients
  end
end
P
  P =
     1     0     0     0     0     0     0
     1     1     0     0     0     0     0
     1     2     1     0     0     0     0
     1     3     3     1     0     0     0
     1     4     6     4     1     0     0
     1     5    10    10     5     1     0
     1     6    15    20    15     6     1
```

Try also command pascal.

8.

```
A = [1 1; 1 0];
for n = 1:6
  n, A^n
  pause                        % press <Enter> to continue
end
```

Do you recognize the Fibonacci sequence?

9.

```
n = 10;
f = ones(1,n);                                           % a sequence of all ones
for k = 3:n
  f(k) = f(k-2) + f(k-1);                                % apply this recursion to
end                                                      % find 3rd, 4th, ... elements

  f =
     1     1     2     3     5     8    13    21    34    55      % first ten Fibonacci numbers
```

10.

```
n = 20;
H = cumsum(1./(1:n));          % e.g., H(5) = 1 + 1/2 + 1/3 + 1/4 + 1/5
I = find(H>3);                 % I contains indexes i for which H(i) > 3
min(I)
  ans =
    11                         % this is the smallest among them
```

11.

```
A = [-1,7; 2,3; 4,7; 2,4; 4,3];                     % coordinates of points
D = zeros(5,5);
for i = 1:5
  for j = 1:5
    D(i,j) = sqrt(sum((A(i,:)-A(j,:)).^2));         % Euclidean distance
  end
end

D
  D =
         0    5.0000    5.0000    4.2426    6.4031
    5.0000         0    4.4721    1.0000    2.0000
    5.0000    4.4721         0    3.6056    4.0000
    4.2426    1.0000    3.6056         0    2.2361
    6.4031    2.0000    4.0000    2.2361         0

max(max(D-D'))
  ans =
     0                                              % therefore it is symmetric

max(max(D))
  ans =
    6.4031                                          % maximum distance

[i,j] = find(D==max(max(D)))
  i =
     5
     1
  j =
     1
     5                                              % the most distant are 1 and 5

D(find(D==0)) = 1000*ones(size(find(D==0)));        % substitute all zeros by 1000's

min(min(D))
  ans =
    1                                               % minimum distance

[i,j] = find(D==min(min(D)))
  i =
     4
     2
  j =
     2
     4                                              % the closest are points 2 and 4
```

12.

```
A = [ 2 -1 1 -1;
      1  1 1  1;
      1 -1 1  1;
      3 -1 2 -1];

B = [-3; 10; 4; 0];           % equation is A*X=B, with X = [x y z w]'

X = inv(A)*B                  % solution
  X =
     1.0000
     3.0000
     2.0000
     4.0000
```

13. If the system is underdetermined, pinv gives us the solution with minimum Euclidean norm: $x_0 = A^T(AA^T)^{-1}b$.

```
A = [ 1  1 1  1;
      1 -1 1  1;
      3 -1 2 -1];

B = [10; 4; 0];

X = pinv(A)*B          % minimum norm solution
  X =
     1.1154
     3.0000
     1.8462
     4.0385
```

If the system is overdetermined, pinv gives us the solution which minimizes the Euclidean norm of the error $B - AX$: $x_0 = (A^TA)^{-1}A^Tb$.

```
A = [ 2 -1 1;
      1  1 1;
      1 -1 1;
      3 -1 2];

B = [1; 6; 0; 5];

X = pinv(A)*B          % this solution minimizes the norm of the error
  X =
     1.2000
     3.0000
     2.0000
```

14.

```
% begin by using only first few primes (2,3,5,7) to find all primes < 100

M = 10;          % should be < 11, because at start we use only 2,3,5,7
N = M^2;
small_seq = 1:N;

small_seq(1) = 0;                            % 1 is not a prime
small_seq(4:2:N) = zeros(size(4:2:N));       % eliminate all divisible by 2
small_seq(6:3:N) = zeros(size(6:3:N));       % eliminate all divisible by 3
small_seq(10:5:N) = zeros(size(10:5:N));     % eliminate all divisible by 5
small_seq(14:7:N) = zeros(size(14:7:N));     % eliminate all divisible by 7

small_primes = find(small_seq > 0);

% now use the primes in small_primes to generate primes < 1000

large_seq = 1:N^2;
large_seq(1) = 0;                                      % 1 is not a prime
for j = 1:size(small_primes,2)
  eliminate = 2*small_primes(j):small_primes(j):N^2;   % eliminate all
  large_seq(eliminate) = zeros(size(eliminate));       % divisible by
end                                                    % small_primes(j)

primes = find(large_seq > 0);
  primes =
      2  3  5  7 11 13 17 19 23 29 31 37 41 43 47 ... 9967 9973

size(primes)
  ans =
          1    1229                  % there are 1229 primes < 10000
```

We see there are 1229 primes < 10000. This agrees well with the Lagrange's approximation for $\pi(x)$, the number of primes $< x$:

$$\pi(x) \approx \frac{x}{\ln x - 1.08366} = 1230.51$$

15.

```
%
% euclid.m
%
% This file implements function d = euclid(a,b)
% which can be called from other programs. In
% particular, it recursively calls itself, until
% the result is 0.
%

function d = euclid(a,b)

if (a==0) | (b==0)
  d = a+b;                  % e.g., GCD(5,0) = 5
end

if a==b
  d = a;                    % e.g., GCD(5,5) = 5
end

if (a>b) & ~(a*b == 0)
  r = a - floor(a/b)*b;
  d = euclid(b,r);
end

if (a<b) & ~(a*b == 0)
  r = b - floor(b/a)*a;
  d = euclid(a,r);
end

end
```

MATLAB has a built-in function **gcd** which does the same. Furthermore, it can be used for the extended Euclid's algorithm, i.e., not only to find $d = \text{GCD}(a, b)$, but also integers α and β such that $\alpha a + \beta b = d$:

```
a = 543312;
b = 65340;
[d,A,B] = gcd(a,b)
  d =
      396
  A =
       73
  B =
     -607              % indeed: 73*543312 - 607*65340 = 396
```

Appendix B

Mathematical preliminaries

B.1 Introduction

This Appendix has a twofold purpose: first, it is a mathematical refresher for the tools used in the rest of the book; secondly, it reviews the notation we use with these tools. The presentation is neither complete nor tutorial, hence the readers not already familiar with the ideas and concepts presented in this Appendix should get better mathematically prepared. We shall discuss the following topics:

- Differential and difference equations
- Laplace and z-transforms
- Matrices and determinants

B.2 Differential and difference equations

The dynamic behavior of many natural phenomena, mechanical systems, or electronic circuits, can be accurately modeled using differential equations. They are, therefore, an essential mathematical tool in physical sciences and engineering. If the sampling interval is properly chosen, difference equations can be used instead of differential equations, allowing for the use of digital computers in modeling, analysis, and control.

Historical background

Calculus. Geometrical problems, such as calculation of areas and volumes and construction of tangents, were the primary source of inspiration for what are now the basic methods of calculus. The first such methods were developed by the Ancient Greek geometers Eudoxus, Euclid, and Archimedes. In the early seventeenth century Kepler, Cavalieri, Torricelli, Descartes, Fermat, Roberval, and Wallis contributed many new ideas and discovered important pieces of what would soon become calculus.

The first to put all these ideas together, to unify the notation, and to apply them to problems in kinematics, dynamics, and celestial mechanics was Newton in 1665. This work wasn't published until much later, in 1736 (nine years after his death), but some of his contemporaries were aware of it. It is quite likely that Newton had discovered most of his revolutionary results published in *Philosophiae Naturalis Principia Mathematica* in 1687 using his method of "fluents" and "fluxions" but in that book he proved them using the traditional Greek geometry[1]. Independently from Newton, in the late 1670's, Leibniz succeeded in unifying the previous knowledge, created his own notation, and formulated algorithms using the symbols d and $\int$. He published his findings in 1684 in the article entitled *A New Method for Maxima and Minima as Well as Tangents, Which Is Impeded Neither by Fractional nor by Irrational Quantities, and a Remarkable Type of Calculus for This.*

The eighteenth century was the century of great discoveries in this field. Calculus has been applied to various problems in geometry and mechanics with great success. The main contributors during this period were Jakob and Johann Bernoulli, Taylor, Maclaurin, Euler, d'Alembert, Laplace, and Lagrange. But it was only in the nineteenth century that the rigor was brought into calculus, mostly through the work of Cauchy and later Weierstrass and Cantor.

Today the calculus comprises the following disciplines: differential and integral calculus of real and complex variables, theory of infinite series, theory of differential equations (ordinary and partial), theory of integral equations, and calculus of variations.

[1] Similarly, Archimedes had a method of calculating areas and volumes (basically it was integration; he called it the "mechanical method") but he didn't consider it rigorous enough for the actual proofs, so he used his method to discover new results, but proved them using the standard geometry.

Differential equations. Almost as soon as the new calculus was invented, the first differential equations appeared. Newton called them "fluxional equations" and solved them using power series with indeterminate coefficients. Since his work wasn't published until more than seventy years later, it was the work of Leibniz and his students and followers that laid the foundation of modern theory of differential equations.

For example, in 1690, Jakob Bernoulli reduced the problem of determining the isochrone (the curve in a vertical plane down which a particle, starting at any initial point on the curve, will descend to the lowest point in the same amount of time) to a first-order nonlinear differential equation

$$(b^2y - a^3)^{1/2}y' = a$$

where the prime denotes the derivative with respect to x. He solved it by what is now called the method of separation of variables.

By the end of the seventeenth century Leibniz and the Bernoulli brothers discovered most of the methods for solving first-order ordinary differential equations. By the time Euler entered the scene, several classes of ordinary differential equations were already investigated: linear, Bernoulli, Riccati, and Clairaut differential equations.

The eighteenth century developments were marked by the work of Euler, who made significant contributions: new methods for lowering the order of an equation, the concept of an integrating factor, the theory of higher-order linear equations, early developments of the theory of elliptic functions, and application to a wide variety of mechanical problems.

All these discoveries were finally mathematically justified in the 1820's when Cauchy put calculus on firm foundations. In the theory of differential equations he established the sufficient conditions for existence and uniqueness of a solution of a first-order differential equation

$$y' = f(x, y)$$

Since that time many mathematicians contributed to the further development of the theory and applications of differential equations, to mention just a few: Lie, Poincare, Picard, Lyapunov, Volterra. In 1926 Schrödinger discovered his famous wave equation, which is a fundamental equation of quantum physics.

Difference equations. Difference equations, or the calculus of finite differences, as this branch of mathematics is also called, were first investigated by Gregory, Newton, and Taylor. They were never as important in theoretical developments as in numerical calculations, where differentials are substituted by finite differences. For example, in 1822 Babbage built a prototype of his Difference Engine, intended to solve differential equations based on the method of finite differences. They came to prominence with the development of digital computers and discrete-time control and communications systems.

Differential equations at a glance

A note on notation. In this general discussion about differential equations we shall use the notation introduced by Leibniz, which is still in widespread use. For example, the second-order linear differential equation in this notation is

$$y''(x) + f(x)y'(x) + g(x)y(x) + h(x) = 0$$

where $y'' = \frac{d^2y}{dx^2}$ and $y' = \frac{dy}{dx}$ are the second and the first derivatives of $y(x)$, respectively.

Later, in the main body of this book, our functions will be functions of time. This presents us with two choices: we can simply substitute t for x and write

$$y''(t) + f(t)y'(t) + g(t)y(t) + h(t) = 0$$

or we can use the notation used by Newton, in which a dot denotes the time derivative:

$$\ddot{y}(t) + f(t)\dot{y}(t) + g(t)y(t) + h(t) = 0$$

In order to avoid the confusion with matrix transposition, which is also denoted using a prime, in the main body of this book we shall use the Newton's notation.

Cauchy's theorem. The following theorem due to Cauchy provides sufficient conditions for a first-order differential equation

$$y' = f(x, y), \qquad y(x_0) = y_0$$

to have a solution. Furthermore, if the conditions of the theorem are satisfied, the solution is unique. We shall also see that this theorem can be generalized to give the sufficient conditions for the existence and uniqueness of solutions of higher-order differential equations. Unfortunately, this theorem doesn't offer much help in finding the actual solution, but knowing that the solution exists and that it is unique is often enough, because then we can have greater confidence in the numerical solutions obtained using a computer. We give this important theorem here without a proof. Interested reader should consult any book on differential equations.

Theorem B.2.1 *Assume that a function $f(x, y)$ satisfies the following two conditions:*

1. *It is continuous in a closed region D of the x-y plane containing the point (x_0, y_0).*
2. *In D this function satisfies the Lipshitz condition with respect to y:*

$$|f(x, y_2) - f(x, y_1)| \le K|y_2 - y_1|$$

where (x, y_1) and (x, y_2) are in D and $K > 0$.

Then the first-order differential equation

$$y' = f(x, y)$$

has a unique solution $y = Y(x)$ satisfying the initial condition $Y(x_0) = y_0$. This solution is differentiable in the neighborhood of x_0.

This theorem can be generalized to a system of n first-order differential equations. This is important, because a differential equation of order n

$$z^{(n)} = f(x, z, z', \ldots, z^{(n-1)}) \qquad \text{with } z^{(j)}(x_0) = z_{j0} \quad (j = 0, 1, \ldots, n-1)$$

can be written as a system of n first-order equations (note that $y_1 = z$):

$$y_1' = y_2 \quad \ldots \quad y_{n-1}' = y_n \qquad y_n' = f(x, y_1, y_2, \ldots, y_n)$$

with initial conditions $y_{j+1}(x_0) = z_{j0}$ $(j = 0, \ldots, n-1)$.

Theorem B.2.2 *Assume that the functions $f_1, \ldots, f_n$ satisfy the following two conditions:*

1. *They are continuous in a closed region D containing $(x_0, y_{10}, \ldots, y_{n0})$.*
2. *In D these functions satisfy the Lipshitz condition:*

$$|f_k(x, y_{12}, \ldots, y_{n2}) - f_k(x, y_{11}, \ldots, y_{n1})| \leq K \sum_{i=1}^{n} |y_{i2} - y_{i1}| \quad (k = 1, \ldots, n)$$

where $(x, y_{12}, \ldots, y_{n2})$ and $(x, y_{11}, \ldots, y_{n1})$ are in D and $K > 0$. Then the system of first-order differential equations

$$\begin{aligned} y_1' &= f_1(x, y_1, \ldots, y_n) \\ &\vdots \\ y_n' &= f_n(x, y_1, \ldots, y_n) \end{aligned}$$

has a unique solution $y_k = Y_k(x)$ $(k = 1, \ldots, n)$ satisfying the initial conditions $Y_k(x_0) = y_{k0}$ $(k = 1, \ldots, n)$. This solution is differentiable in the neighborhood of x_0.

As a special case which is of particular interest to us, consider a homogeneous linear differential equation of order n

$$y^{(n)} + a_1(x) y^{(n-1)} + \ldots + a_{n-1}(x) y' + a_n(x) y = 0$$

In the notation of the generalized Cauchy's theorem we have

$$f_1 = y_2 \quad \dots \quad f_{n-1} = y_n \qquad f_n = -(a_1(x)y_n + \dots + a_{n-1}(x)y_2 + a_n(x)y_1)$$

and it is easy to see that as long as functions $a_k(x)$ $(k = 1, \dots, n)$ are continuous, the functions $f_1, \dots, f_n$ satisfy the conditions of Theorem B.2.2. In this book we shall consider only the cases in which $a_k(x)$'s are constants, which further simplifies the analysis.

For further results we are going to need in this book, the reader should take a look at problems in Section 3.1.

Difference equations at a glance

Similarities with differential equations. A detailed comparison of methods used for solution of differential and difference equations shows many similarities. For example, in differential and integral calculus we often use the following identities:

$$\frac{d}{dx}x^n = nx^{n-1} \qquad \text{and} \qquad \int x^n\,dx = \frac{1}{n+1}x^{n+1} + C$$

The following function and associated identities are equally important in the calculus of finite differences:

Definition B.2.1 *The falling factorial power of k is*

$$k^{\underline{n}} = k(k-1)\dots(k-n+1)$$

In the special case when $n = 0$ it is defined to be 1. The symbol $k^{\underline{n}}$ is pronounced "k to the n falling."

It is easy to show that the difference of the falling factorial power is

$$\Delta k^{\underline{n}} = (k+1)^{\underline{n}} - k^{\underline{n}} = nk^{\underline{n-1}}$$

while the indefinite sum of the falling factorial power is

$$\sum k^{\underline{n}} = \frac{k^{\underline{n+1}}}{n+1} + C$$

Similar identities hold for the rising factorial power:

Definition B.2.2 *The rising factorial power of k is*

$$k^{\overline{n}} = k(k+1)\dots(k+n-1)$$

In the special case when $n = 0$ it is defined to be 1. The symbol $k^{\overline{n}}$ is pronounced "k to the n rising."

Another important similarity is the analogy between the function e^{ax} in the "continuous" calculus and the sequence a^k in the "discrete" calculus:

$$f'(x) = af(x) \quad \Rightarrow \quad f(x) = Ce^{ax}$$

while

$$g[k+1] = ag[k] \quad \Rightarrow \quad g[k] = Ca^k$$

With all these and other similarities between the differential calculus and the calculus of finite differences, it should be no surprise that the general solution of the order-n homogeneous linear difference equation with constant coefficients

$$y[k] + a_1 y[k-1] + \ldots + a_n y[k-n] = 0$$

can be written as soon as we determine the roots of its characteristic equation (these roots are called characteristic values or eigenvalues of the difference equation):

$$r^n + a_1 r^{n-1} + \ldots + a_n = 0$$

To learn more about solving difference equations the reader should refer to Section 4.1.

B.3 Laplace and z-transforms

Laplace transform is the most popular integral transform used for solving linear differential equations with constant coefficients. It transforms them into algebraic equations, which are easier to solve. The z-transform takes the place of the Laplace transform in the "discrete" world.

Historical background

Fourier transform. Fourier was the first to use an integral transform to solve differential equations. In 1807, Fourier discovered that periodic functions[2] can be represented using a trigonometric series[3]

$$f(x) = \frac{1}{2}a_0 + \sum_{k=1}^{\infty}(a_k \cos kx + b_k \sin kx)$$

where coefficients a_k and b_k ($k = 0, 1, 2, \ldots$) can be determined from the *Euler-Fourier formulas*

$$a_k = \frac{1}{\pi}\int_{-\pi}^{\pi} f(x) \cos kx \, dx$$

and

$$b_k = \frac{1}{\pi}\int_{-\pi}^{\pi} f(x) \sin kx \, dx$$

In his *Analytical Theory of Heat* published in 1822, Fourier solved the following partial differential equation[4]

$$\frac{\partial u}{\partial t} = k\left(\frac{\partial^2 u}{\partial x^2} + \frac{\partial^2 u}{\partial y^2} + \frac{\partial^2 u}{\partial z^2}\right)$$

where $u = u(x, y, z, t)$ was the temperature and k was a constant dependent on the properties of the medium. In Chapter III of that book he presented "The first example of the use of trigonometric series in the theory of heat." There he solved the heat equation by showing that it was satisfied by sinusoidal functions of various frequencies. He then used the linearity of the equation to combine them into a trigonometric series with coefficients chosen so that the boundary conditions were satisfied as in Example B.3.1:

---

[2]To simplify the notation here we assume that the period is equal to 2π.

[3]This is true under certain conditions, the so-called Dirichlet conditions (1829). This representation was first used by D. Bernoulli and Euler in their work on the problem of an oscillating chord in the 18th century, but Fourier discovered their real importance.

[4]This is the so-called heat equation, also known as the diffusion equation. Many unrelated physical phenomena can be described by it.

Example B.3.1 *Consider a homogeneous solid bounded from the left and right by vertical planes, and from above by a horizontal plane (see Figure B.1). Let the boundary conditions be (for any $x \in (-d,d)$ and $y > 0$)*

$$u(x,0)=1, \quad u(-d,y)=0, \quad u(d,y)=0, \quad \text{and} \quad \lim_{y\to\infty} u(x,y)=0$$

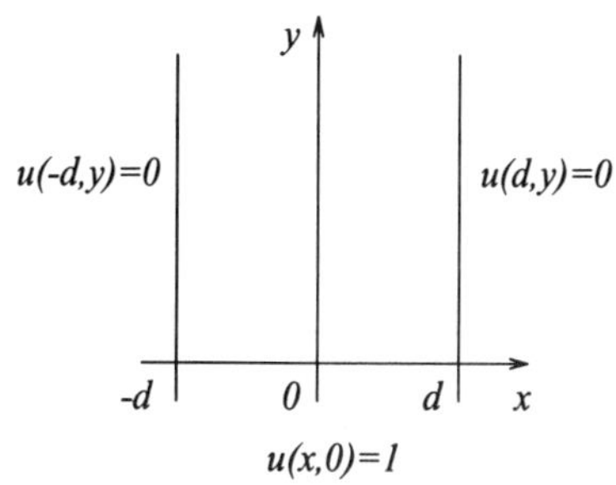

Figure B.1: The boundary conditions in Fourier's first example.

The heat equation in this case becomes

$$\frac{\partial u}{\partial t} = k\left(\frac{\partial^2 u}{\partial x^2} + \frac{\partial^2 u}{\partial y^2}\right)$$

If we are interested in the stationary solution alone, then $\partial u/\partial t = 0$, so we have

$$\frac{\partial^2 u}{\partial x^2} + \frac{\partial^2 u}{\partial y^2} = 0 \tag{B.1}$$

At this point, Fourier wrote: "In order to consider the problem in its elements, we shall in the first place seek for the simplest functions of x and y, which satisfy equation (B.1); we shall then generalize the value of u in order to satisfy all the stated conditions."

If the solution is assumed to be of the form $u(x,y) = f(x)g(y)$, then from (B.1)

$$\frac{f''(x)}{f(x)} = -\frac{g''(y)}{g(y)}$$

which means that both sides are equal to some real constant m. Therefore,

$$u(x,y) = e^{-my}\cos mx$$

Since $u(x,y)$ must be bounded, $m \geq 0$. In order to simplify further analysis, set $d = \pi/2$. Boundary conditions $u(\pm\pi/2, y) = 0$ imply that m can be an odd integer only.

Since the equation (B.1) is linear, before imposing the boundary condition $u(x,0) = 1$, we can say that, in its most general form, the solution is a linear combination of the solutions we have obtained earlier:

$$u(x,y) = ae^{-y}\cos x + be^{-3y}\cos 3x + ce^{-5y}\cos 5x + \ldots$$

Now we use $u(x,0)=1$ to obtain

$$a\cos x + b\cos 3x + c\cos 5x + \ldots \;=\; 1, \qquad -\frac{\pi}{2} \le x \le \frac{\pi}{2} \tag{B.2}$$

The unknown constants $a, b, c, \ldots$ can be determined by multiplying (B.2) by $\cos x$, $\cos 3x$, $\cos 5x$, ... respectively, and integrating from $-\pi/2$ to $\pi/2$ (the Euler-Fourier formulas). In fact, that is the way Fourier did that later in the book. But here he considered new equations, obtained from (B.2) by successive differentiations, at $x=0$

$$\begin{array}{ccccccc} a + & b + & c & + & \ldots & = & 1 \\ a + & 3b + & 5c & + & \ldots & = & 0 \\ a + & 3^2 b + & 5^2 c & + & \ldots & = & 0 \\ & & \ldots & & & & \end{array}$$

Using Wallis' formula

$$\frac{2}{1}\cdot\frac{2}{3}\cdot\frac{4}{3}\cdot\frac{4}{5}\cdot\frac{6}{5}\cdot\frac{6}{7}\cdots = \frac{\pi}{2}$$

this system can be seen to have the following solution

$$a = \frac{4}{\pi}, \quad b = -\frac{4}{3\pi}, \quad c = \frac{4}{5\pi}, \quad \ldots$$

thus, the solution $u(x,y)$ is given by

$$u(x,y) = \frac{4}{\pi}\sum_{k=1}^{\infty}\frac{(-1)^{k-1}}{2k-1}e^{-(2k-1)y}\cos(2k-1)x \qquad \square$$

Laplace transform. Laplace was the first to use the following integral transform

$$F(s) = \int_0^\infty f(t)e^{-st}\,dt$$

today called the one-sided or unilateral Laplace transform, to solve differential equations. Most of the early work on this transform was done by Petzval[5]. In the early 1900s Bromwich discovered the inversion formula

$$f(t) = \frac{1}{2\pi j}\int_{r-j\infty}^{r+j\infty} F(s)e^{st}\,ds \qquad \text{(for any } r \text{ in the region of convergence)}$$

Unaware of these mathematical developments, in 1892 Heaviside introduced his operational calculus to solve differential equations arising in electrical circuits and problems in electrical transmission and telegraphy. Heaviside wrote p instead of d/dt, thus obtaining algebraic equations and did the inversion using the tables. For a while Heaviside's operational calculus was disputed as having no mathematical foundations, but in 1926 Carson recognized the connection between the Heaviside's operational calculus and the Laplace transform. Like Heaviside, he used an extra p in the definition. In our notation they used

$$F(s) = s\int_0^\infty f(t)e^{-st}\,dt \quad \text{instead of} \quad F(s) = \int_0^\infty f(t)e^{-st}\,dt$$

[5] After a public quarrel with a student who unjustly accused Petzval of plagiarizing Laplace, mathematicians started calling this transform the Laplace transform.

***z*-transform.** What the electrical engineers call the z-transform the mathematicians call the generating functions. Historically, de Moivre was the first to use this method in 1718. He used it to determine the explicit formula for the members of the famous Fibonacci sequence, as in the following.

Example B.3.2 *The Fibonacci sequence* $0, 1, 1, 2, 3, 5, 8, 13, 21, 34, 55, \ldots$ *is defined by the following recursion*

$$f_n = f_{n-1} + f_{n-2} \qquad f_0 = 0, f_1 = 1$$

By definition, the generating function for the sequence of Fibonacci numbers is

$$G(x) = f_0 + f_1 x + f_2 x^2 + \ldots$$

Note that its z*-transform is given by*[6] $F(z) = \mathcal{Z}\{f_n\} = f_0 + f_1 z^{-1} + f_2 z^{-2} + \ldots$ *From the recursion and the initial conditions de Moivre found the following algebraic equation for* $G(x)$

$$G(x) - xG(x) - x^2 G(x) = x$$

which implies

$$G(x) = \frac{x}{1 - x - x^2}$$

This expression can be written using the partial fraction expansion as

$$G(x) = \frac{\sqrt{5}}{5}\frac{1}{1 - \phi x} - \frac{\sqrt{5}}{5}\frac{1}{1 - \hat{\phi} x}$$

where $\phi = \frac{1+\sqrt{5}}{2}$ *and* $\hat{\phi} = \frac{1-\sqrt{5}}{2}$. *Note that* $\phi = 1.61803\ldots$ *is the so-called golden section, a very important mathematical constant which appears not only in mathematics, but also in astronomy, biology, psychology, art, and architecture. Now* $G(x)$ *can be rewritten as*

$$G(x) = \frac{\sqrt{5}}{5}(1 + \phi x + \phi^2 x^2 + \ldots) - \frac{\sqrt{5}}{5}(1 + \hat{\phi} x + \hat{\phi}^2 x^2 + \ldots)$$

which finally yields

$$G(x) = 0 + \frac{\sqrt{5}}{5}(\phi - \hat{\phi})x + \frac{\sqrt{5}}{5}(\phi^2 - \hat{\phi}^2)x^2 + \ldots$$

Therefore, we can immediately write

$$f_n = \frac{\sqrt{5}}{5}\left[\left(\frac{1+\sqrt{5}}{2}\right)^n - \left(\frac{1-\sqrt{5}}{2}\right)^n\right] \qquad \square$$

Generating functions were then used by Laplace, who applied them in the theory of probability. The first to use them in engineering related problems were Hurewicz, Zadeh, and Ragazzini (see Problem 4.4.12).

[6]For further details on this notational difference see Problem 4.4.12.

Laplace transform at a glance

Useful properties. In the following table we summarize the most commonly used properties of the Laplace transform (see also Problem 3.1.5):

$$\begin{aligned} F(s) &= \mathcal{L}\{f(t)\} = \int_{0-}^{\infty} f(t)e^{-st}\,dt \\ f(t) &= \mathcal{L}^{-1}\{F(s)\} = \frac{1}{2\pi j}\int_{r-j\infty}^{r+j\infty} F(s)e^{st}\,ds \end{aligned}$$

for any r in the region of convergence, i.e., to the right from all poles of $F(s)$.

LAPLACE TRANSFORM – PROPERTIES AND PAIRS

original	**transform**	**property**
$\alpha f(t) + \beta g(t)$	$\alpha F(s) + \beta G(s)$	linearity
$e^{at} f(t)$	$F(s-a)$	s-domain shift
$t^n f(t)$	$(-1)^n F^{(n)}(s)$	s-domain deriv.
$f(t-a)\ (a>0)$	$e^{-as}F(s)$	t-domain shift
$f^{(n)}(t)$	$s^n F(s) - \sum_{i=1}^{n} s^{n-i} f^{(i-1)}(0)$	t-domain derivative
$\int_0^t f(\tau)\,d\tau$	$F(s)/s$	time integral
$\int_0^t f(t-\tau)g(\tau)\,d\tau$	$F(s)G(s)$	time convolution
$\delta(t)$	1	Dirac's delta impulse
1	$\frac{1}{s}$	Heaviside's unit step
$\frac{t^n e^{at}}{n!}$	$\frac{1}{(s-a)^{n+1}}$	
$e^{-at}\cos\omega t$	$\frac{(s+a)}{(s+a)^2+\omega^2}$	
$e^{-at}\sin\omega t$	$\frac{\omega}{(s+a)^2+\omega^2}$	

z-transform at a glance

Useful properties. The following table summarizes the properties of the z-transform:

$$F(z) = \mathcal{Z}\{f[k]\} = \sum_{k=0}^{\infty} f[k]z^{-k}$$

$$f[k] = \mathcal{Z}^{-1}\{F(z)\} = \frac{1}{2\pi j}\oint_C F(z)z^{k-1}\,dz$$

where C is any circle centered at the origin such that all poles of $F(z)z^{k-1}$ are in its interior.

Z-TRANSFORM – PROPERTIES AND PAIRS

original	**transform**	**property**
$\alpha f[k] + \beta g[k]$	$\alpha F(z) + \beta G(z)$	linearity
$a^k f[k]$	$F(z/a)$	z-domain scaling
$(k+1)^{\overline{m}} f[k+m]$	$\frac{d^m}{(dz^{-1})^m} F(z)$	z-domain derivative
$f[k-n] \quad (n \geq 0)$	$z^{-n}F(z)$	k-domain shift
$f[k+n] \quad (n \geq 0)$	$z^n \left(F(z) - \sum_{i=0}^{n-1} f[i]z^{-i}\right)$	k-domain shift
$\sum_{i=0}^{k} f[k-i]g[i]$	$F(z)G(z)$	k-domain convolution
$\delta[k]$	1	Kronecker's delta
a^k	$\frac{1}{1-az^{-1}}$	
$\frac{(k+1)^{\overline{m}} a^k}{m!}$	$\frac{1}{(1-az^{-1})^{m+1}}$	
$a^k \cos k\omega$	$\frac{1-az^{-1}\cos\omega}{1-2az^{-1}\cos\omega+a^2z^{-2}}$	
$a^k \sin k\omega$	$\frac{az^{-1}\sin\omega}{1-2az^{-1}\cos\omega+a^2z^{-2}}$	

B.4 Matrices and determinants

Matrix notation and methods are the most important mathematical tool that we use in this book. Therefore it should be no surprise that besides this brief historical and theoretical introduction and problems in Sections 3.2 and 4.2, we have also dedicated Appendix C to matrices. Their usefulness stems from the compact notation they offer for many classes of problems, especially for systems of equations and for quadratic forms.

Historical background

The need for matrices and determinants arose first in the context of systems of linear equations, and later with investigations of quadratic forms. This same notation became useful in other areas, e.g., in analytic geometry, functional analysis, probability, physics, and engineering.

Systems of linear equations. A Babylonian clay tablet from around 300 BC contains the following problem:

> There are two fields whose total area is 1800 square yards. One produces 2/3 of a bushel of grain per square yard while the other produces 1/2 a bushel of grain per square yard. If the total yield is 1100 bushels, what is the size of each field?

This is the oldest known problem which reduces to a system of simultaneous equations. The oldest known use of matrix methods is found in the Chinese mathematical text *Nine Chapters on the Mathematical Procedures*, which was probably compiled in the 1st century AD. It contained the following problem, whose solution used all but modern notation: the reader will recognize the Gaussian elimination and rules to transform matrices.

Example B.4.1 *There are three types of corn. Three bundles of the first, two of the second, and one of the third make 39 measures. Two of the first, three of the second and one of the third make 34 measures. And one of the first, two of the second and three of the third make 26 measures. How many measures of corn are contained in one bundle of each type?*

In our modern notation we would write this as

$$\begin{aligned} 3x + 2y + z &= 39 \\ 2x + 3y + z &= 34 \\ x + 2y + 3z &= 26 \end{aligned}$$

The ancient text proceeds by writing the coefficients in a form of a table

$$\begin{array}{ccc} 1 & 2 & 3 \\ 2 & 3 & 2 \\ 3 & 1 & 1 \\ 26 & 34 & 39 \end{array}$$

which is then transformed using the following rule: multiply the middle column by 3 and subtract the right column from it as many times as possible. Then similar is done with the left and the right column, which yields

0	0	3
4	5	2
8	1	1
39	24	39

Finally, if the left column is multiplied by 5 and the middle column is subtracted from it as many times as possible, the following new equivalent system is obtained

0	0	3
0	5	2
36	1	1
99	24	39

This same method was used centuries later by Gauss when he calculated the six orbital elements of the asteroid Pallas. □

The rule for solving a 2×2 system was first given by Cardano in his *Ars Magna* in 1545. Today we recognize it as the first instance of the Cramer's rule.

Determinants were first defined by the Japanese mathematician Seki in 1683. He was able to calculate determinants up to 5×5 and to demonstrate the general rules for their evaluation through examples. That same year, Leibniz was the first to introduce determinants in Europe. He proved various results about 3×3 determinants, including what we now call the Cramer's rule and Laplace expansion. In 1750 Cramer gave the general rule for systems $n \times n$. The first to use the compactness of determinants to simplify the discussion was Laplace in 1772 in a paper about the orbits of the inner planets. He also proved the general case of the expansion rule now named after him. Gauss' motivation for inventing an efficient method for solving simultaneous equations was also coming from the celestial mechanics. In 1809 he introduced the elimination algorithm, now named after him, in his work on orbital elements of the asteroid Pallas, where he dealt with six linear equations with six unknowns.

Quadratic forms. Gauss was the first to use the term "determinant" in his *Disquisitiones Arithmeticae* in 1801. He used that name because these objects determined the properties of the quadratic forms he was studying. In the same context he described matrix multiplication and inversion.

In 1812 Cauchy and Binet found the rules for determinant multiplication. In 1826 Cauchy worked on quadratic forms and in that context he calculated the eigenvalues of the corresponding matrices, and showed that real symmetric matrices are diagonalizable. In 1846 Finck published the rule for evaluation of 3×3 determinants and credited Sarrus for it.

The modern notation for determinants (two vertical lines) was first introduced by Cayley in 1841, while in 1850 Sylvester was the first to coin the term "matrix."

Later developments. In 1858 Cayley gave the abstract definition of a matrix, thus generalizing the rectangular arrays of numbers encountered in various mathematical investigations. He proved that 2×2 and 3×3 matrices satisfy their own characteristic equations. Hamilton proved the same result for matrices 4×4 in his work on quaternions. The general case was proved by Frobenius in 1878, in the same paper in which he introduced the notion of the rank. When in 1896 he became aware of Cayley's work, he generously attributed this important theorem to him. Important work on matrices and determinants was also done by Weierstrass and Kronecker.

In 1925 Heisenberg formulated his quantum theory using arrays of numbers describing probabilities of transitions between different quantum states. It was Born who first recognized the matrices in Heisenberg's work. In 1927 Schrödinger proved the equivalence of his and Heisenberg's approach. Today matrices are useful in many areas of science and engineering, such as signal and image processing and control theory.

Matrices and determinants at a glance

Matrix operations. Consider the following system of m equations in n unknowns $x_1, x_2, \ldots, x_n$:

$$\begin{array}{ccccccccc} a_{11}x_1 & + & a_{12}x_2 & + & \ldots & + & a_{1n}x_n & = & b_1 \\ a_{21}x_1 & + & a_{22}x_2 & + & \ldots & + & a_{2n}x_n & = & b_2 \\ & & & & & & & \vdots & \\ a_{m1}x_1 & + & a_{m2}x_2 & + & \ldots & + & a_{mn}x_n & = & b_m \end{array} \tag{B.3}$$

If we use the usual definition of multiplication of a matrix and a vector, we can write this system as follows:

$$Ax = b$$

where

$$A = \begin{bmatrix} a_{11} & a_{12} & \ldots & a_{1n} \\ a_{21} & a_{22} & \ldots & a_{2n} \\ \vdots & \vdots & \ddots & \vdots \\ a_{m1} & a_{m2} & \ldots & a_{mn} \end{bmatrix} \qquad x = \begin{bmatrix} x_1 \\ x_2 \\ \vdots \\ x_n \end{bmatrix} \qquad b = \begin{bmatrix} b_1 \\ b_2 \\ \vdots \\ b_m \end{bmatrix}$$

Similarly, all other basic matrix operations have interpretation, probably even the origin, in the world of systems of linear equations. For example, to see why the standard[7] definition of matrix multiplication makes so much sense, consider the change of variables in the system (B.3):

$$\begin{array}{lllllllll} x_1 & = & p_{11}w_1 & + & p_{12}w_2 & + & \ldots & + & p_{1n}w_n \\ x_2 & = & p_{21}w_1 & + & p_{22}w_2 & + & \ldots & + & p_{2n}w_n \\ \vdots & & & & & & & & \\ x_n & = & p_{n1}w_1 & + & p_{n2}w_2 & + & \ldots & + & p_{nn}w_n \end{array} \tag{B.4}$$

Then the system (B.3) becomes

$$\begin{array}{lllllll} (a_{11}p_{11} + \ldots + a_{1n}p_{n1})w_1 & + & \ldots & + & (a_{11}p_{1n} + \ldots + a_{1n}p_{nn})w_n & = & b_1 \\ (a_{21}p_{11} + \ldots + a_{2n}p_{n1})w_1 & + & \ldots & + & (a_{21}p_{1n} + \ldots + a_{2n}p_{nn})w_n & = & b_2 \\ & & & & & \vdots & \\ (a_{m1}p_{11} + \ldots + a_{mn}p_{n1})w_1 & + & \ldots & + & (a_{m1}p_{1n} + \ldots + a_{mn}p_{nn})w_n & = & b_m \end{array}$$

which is consistent with the definition of matrix multiplication and the matrix form of this system

$$APw = b$$

where

$$P = \begin{bmatrix} p_{11} & p_{12} & \ldots & p_{1n} \\ p_{21} & p_{22} & \ldots & p_{2n} \\ \vdots & \vdots & \ddots & \vdots \\ p_{m1} & p_{m2} & \ldots & p_{mn} \end{bmatrix} \qquad w = \begin{bmatrix} w_1 \\ w_2 \\ \vdots \\ w_n \end{bmatrix} \qquad b = \begin{bmatrix} b_1 \\ b_2 \\ \vdots \\ b_m \end{bmatrix}$$

All this is also consistent with writing the transformation equations (B.4) as $x = Pw$.

---

[7] The standard matrix multiplication is named after Cayley. There are other types of matrix multiplication, such as Kronecker (also called tensor product or direct product of matrices), Hadamard, inner, outer, cojoint, Lie, and others (see [15]). For example, the Kronecker product of two square matrices is defined as

$$U \otimes V \stackrel{\text{def}}{=} \begin{bmatrix} u_{11}V & u_{12}V & \ldots & u_{1r}V \\ u_{21}V & u_{22}V & \ldots & u_{2r}V \\ \vdots & \vdots & \ddots & \vdots \\ u_{r1}V & u_{r2}V & \ldots & u_{rr}V \end{bmatrix}$$

therefore, if the orders of U and V are r and s, respectively, the order of their Kronecker product $U \otimes V$ is rs.

A note on notation. In this book we use the following notation:

- In A^T the operator T denotes matrix transposition.
- In $\bar{A}$ the bar denotes the complex conjugation.
- In A^H the operator H denotes the Hermitian operator, i.e., $A^H = (\bar{A})^T$.

Obviously, if A is real, then $A^H = A^T$. We also use the prime to denote the Hermitian operator: $A' \equiv A^H$. Hence, for real matrices, the prime denotes the transpose.

The reason for this double notation is that one notation is sometimes easier to read or use than the other. For example, MATLAB adopted the prime because it is easier to use when typing programs. On the other hand, writing A^{-T} is more elegant than the cumbersome $(A')^{-1}$ or $(A^{-1})'$.

Operations on determinants. In order to illustrate the operations on determinants we shall investigate one important class of determinants, the so-called Vandermonde determinants.

Example B.4.2 *The Vandermonde determinant of order n is defined by*

$$V_n(a_1, \ldots, a_n) = \begin{vmatrix} 1 & a_1 & \ldots & a_1^{n-1} \\ 1 & a_2 & \ldots & a_2^{n-1} \\ \vdots & \vdots & \ddots & \vdots \\ 1 & a_n & \ldots & a_n^{n-1} \end{vmatrix}$$

We shall use the induction to prove that for $n \geq 2$

$$V_n(a_1, \ldots, a_n) = \prod_{1 \leq i < j \leq n} (a_j - a_i)$$

For example, for $n = 3$ *we shall find*

$$V_3(a, b, c) = \begin{vmatrix} 1 & a & a^2 \\ 1 & b & b^2 \\ 1 & c & c^2 \end{vmatrix} = (c-a)(c-b)(b-a)$$

First step: For $n = 2$ *we have*

$$\begin{vmatrix} 1 & a_1 \\ 1 & a_2 \end{vmatrix} = a_2 - a_1$$

Second step: Let

$$V_k(a_1, \ldots, a_k) = \prod_{1 \leq i < j \leq k} (a_j - a_i)$$

Third step: If in the determinant

$$V_{k+1}(a_1,\ldots,a_{k+1}) = \begin{vmatrix} 1 & a_1 & a_1^2 & \ldots & a_1^k \\ 1 & a_2 & a_2^2 & \ldots & a_2^k \\ \vdots & \vdots & \vdots & & \vdots \\ 1 & a_k & a_k^2 & \ldots & a_k^k \\ 1 & a_{k+1} & a_{k+1}^2 & \ldots & a_{k+1}^k \end{vmatrix}$$

from the j-th column we subtract the $(j-1)$th column multiplied by a_{k+1}, for all $j = 2,3,\ldots,(k+1)$, and then extract $(a_i - a_{k+1})$ from the i-th row, for every $i = 1,2,\ldots,k$, we find

$$V_{k+1}(a_1,\ldots,a_{k+1}) = \begin{vmatrix} 1 & 1 & a_1 & \ldots & a_1^{k-1} \\ 1 & 1 & a_2 & \ldots & a_2^{k-1} \\ \vdots & \vdots & \vdots & & \vdots \\ 1 & 1 & a_k & \ldots & a_k^{k-1} \\ 1 & 0 & 0 & \ldots & 0 \end{vmatrix} \cdot (-1)^k \cdot \prod_{i=1}^{k}(a_{k+1} - a_i)$$

Using the Laplace's determinant expansion, we find

$$V_{k+1}(a_1,\ldots,a_{k+1}) = (-1)^k \cdot 1 \cdot V_k(a_1,\ldots,a_k) \cdot (-1)^k \cdot \prod_{j=1}^{k}(a_{k+1} - a_j)$$

hence, using the inductive hypothesis,

$$V_{k+1}(a_1,\ldots,a_{k+1}) = \prod_{1\le i<j\le k+1} (a_j - a_i) \qquad \square$$

Appendix C

Results from advanced matrix theory

C.1 Eigenvectors and eigenvalues

If A is a square complex matrix of order n, a function $\mathcal{A} : C^n \mapsto C^n$ defined by

$$y = \mathcal{A}(x) = Ax$$

is linear (i.e., it is additive and homogeneous). Function $\mathcal{A}(x)$ is usually called a linear transformation.

Very often it is important to determine those vectors $r \neq 0$ transformed by $\mathcal{A}$ into vectors y, such that $y = \lambda r$, for some complex scalar λ, i.e.,

$$Ar = \lambda r, \quad r \neq 0, \ \lambda \in C$$

Such vectors are called the eigenvectors (or the characteristic vectors) of matrix A.

Definition C.1.1 (Eigenvectors and eigenvalues) *Let A be an order n complex square matrix. Every vector $r \in C^n$ satisfying*

$$Ar = \lambda r, \quad r \neq 0 \tag{C.1}$$

is an eigenvector of the matrix A, and scalar $\lambda \in C$ is the corresponding eigenvalue.

We can write (C.1) as

$$(\lambda I - A)r = 0, \quad r \neq 0 \tag{C.2}$$

Since eigenvectors must be non-zero, i.e., nontrivial solutions of (C.2), we see that the eigenvectors can be found if and only if

$$\det(\lambda I - A) = 0 \tag{C.3}$$

We just proved the following theorem:

Theorem C.1.1 *A complex number λ is an eigenvalue of A if and only if it satisfies the equation* $\det(\lambda I - A) = 0$.

Equation (C.3) is called the characteristic equation of A. Since $\det(\lambda I - A)$ is a polynomial in λ, and its degree is n, the Equation (C.3) has n solutions, with some of them possibly equal to each other.

Note that, since for each eigenvalue λ_k $(k = 1, 2, \ldots, n)$ we have $\det(\lambda_k I - A) = 0$, we can find at least one eigenvector for each distinct λ_k. Therefore, if A has $q \leq n$ distinct eigenvalues it has at least q eigenvectors. The following theorem tells us that these eigenvectors are linearly independent:

Theorem C.1.2 *If a matrix of order n has $q \leq n$ distinct eigenvalues, then it has at least q linearly independent eigenvectors, at least one corresponding to each distinct eigenvalue.*

Proof. Denote by r_k the eigenvector corresponding to the eigenvalue λ_k $(k = 1, 2, \ldots, q)$, and suppose that the theorem is not true. If eigenvectors r_k are not linearly independent, then

$$\alpha_1 r_1 + \alpha_2 r_2 + \ldots + \alpha_q r_q = 0 \tag{C.4}$$

where at least one of the scalar coefficients is non-zero, for example $\alpha_m \neq 0$.

From $Ar_k = \lambda_k r_k$ it follows that $A^p r_k = \lambda_k^p r_k$, for any nonnegative integer p, so for any polynomial g, it is true that

$$g(A) r_k = g(\lambda_k) r_k$$

We will pick polynomial g such that

$$g(\lambda_k) = \delta_{mk} \quad (k = 1, 2, \ldots, q)$$

Note that there is exactly one such polynomial of degree $(q - 1)$ (recall the Lagrange's method of interpolation), and infinitely many such polynomials of higher orders.

Now premultiply Equation (C.4) by $g(A)$. We get

$$g(A) \sum_{k=1}^{q} \alpha_k r_k = 0 \quad \Rightarrow \quad \sum_{k=1}^{q} \alpha_k g(A) r_k = 0 \quad \Rightarrow \quad \sum_{k=1}^{q} \alpha_k g(\lambda_k) r_k = 0$$

Since we picked g so that $g(\lambda_k) = \delta_{mk}$, we have $\alpha_m r_m = 0$, but since r_m is an eigenvector, it must be non-zero, therefore $\alpha_m = 0$, which contradicts our initial assumption $\alpha_m \neq 0$. This proves the theorem. □

Corollary C.1.1 *If all eigenvectors of a matrix A are distinct, i.e., if*

$$\lambda_i \neq \lambda_j \quad (i \neq j)$$

then the eigenvectors of A corresponding to λ_k $(k = 1, 2, \ldots, n)$ are linearly independent.

If A has repeated eigenvalues, it has $\leq n$ (but $\geq q$) linearly independent eigenvectors[1].

Example C.1.1 *Let*

$$A = \begin{bmatrix} -1 & 1 & 1 \\ 0 & 0 & 1 \\ 0 & -2 & -3 \end{bmatrix}$$

It can be seen that the eigenvalues of A are $\lambda_{1,2} = -1$ and $\lambda_3 = -2$, and that the eigenvectors corresponding to them are of the form

$$r(\lambda_{1,2}) = \begin{bmatrix} a \\ b \\ -b \end{bmatrix} \quad \textit{and} \quad r(\lambda_3) = \begin{bmatrix} c \\ c \\ -2c \end{bmatrix}$$

Obviously, we can pick two linearly independent eigenvectors corresponding to $\lambda_{1,2} = -1$, and one corresponding to $\lambda_3 = -2$, for example

$$r_1 = \begin{bmatrix} 1 \\ 0 \\ 0 \end{bmatrix}, \quad r_2 = \begin{bmatrix} 0 \\ 1 \\ -1 \end{bmatrix}, \quad \textit{and} \quad r_3 = \begin{bmatrix} 1 \\ 1 \\ -2 \end{bmatrix} \qquad \square$$

Example C.1.2 *Let*

$$A = \begin{bmatrix} 1 & 1 & 0 \\ 0 & 0 & 1 \\ 0 & 0 & 1 \end{bmatrix}$$

The eigenvalues of A are $\lambda_1 = 0$ and $\lambda_{2,3} = 1$. In this example, all eigenvectors corresponding to the double eigenvalue $\lambda_{2,3} = 1$ are of the form $r(\lambda_{2,3}) = [b \ 0 \ 0]^T$. Therefore, this matrix has only two linearly independent eigenvectors, the minimum guaranteed by Theorem C.1.2. □

It is easy to see that the number of linearly independent eigenvectors corresponding to the eigenvalue λ_k of A is in general equal to the nullity of $(\lambda_k I - A)$, $\nu(\lambda_k I - A) = n - \rho(\lambda_k I - A)$, where ρ denotes the matrix rank. This is so because the nullity determines the number of linearly independent nontrivial solutions of $(\lambda_k I - A) r(\lambda_k) = 0$.

More details about the number of linearly independent eigenvectors corresponding to each of the distinct eigenvalues can be found in Problem C.8.4.

[1]For example, normal matrices (which include real symmetric and Hermitian matrices) have n linearly independent eigenvectors even if they have repeated eigenvalues. Furthermore, these eigenvectors are mutually orthogonal.

Now we shall consider several important properties of eigenvalues:

Theorem C.1.3 *If $\lambda_1, \lambda_2, \ldots, \lambda_n$ are the eigenvalues of A, then*

$$\lambda_1 + \lambda_2 + \ldots + \lambda_n = \operatorname{tr}(A) \tag{C.5}$$

$$\lambda_1 \lambda_2 \ldots \lambda_n = \det(A) \tag{C.6}$$

where $\operatorname{tr}(A) = a_{11} + a_{22} + \ldots + a_{nn}$ *is the trace of* A.

Proof. Consider the characteristic polynomial of A

$$\det(\lambda I - A) = (\lambda - \lambda_1)(\lambda - \lambda_2)\ldots(\lambda - \lambda_n) \tag{C.7}$$

The coefficient next to λ^{n-1} on the right-hand side of (C.7) is equal to $-(\lambda_1 + \lambda_2 + \ldots + \lambda_n)$. On the left-hand side of (C.7) the coefficient with λ^{n-1} comes only from the product of the elements on the main diagonal, i.e., from $(\lambda - a_{11})(\lambda - a_{22})\ldots(\lambda - a_{nn})$. Therefore it is equal to $-(a_{11} + a_{22} + \ldots + a_{nn})$, and we see that (C.5) is true.

To prove (C.6) consider the value of both sides of (C.7) when $\lambda = 0$. The left-hand side is equal to $\det(-A) = (-1)^n \det(A)$, while the right-hand side is equal to $(-1)^n \lambda_1 \lambda_2 \ldots \lambda_n$. Hence (C.6) is true. □

Theorem C.1.4 *If λ is an eigenvalue of A, then, if A^{-1} exists, one of its eigenvalues is λ^{-1}.*

Proof. First note that if A^{-1} exists, than $\det(A) \neq 0$, and from Theorem C.1.3 we see that none of the eigenvalues of A can be zero.

If λ is an eigenvalue of A, then it satisfies the characteristic equation of A. Since

$$\det(\lambda I - A) = 0 \quad \Leftrightarrow \quad \lambda \det(A^{-1} - \lambda^{-1} I) \det(A) = 0$$

from $\lambda \neq 0$ and $\det(A) \neq 0$, we see that λ^{-1} satisfies the characteristic equation of A^{-1}, which proves the theorem. □

We noted earlier that for an arbitrary square matrix A

$$Ar = \lambda r \quad \Rightarrow \quad A^p r = \lambda^p r \quad \text{(for all nonnegative integers } p\text{)}$$

i.e., if λ is an eigenvalue of A, then for any nonnegative integer p, λ^p is an eigenvalue of A^p.

Using the previous theorem, we can say that, if A is nonsingular, then if λ is an eigenvalue of A, λ^p is an eigenvalue of A^p, for any integer p.

Thus we proved the following theorem:

Theorem C.1.5 *If A is a nonsingular matrix, then if λ is an eigenvalue of A, λ^p is an eigenvalue of A^p for any integer p. If A is singular, the same is true, but for nonnegative integers p only.*

At the end of this section, we shall prove one of the most important theorems in matrix theory, the Cayley-Hamilton (C-H) theorem:

Theorem C.1.6 (Cayley-Hamilton) *Let* $a(\lambda) = \lambda^n + a_1\lambda^{n-1} + \ldots + a_n$ *be the characteristic polynomial of* A, *i.e.,* $a(\lambda) = \det(\lambda I - A)$. *Then*

$$a(A) = 0$$

Proof. Recall that for any matrix U we can find a matrix $\mathrm{adj}(U)$, the adjoint matrix of matrix U, such that

$$U\mathrm{adj}(U) = \det(U)I$$

Let us also mention that if $\det(U) \neq 0$, then $U\mathrm{adj}(U)/\det(U) = I$, therefore

$$\det(U) \neq 0 \quad \Rightarrow \quad U^{-1} = \frac{\mathrm{adj}(U)}{\det(U)}$$

Let B be the adjoint matrix of $\lambda I - A$:

$$B = \mathrm{adj}(\lambda I - A)$$

All elements of B are polynomials in λ with degree less than n, therefore we can write

$$B = B_0\lambda^{n-1} + B_1\lambda^{n-2} + \ldots + B_{n-1}$$

where matrices B_k $(k = 0, 1, \ldots, n-1)$ do not depend on λ.

Since

$$(\lambda I - A)B = \det(\lambda I - A)I$$

we have

$$\begin{array}{rcl} B_0 & = & I \\ -AB_0 \quad + B_1 & = & a_1 I \\ & \cdots & \\ -AB_{n-2} + B_{n-1} & = & a_{n-1}I \\ -AB_{n-1} \qquad & = & a_n I \end{array}$$

If we multiply the first of these equations by A^n, the second by A^{n-1}, etc., and the last by I, and add them together, we get

$$A^n + a_1A^{n-1} + \ldots + a_{n-1}A + a_nI = 0$$

i.e., $a(A) = 0$. □

C.2 Diagonal and Jordan forms

In many cases, properties of a square matrix are the same as those of some diagonal matrix. For example, their eigenvalues are the same. But, in general, we can not always find the appropriate diagonal matrix, because not all matrices are diagonalizable. Fortunately, to cover all cases, we don't have to go far in generalizing the diagonal matrices, because every matrix can be transformed into a Jordan form.

First, we shall use Corollary C.1.1 to show that if A has distinct eigenvalues, then it can be diagonalized. If R is a matrix whose columns are the right eigenvectors corresponding to n distinct eigenvalues of A, i.e.,

$$R = \begin{bmatrix} r_1 & r_2 & \dots & r_n \end{bmatrix}$$

than the equations $Ar_k = \lambda_k r_k \quad (k = 1, 2, \dots, n)$ can be written as

$$AR = RD$$

where $D = \text{diag}(\lambda_1, \lambda_2, \dots, \lambda_n)$.

Since r_k $(k = 1, 2, \dots, n)$ are linearly independent vectors, matrix R is non-singular, and we can write

$$R^{-1}AR = D$$

Before stating this result in the form of a theorem, let us define the notion of similarity. Section C.3 is exploring matrix similarity in more details.

Definition C.2.1 *Matrix F is said to be similar*[2] *to a matrix G if there exists a nonsingular matrix S (the similarity transformation matrix) such that*

$$F = SGS^{-1}$$

Theorem C.2.1 *If A has distinct eigenvalues, then it is similar to a diagonal matrix of its eigenvalues. The similarity transformation matrix in that case is the matrix of right eigenvectors of A.*

Theorem C.2.1 gives us a sufficient condition for a matrix to be diagonalizable, but this condition is not necessary[3]. In the following theorem we give the condition which is both necessary and sufficient for a matrix to be similar to a diagonal matrix.

[2]Similarity is an equivalence relation (i.e., it is reflexive, symmetric, and transitive). Similar matrices have many common properties, for example their eigenvalues are the same. To see that, we can use the fact that $\det(S) \neq 0$ to show that their characteristic equations are the same: $\det(\lambda I - F) = 0 \Leftrightarrow \det(\lambda I - SGS^{-1}) = 0 \Leftrightarrow \det(S(S^{-1}\lambda IS - G)S^{-1}) = 0 \Leftrightarrow \det(S)\det(\lambda I - G)\det(S^{-1}) = 0 \Leftrightarrow \det(\lambda I - G) = 0$.

[3]A unity matrix, which is diagonal, has repeated eigenvalues (see Example C.1.1).

Theorem C.2.2 *Arbitrary matrix A of order n is similar to a diagonal matrix if and only if it has n linearly independent eigenvectors.*

Proof. If A has n linearly independent eigenvectors[4] $r_1, r_2, \ldots, r_n$ then, as in the derivation of Theorem C.2.1, $R^{-1}AR = \mathrm{diag}(\lambda_1, \lambda_2, \ldots, \lambda_n)$, where $R = [r_1\ r_2\ \ldots\ r_n]$.

Now suppose that A is similar to some diagonal matrix D. As we noted earlier, similar matrices have the same eigenvalues, so D must be of the form $D = \mathrm{diag}(\lambda_{i_1}, \lambda_{i_2}, \ldots, \lambda_{i_n})$, where λ_k $(k = 1, 2, \ldots, n)$ are the eigenvalues of A, and $(i_1, i_2, \ldots, i_n)$ is some permutation of the set $\{1, 2, \ldots, n\}$. From the definition of similarity, there exists a nonsingular matrix S such that

$$S^{-1}AS = D, \quad \text{i.e.,} \quad AS = SD$$

From the last relation we see that the columns of S are the eigenvectors of A. Since S is nonsingular, A has n linearly independent eigenvectors. □

As we mentioned earlier, if A has repeated eigenvalues, it may have less than n linearly independent eigenvectors (see Example C.1.2). If that is the case, then by Theorem C.2.2, this means that A is not diagonalizable. In such cases the best we can do is to transform A into a matrix in Jordan (canonical, normal) form, which we define via Jordan blocks:

Definition C.2.2 (Jordan block) *A Jordan block of order m is*

$$J_1(\lambda) = [\lambda] \qquad (m = 1)$$

$$J_m(\lambda) = \begin{bmatrix} \lambda & 1 & 0 & 0 & \ldots & 0 & 0 \\ 0 & \lambda & 1 & 0 & \ldots & 0 & 0 \\ 0 & 0 & \lambda & 1 & \ldots & 0 & 0 \\ & & & & \ldots & & \\ 0 & 0 & 0 & 0 & \ldots & \lambda & 1 \\ 0 & 0 & 0 & 0 & \ldots & 0 & \lambda \end{bmatrix} \qquad (m > 1)$$

Definition C.2.3 (Jordan form) *Matrix J of order n is in Jordan form if*

$$J = \mathrm{diag}(J_{m_1}(\lambda_1), J_{m_2}(\lambda_2), \ldots, J_{m_k}(\lambda_k))$$

where $m_1 + m_2 + \ldots + m_k = n$, and it is possible that some of the numbers λ_i $(i = 1, 2, \ldots, k)$ are equal to each other.

---

[4] As we shall see later, normal matrices (including real symmetric and Hermitian matrices) have n mutually orthogonal (and therefore linearly independent) eigenvectors even when they have repeated eigenvalues.

Example C.2.1 *Matrix*

$$J = \begin{bmatrix} 5 & 0 & 0 & 0 & 0 & 0 \\ 0 & 5 & 1 & 0 & 0 & 0 \\ 0 & 0 & 5 & 1 & 0 & 0 \\ 0 & 0 & 0 & 5 & 0 & 0 \\ 0 & 0 & 0 & 0 & 9 & 1 \\ 0 & 0 & 0 & 0 & 0 & 9 \end{bmatrix}$$

is in Jordan form, with

$$J = \text{diag}(J_1(5), J_3(5), J_2(9)) \quad \square$$

The following important theorem is given without a proof:

Theorem C.2.3 *Every complex square matrix is similar to some matrix in Jordan form.*

Example C.2.2 *Matrix*

$$A = \begin{bmatrix} 5 & -4 & -3 & 6 \\ 1 & 3 & -3 & 3 \\ 1 & -3 & 4 & 2 \\ 1 & -4 & -3 & 10 \end{bmatrix}$$

is similar to

$$J = \begin{bmatrix} 4 & 1 & 0 & 0 \\ 0 & 4 & 0 & 0 \\ 0 & 0 & 7 & 1 \\ 0 & 0 & 0 & 7 \end{bmatrix}$$

because

$$J = S^{-1}AS, \quad \text{where } S = \begin{bmatrix} 1 & 1 & 1 & 1 \\ 1 & 0 & 1 & 0 \\ 1 & 0 & 0 & 1 \\ 1 & 0 & 1 & 1 \end{bmatrix}$$

Note that the columns of S are the eigenvectors and the generalized eigenvectors of A. In the next example we show how to determine the generalized eigenvectors of a matrix. Together, eigenvectors and generalized eigenvectors are called principal vectors of a matrix. Also, note that the diagonal elements of J are the eigenvalues of both J and A. $\square$

Example C.2.3 *Consider again*

$$A = \begin{bmatrix} 1 & 1 & 0 \\ 0 & 0 & 1 \\ 0 & 0 & 1 \end{bmatrix}$$

Recall that the eigenvalues of A are $\lambda_1 = 0$ and $\lambda_{2,3} = 1$. All eigenvectors corresponding to $\lambda_1 = 0$ are of the form

$$r(\lambda_1) = \begin{bmatrix} a \\ -a \\ 0 \end{bmatrix}$$

while all eigenvectors corresponding to $\lambda_{2,3} = 1$ *are of the form*

$$r(\lambda_{2,3}) = \begin{bmatrix} b \\ 0 \\ 0 \end{bmatrix}$$

Therefore, A does not have a full set of linearly independent eigenvectors, i.e., it is not diagonalizable. But every matrix is similar to some matrix in Jordan form, the generalization of the diagonal form. In order to transform A into its Jordan form, we have to find the generalized eigenvector corresponding to $\lambda_{2,3} = 1$.

The generalized eigenvector $r_g(\lambda_{2,3})$ *can be found from*

$$(A - \lambda_{2,3}I)r_g(\lambda_{2,3}) = r(\lambda_{2,3}) \qquad (or\ \ (A - \lambda_{2,3}I)^2 r_g(\lambda_{2,3}) = 0)$$

hence

$$r_g(\lambda_{2,3}) = \begin{bmatrix} c \\ b \\ b \end{bmatrix}$$

If we put $a = b = c = 1$, *we have*

$$r_1 = \begin{bmatrix} 1 \\ -1 \\ 0 \end{bmatrix}, \quad r_2 = \begin{bmatrix} 1 \\ 0 \\ 0 \end{bmatrix}, \quad and \ \ r_3 = \begin{bmatrix} 1 \\ 1 \\ 1 \end{bmatrix}$$

Note that r_3 *is not an eigenvector of A, it is a generalized eigenvector of A. The purpose of introducing the generalized eigenvector was to enable us to find the Jordan form of A:*

$$R = \begin{bmatrix} 1 & 1 & 1 \\ -1 & 0 & 1 \\ 0 & 0 & 1 \end{bmatrix} \quad \Rightarrow \quad R^{-1}AR = \begin{bmatrix} 0 & 0 & 0 \\ 0 & 1 & 1 \\ 0 & 0 & 1 \end{bmatrix} = \begin{bmatrix} \lambda_1 & 0 & 0 \\ 0 & \lambda_{2,3} & 1 \\ 0 & 0 & \lambda_{2,3} \end{bmatrix} \quad \square$$

C.3 Similarity of matrices

In this Section we consider the properties shared by similar matrices. Similarity occurs, for example, when we consider a linear system and make a change of variables describing it. Also, since similarity of matrices is an equivalence relation, we can use it to simplify the study of arbitrary matrices. For example, companion matrices are very useful in control theory, so we investigate the conditions under which an arbitrary matrix is similar to a companion matrix.

A linear system is described by its states $x_1(t), x_2(t), \ldots, x_n(t)$ which satisfy the system of equations

$$\begin{aligned}\dot{x}(t) &= Ax(t) + Bu(t)\\ y(t) &= Cx(t)\end{aligned}$$

where $u(t)$ is the input to the system, and $y(t)$ is the system's output.

If we decide to define the states in some other way (so that some property of the system becomes more apparent), we may use the following change of variables:

$$x_{new}(t) = Sx(t)$$

where S is some nonsingular matrix.

Then the equations become

$$\begin{aligned}S^{-1}\dot{x}_{new}(t) &= AS^{-1}x_{new}(t) + Bu(t)\\ y(t) &= CS^{-1}x_{new}(t)\end{aligned}$$

or, after premultuplying the first equation by S,

$$\begin{aligned}\dot{x}_{new}(t) &= SAS^{-1}x_{new}(t) + SBu(t)\\ y(t) &= CS^{-1}x_{new}(t)\end{aligned}$$

We see that the new system matrix of the system is $A_{new} = SAS^{-1}$.

More generally, consider a linear transformation $w = Av$ in standard basis $\{e^{(i)}\}_{i=1,\ldots,n}$, where $e^{(i)} = [0\ \ldots\ 0\ 1\ 0\ \ldots\ 0]^T$, with 1 at the i-th position. If we wish to look at the same linear transformation, but using some other basis $\{\sigma^{(i)}\}_{i=1,\ldots,n}$, we can see that $w_{new} = Sw$ and $v_{new} = Sv$, where $S = [\sigma^{(1)}\ \sigma^{(2)}\ \ldots\ \sigma^{(n)}]$. Note that S is nonsingular because $\{\sigma^{(i)}\}_{i=1,\ldots,n}$ is a basis. Now we have

$$w_{new} = SAS^{-1}v_{new}$$

Definition C.3.1 *Matrix F is said to be similar to a matrix G if there exists a nonsingular matrix S (the similarity transformation matrix) such that*

$$F = S^{-1}GS$$

In that case we write $F \sim G$.

Theorem C.3.1 *Similarity of matrices is an equivalence relation.*

Proof. Recall that a relation is an equivalence relation if it is reflexive, symmetric and transitive.

- Reflexivity: Every matrix is similar to itself, because $F = I^{-1}FI$, and $\det(I) \neq 0$.
- Symmetry: $F \sim G \Rightarrow (\exists S \mid \det(S) \neq 0)\ F = S^{-1}GS \Rightarrow (\exists T = S^{-1})\ G = T^{-1}FT \Rightarrow G \sim F$.
- Transitivity: $(F \sim G) \wedge (G \sim H) \Rightarrow (\exists S, T \mid \det(S) \neq 0,\ \det(T) \neq 0)\ (F = S^{-1}GS) \wedge (G = T^{-1}HT) \Rightarrow F = (TS)^{-1}HTS \Rightarrow F \sim H$, because $\det(TS) = \det(T)\det(S) \neq 0$. □

Theorem C.3.2 *The eigenvalues of a matrix are invariant under the similarity transformation.*

Proof. We shall show that the characteristic equations of similar matrices are the same. If $F = S^{-1}GS$, where $\det(S) \neq 0$, then

$$\det(\lambda I - F) = 0 \Leftrightarrow \det(\lambda I - S^{-1}GS) = 0 \Leftrightarrow \det(S^{-1}(S\lambda IS^{-1} - G)S) = 0 \Leftrightarrow$$

$$\Leftrightarrow \det(S^{-1})\det(\lambda I - G)\det(S) = 0 \Leftrightarrow \det(\lambda I - G) = 0$$

Since F and G have the same characteristic equations, they have the same eigenvalues. □

Corollary C.3.1 *The trace and the determinant of a matrix are invariant under the similarity transformation.*

Proof. This is a direct consequence of Theorems C.1.3 and C.3.2, because $\mathrm{tr}(F)$ and $\det(F)$ depend on the eigenvalues of F only. □

Corollary C.3.2 *If F is (non)singular, so are all matrices in its similarity class.*

Proof. This Corollary is a consequence of the previous Corollary, because F is nonsingular if and only if $\det(F) \neq 0$. □

Note that although similar matrices have the same eigenvalues, it doesn't mean that all matrices with the same eigenvalues are similar.

Example C.3.1 *Both matrices*

$$F = \begin{bmatrix} 5 & 1 \\ 0 & 5 \end{bmatrix} \quad \text{and} \quad G = \begin{bmatrix} 5 & 0 \\ 0 & 5 \end{bmatrix}$$

have the same eigenvalues $\lambda_{1,2} = 5$, *but if we assume that there exists a nonsingular matrix* S *such that* $F = S^{-1}GS$, *we will find that*

$$F = \begin{bmatrix} 5 & 1 \\ 0 & 5 \end{bmatrix} = S^{-1} \begin{bmatrix} 5 & 0 \\ 0 & 5 \end{bmatrix} S = 5 \cdot S^{-1} I S = 5I = G$$

which is a contradiction. □

Important applications of similarity of matrices are based on the following theorem (already stated as Theorem C.2.3). For more details, see Section C.2.

Theorem C.3.3 *Every complex square matrix is similar to some matrix in Jordan form.*

Another group of applications of similarity transformations is based on the following theorem. For more details, see Section C.4.

Theorem C.3.4 *Every Hermitian matrix is similar to a diagonal matrix of its eigenvalues.*

Still another group of applications of matrix similarity is based on the properties of the companion matrices.

Definition C.3.2 *Matrix* A *is a companion matrix if*

$$A = \begin{bmatrix} -a_1 & -a_2 & -a_3 & \dots & -a_{n-1} & -a_n \\ 1 & 0 & 0 & \dots & 0 & 0 \\ 0 & 1 & 0 & \dots & 0 & 0 \\ 0 & 0 & 1 & \dots & 0 & 0 \\ & & & \dots & & \\ 0 & 0 & 0 & \dots & 1 & 0 \end{bmatrix} \tag{C.8}$$

Theorem C.3.5 *The characteristic polynomial of a companion matrix given by* (C.8) *is*

$$a(\lambda) = \lambda^n + a_1\lambda^{n-1} + \dots + a_n$$

Proof. From the definitions we have

$$a(\lambda) \quad = \quad \det(\lambda I - A) \quad = \quad \begin{vmatrix} \lambda + a_1 & a_2 & a_3 & \dots & a_{n-1} & a_n \\ -1 & \lambda & 0 & \dots & 0 & 0 \\ 0 & -1 & \lambda & \dots & 0 & 0 \\ 0 & 0 & -1 & \dots & 0 & 0 \\ & & & \dots & & \\ 0 & 0 & 0 & \dots & -1 & \lambda \end{vmatrix}$$

$$
\begin{aligned}
&= (\lambda + a_1)\begin{vmatrix} \lambda & 0 & \dots & 0 & 0 \\ -1 & \lambda & \dots & 0 & 0 \\ 0 & -1 & \dots & 0 & 0 \\ & & \dots & & \\ 0 & 0 & \dots & -1 & \lambda \end{vmatrix} + \begin{vmatrix} a_2 & a_3 & \dots & a_{n-1} & a_n \\ -1 & \lambda & \dots & 0 & 0 \\ 0 & -1 & \dots & 0 & 0 \\ & & \dots & & \\ 0 & 0 & \dots & -1 & \lambda \end{vmatrix} \\
&= (\lambda + a_1)\lambda^{n-1} + a_2\lambda^{n-2} + \begin{vmatrix} a_3 & a_4 & \dots & a_{n-1} & a_n \\ -1 & \lambda & \dots & 0 & 0 \\ 0 & -1 & \dots & 0 & 0 \\ & & \dots & & \\ 0 & 0 & \dots & -1 & \lambda \end{vmatrix} \\
&= \lambda^n + a_1\lambda^{n-1} + a_2\lambda^{n-2} + \dots + a_{n-2}\lambda^2 + \begin{vmatrix} a_{n-1} & a_n \\ -1 & \lambda \end{vmatrix} \\
&= \lambda^n + a_1\lambda^{n-1} + a_2\lambda^{n-2} + \dots + a_{n-2}\lambda^2 + a_{n-1}\lambda + a_n
\end{aligned}
$$

□

Theorem C.3.6 *If λ is an eigenvalue of a companion matrix A then*

$$\rho(\lambda I - A) = n - 1$$

Proof. The rank of the characteristic matrix is in this case

$$\rho(\lambda I - A) = \rho \begin{bmatrix} \lambda + a_1 & a_2 & a_3 & \dots & a_{n-1} & a_n \\ -1 & \lambda & 0 & \dots & 0 & 0 \\ 0 & -1 & \lambda & \dots & 0 & 0 \\ 0 & 0 & -1 & \dots & 0 & 0 \\ & & & \dots & & \\ 0 & 0 & 0 & \dots & -1 & \lambda \end{bmatrix}$$

Now let's multiply the first column by λ, and add it to the second column, then multiply so obtained second column by λ and add it to the third column, and so on, until we finally get to the last column. The first element of the last column now equals zero, because we made it equal to the characteristic polynomial in Horner's form:

$$(\dots((\lambda + a_1)\lambda + a_2)\lambda + \dots + a_{n-1})\lambda + a_n = 0$$

The last element of the last column also became zero in this process. Because of the specific positions of -1's below the main diagonal, all other columns of $\lambda I - A$ are linearly independent, so that $\rho(\lambda I - A) = n - 1$. □

Corollary C.3.3 *The number of linearly independent eigenvectors corresponding to λ, an eigenvalue of a companion matrix, is $\nu(\lambda I - A) = 1$, no matter what the multiplicity of λ might be (see Example* C.3.3*).*

Corollary C.3.4 *All companion matrices are similar to Jordan matrices made of Jordan blocks having distinct eigenvalues.*

Corollary C.3.5 *An arbitrary matrix is similar to a companion matrix if and only if it is similar to a Jordan matrix whose Jordan blocks have distinct eigenvalues.*

Example C.3.2 *No matrix similar to* $J = \text{diag}(J_1(4), J_2(4), J_2(5))$ *can be similar to a companion matrix. On the other hand, any matrix similar to* $J = \text{diag}(J_3(4), J_2(5))$ *is similar to*

$$\begin{bmatrix} 22 & -193 & 844 & -1840 & 1600 \\ 1 & 0 & 0 & 0 & 0 \\ 0 & 1 & 0 & 0 & 0 \\ 0 & 0 & 1 & 0 & 0 \\ 0 & 0 & 0 & 1 & 0 \end{bmatrix}$$

because $(\lambda - 4)^3(\lambda - 5)^2 = \lambda^5 - 22\lambda^4 + 193\lambda^3 - 844\lambda^2 + 1840\lambda - 1600.$ □

Corollary C.3.6 *A companion matrix is diagonalizable if and only if it has no repeated eigenvalues.*

Here we considered only the so-called top companion matrices, because the other three varieties (bottom, left, and right) are always similar to the "top" variety. If we denote

$$A_t = \begin{bmatrix} -a_1 & -a_2 & \dots & -a_{n-1} & -a_n \\ 1 & 0 & \dots & 0 & 0 \\ 0 & 1 & \dots & 0 & 0 \\ & & \dots & & \\ 0 & 0 & \dots & 1 & 0 \end{bmatrix}, \quad A_l = \begin{bmatrix} -a_1 & 1 & 0 & \dots & 0 \\ -a_2 & 0 & 1 & \dots & 0 \\ & & & \dots & \\ -a_{n-1} & 0 & 0 & \dots & 1 \\ -a_n & 0 & 0 & \dots & 0 \end{bmatrix}$$

$$A_b = \begin{bmatrix} 0 & 1 & \dots & 0 & 0 \\ & & \dots & & \\ 0 & 0 & \dots & 1 & 0 \\ 0 & 0 & \dots & 0 & 1 \\ -a_n & -a_{n-1} & \dots & -a_2 & -a_1 \end{bmatrix}, \quad A_r = \begin{bmatrix} 0 & \dots & 0 & 0 & -a_n \\ 1 & \dots & 0 & 0 & -a_{n-1} \\ & \dots & & & \\ 0 & \dots & 1 & 0 & -a_2 \\ 0 & \dots & 0 & 1 & -a_1 \end{bmatrix}$$

the similarity transformation between any two of these four varieties of companion matrices can be deduced from the following three relations[5]

$$A_t = \tilde{I}^{-1} A_b \tilde{I} \qquad A_t = \mathsf{a}_-^{-T} A_r \mathsf{a}_-^T \qquad A_r = \tilde{I}^{-1} A_l \tilde{I}$$

where

$$\tilde{I} = \tilde{I}^{-1} = \begin{bmatrix} 0 & 0 & \dots & 0 & 1 \\ 0 & 0 & \dots & 1 & 0 \\ & & \dots & & \\ 0 & 1 & \dots & 0 & 0 \\ 1 & 0 & \dots & 0 & 0 \end{bmatrix}$$

[5]Operators T and $-T$ denote the transpose and the transpose inverse, respectively.

and

$$a_{-}^{T} = \begin{bmatrix} 1 & a_1 & a_2 & \dots & a_{n-2} & a_{n-1} \\ 0 & 1 & a_1 & \dots & a_{n-3} & a_{n-2} \\ 0 & 0 & 1 & \dots & a_{n-4} & a_{n-3} \\ & & & \dots & & \\ 0 & 0 & 0 & \dots & 1 & a_{n-1} \\ 0 & 0 & 0 & \dots & 0 & 1 \end{bmatrix}$$

Since $\det(a_{-}) = 1$, so a_{-} is always nonsingular.

Note: *In control theory A_t corresponds to the system matrix of the controller realization. Similarly, A_l corresponds to the observer form, while A_b corresponds to the observability form and A_r to the controllability form. As we saw, these four matrices are always similar, but that doesn't mean the corresponding realizations are. This is because the similarity of systems requires additional rclations to hold between other matrices or vectors that describe the system.*

Example C.3.3 If λ is an eigenvalue of a companion matrix A and if its multiplicity is $k \geq 1$, then

$$\begin{bmatrix} \lambda^{n-1} \\ \lambda^{n-2} \\ \vdots \\ \lambda^2 \\ \lambda \\ 1 \end{bmatrix}$$

is its eigenvector, and if $k > 1$

$$\begin{bmatrix} (n-1)\lambda^{n-2} \\ (n-2)\lambda^{n-3} \\ \vdots \\ 2\lambda \\ 1 \\ 0 \end{bmatrix}, \quad \begin{bmatrix} \binom{n-1}{2}\lambda^{n-3} \\ \binom{n-2}{2}\lambda^{n-4} \\ \vdots \\ 1 \\ 0 \\ 0 \end{bmatrix}, \quad \dots, \quad \begin{bmatrix} \binom{n-1}{k-1}\lambda^{n-k} \\ \binom{n-2}{k-1}\lambda^{n-k-1} \\ \vdots \\ 0 \\ 0 \\ 0 \end{bmatrix}$$

are its generalized eigenvectors.

Example C.3.4 The inverse of a companion matrix is another companion matrix. For example

$$\begin{bmatrix} -6 & -7 & 6 & 8 \\ 1 & 0 & 0 & 0 \\ 0 & 1 & 0 & 0 \\ 0 & 0 & 1 & 0 \end{bmatrix}^{-1} = \begin{bmatrix} 0 & 1 & 0 & 0 \\ 0 & 0 & 1 & 0 \\ 0 & 0 & 0 & 1 \\ 0.125 & 0.75 & 0.875 & -0.75 \end{bmatrix}$$

C.4 Symmetric and Hermitian matrices

Matrices appearing in the description of physical or engineering problems are often real and symmetric with respect to the main diagonal.

Such matrices have many interesting and important properties worth exploring. For example, their eigenvalues are real, and their eigenvectors can be chosen so that they form an orthonormal basis. But it is interesting that these properties do not hold for complex symmetric matrices in general. We shall see that the proper generalization are the Hermitian matrices.

Thus, with respect to the properties we are interested in, real symmetric matrices are special cases of Hermitian matrices, which in turn are special cases of the so-called normal matrices. For additional properties of normal matrices see Section C.8.

We begin this Section with definitions of these important classes of matrices, and continue with several important theorems about Hermitian matrices.

Definition C.4.1 *Square matrix A is symmetric if it equals its transpose:*

$$A^T = A$$

Definition C.4.2 *Square matrix A is Hermitian if it equals its conjugate transpose:*

$$A^H = A$$

where $A^H = (\bar{A})^T$ is the conjugate transpose[6] *of A.*

Definition C.4.3 *Square matrix A is normal if*

$$AA^H = A^H A$$

The normal matrices are the most general of these special matrices. We illustrate that in Figure C.1. Another important special case of the set of normal matrices is the set of unitary matrices. Real orthogonal matrices are special cases of unitary matrices.

Definition C.4.4 *Square matrix A is orthogonal if*

$$AA^T = I$$

Definition C.4.5 *Square matrix A is unitary if*

$$AA^H = I$$

[6] In mathematical and technical literature, $*$ and $'$ are often used instead of H.

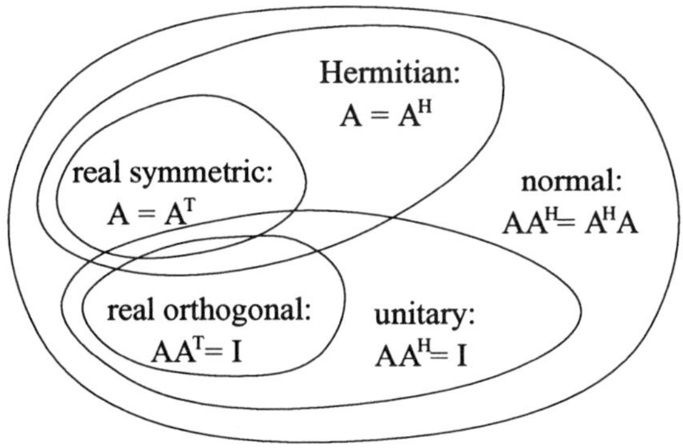

Figure C.1: Some important classes of normal matrices. Operator H (Hermitian operator) denotes the conjugate transpose: $A^H = (\bar{A})^T$. Operators $*$ and $'$ are often used instead of H.

Theorem C.4.1 *All eigenvalues of a Hermitian matrix are real.*

Proof. If λ is an eigenvalue of a Hermitian matrix A, and r is the eigenvector corresponding to λ, then

$$Ar = \lambda r \tag{C.9}$$

From (C.9) it follows

$$r^H A^H = \bar{\lambda} r^H$$

or, since A is Hermitian

$$r^H A = \bar{\lambda} r^H \tag{C.10}$$

Now, multiply (C.9) by r^H from the left to get

$$r^H A r = \lambda r^H r \tag{C.11}$$

and (C.10) by r from the right to get

$$r^H A r = \bar{\lambda} r^H r \tag{C.12}$$

From (C.11) and (C.12), using the fact that $r \neq 0$, we find that

$$\bar{\lambda} = \lambda$$

which means that λ is a real number. □

Corollary C.4.1 *All eigenvalues of a real symmetric matrix are real.*

Example C.4.1 *Hankel matrices are square matrices with constant elements along the anti-diagonals, therefore they are symmetric. If a matrix is Hankel and its elements are real, it is real symmetric, and therefore has real eigenvalues. For example, the eigenvalues of*

$$H = \begin{bmatrix} 0 & 1 & 2 \\ 1 & 2 & 0 \\ 2 & 0 & 1 \end{bmatrix}$$

are 3 *and* $\pm\sqrt{3}$. □

Example C.4.2 *The DFT matrix of order n (DFT stands for "Discrete Fourier Transform") is a matrix defined as*

$$F_n = \begin{bmatrix} 1 & 1 & 1 & \cdots & 1 \\ 1 & \omega_n & \omega_n^2 & \cdots & \omega_n^{n-1} \\ 1 & \omega_n^2 & \omega_n^4 & \cdots & \omega_n^{2(n-1)} \\ & & & \cdots & \\ 1 & \omega_n^{n-1} & \omega_n^{2(n-1)} & \cdots & \omega_n^{(n-1)^2} \end{bmatrix}$$

where $\omega_n = e^{-2\pi j/n}$ is the n-th primitive root of unity.

Obviously, for $n > 2$, the matrix F_n is complex symmetric (not Hermitian), therefore its eigenvalues are not necessarily real. For example, the eigenvalues of F_4 are ± 2 and $\pm 2j$. Can all the eigenvalues of a complex symmetric matrix be real? □

In the rest of this Section we shall prove the following important theorem and consider some of its immediate consequences:

Theorem C.4.2 *Every Hermitian matrix is similar to a diagonal matrix of its eigenvalues via some unitary matrix.*

Proof. First, we shall use induction to prove the following lemma which is important in its own right:

Lemma C.4.1 (Schur) *For any complex square matrix A of order n there exists a unitary matrix U such that*

$$B = U^{-1}AU = U^H AU$$

is an upper triangular matrix.

Proof of Lemma C.4.1. The base case $n = 1$ is trivially true, because any 1×1 matrix can be considered upper triangular. Assume correct for matrices of order $\leq n$. Let λ_1 be an eigenvalue of A and r_1 a normalized eigenvector corresponding to λ_1. Using Gram-Schmidt orthonormalization, we can always construct a matrix

$$V = [r_1 \; p_2 \; p_3 \; \ldots \; p_n] = [r_1 \; P]$$

such that its columns form an orthonormal basis. Note that P is $n \times (n-1)$. By construction, $VV^H = I$, i.e., V is unitary. Now observe that

$$V^{-1}AV = V^H AV = \begin{bmatrix} r_1^H \\ P^H \end{bmatrix} A \, [r_1 \; P] = \begin{bmatrix} \lambda_1 & r_1^H AP \\ 0 & P^H AP \end{bmatrix}$$

because the columns of V form an orthonormal basis, so that $r_1^H r_1 = 1$ and $P^H r_1 = 0$.

Now note that $P^H AP$ is $(n-1) \times (n-1)$, so according to the inductive hypothesis there exists a unitary matrix W such that $W^{-1}(P^H AP)W$ is upper triangular.

If we take

$$U = V \begin{bmatrix} 1 & 0 \\ 0 & W \end{bmatrix}$$

we can easily see that it is unitary and $U^{-1}AU = U^H AU$ is an upper triangular matrix. This proves the lemma. □

Now that we have this powerful lemma, we can easily finish the proof of Theorem C.4.2. Let A be Hermitian, i.e., $A^H = A$. According to Lemma C.4.1, there exists a unitary matrix U such that $B = U^{-1}AU = U^H AU$ is upper triangular. But since A is Hermitian, B must be Hermitian too, because

$$B^H = (U^H AU)^H = U^H A^H U = U^H AU = B$$

Thus, B is both upper triangular and Hermitian, therefore it is diagonal. Since the similarity transformation does not change the eigenvalues and the eigenvalues of a diagonal matrix are its diagonal elements, we see that the theorem is indeed true. □

Corollary C.4.2 *Hermitian matrices of order n have n orthonormal eigenvectors.*

Proof. Since every Hermitian matrix A is diagonalizable via some unitary matrix, the columns of that unitary matrix can be taken as the orthonormal set of eigenvectors of A. □

This is a very important result about Hermitian matrices. Similarly, we can prove the following theorem:

Theorem C.4.3 *Every real symmetric matrix is similar to a diagonal matrix of its eigenvalues via some real orthogonal matrix.*

Corollary C.4.3 *Real symmetric matrices of order n have n real orthonormal eigenvectors.*

Some important examples of Hermitian and symmetric matrices come from controls and signal processing.

Example C.4.3 *Gram matrix of a (not necessarily square) real matrix A is*

$$G(A) = A^T A$$

Since

$$G^T(A) = (A^T A)^T = A^T A = G(A)$$

we see that Gram matrix is real symmetric. If A is complex, Gram matrix is defined as $G(A) = A^H A$, and it is Hermitian. □

Example C.4.4 *Let u be the vector of samples of a signal $u(t)$ at discrete time points $t = n, n-1, n-2, \ldots, n-m+1$. The correlation matrix of u is*

$$R = E[uu^T] = \begin{bmatrix} r_{n,n} & r_{n,n-1} & \cdots & r_{n,n-m+1} \\ r_{n-1,n} & r_{n-1,n-1} & \cdots & r_{n-1,n-m+1} \\ & & \cdots & \\ r_{n-m+1,n} & r_{n-m+1,n-1} & \cdots & r_{n-m+1,n-m+1} \end{bmatrix}$$

where $r_{n-i,n-j} = E[u(n-i)u(n-j)]$.

If $u(t)$ is wide-sense stationary, i.e., if $E[u(k)] = \text{const}$ and $r_{n-i,n-j} = r(j-i)$ then

$$R = \begin{bmatrix} r(0) & r(1) & \cdots & r(m-1) \\ r(-1) & r(0) & \cdots & r(m-2) \\ & & \cdots & \\ r(-(m-1)) & r(-(m-2)) & \cdots & r(0) \end{bmatrix}$$

Since $r(-k) = E[u(n-k)u(n)] = E[u(n)u(n-k)] = r(k)$, we see that $R^T = R$. Thus, the correlation matrix of a wide-sense stationary discrete-time stochastic signal is real symmetric. If $u(t)$ is complex, its correlation matrix is defined by $R = E[uu^H]$. In that case $R^H = R$, i.e., R is Hermitian. □

C.5 Quadratic forms and definiteness

In this Section we shall see that quadratic forms can be written using matrix notation. Also, the question of whether a certain quadratic form is definite, semi-definite or indefinite can be answered using the tests based on the matrix notation. Thus, the notions of definiteness, semi-definiteness and, indefiniteness can be defined for matrices via this connection to quadratic forms. Many important applications of linear algebra use the material contained in this Section, for example, the Lyapunov stability theory applied to linear and linearized systems.

Definition C.5.1 *Quadratic form $q(x_1, \ldots, x_n)$ is any polynomial of order* 2 *in n real variables $x_1, \ldots, x_n$.*

Any quadratic form $q(x_1, \ldots, x_n)$ can be written as

$$q(x) = x^T Q x$$

where $x = [x_1 \; \ldots \; x_n]^T$ and Q is any conveniently chosen matrix. To see that, consider the following: If $Q = [q_{i,j}]_{n \times n}$ then

$$x^T Q x = \sum_{k=1}^{n} q_{kk} x_k^2 + \sum_{i<j} (q_{ij} + q_{ji}) x_i x_j$$

Therefore, if we are given $q(x)$, we can pick Q so that $q(x) = x^T Q x$. In particular, we can pick Q to be real symmetric and therefore diagonalizable.

Example C.5.1 *Consider*

$$q(x) = a x_1^2 + b x_2^2 + c x_3^2 + 2r x_1 x_2 + 2s x_1 x_3 + 2t x_2 x_3$$

where $x = [x_1 \; x_2 \; x_3]^T \in R^3$ and $a, b, c, r, s, t \in R$ (here R denotes the set of real numbers).

The symmetric matrix Q corresponding to this quadratic form is given by

$$Q = \begin{bmatrix} a & r & s \\ r & b & t \\ s & t & c \end{bmatrix}$$

According to Theorem C.4.3, *since Q is real symmetric, it can be diagonalized using the matrix of its normalized eigenvectors U:*

$$U^T Q U = D$$

where $D = \mathrm{diag}(\lambda_1, \lambda_2, \lambda_3)$, $\lambda_1, \lambda_2, \lambda_3$ are the eigenvalues of Q, and $U^T U = U U^T = I$.

If we put $y = U^T x$, we can write

$$\begin{aligned} q(x) &= x^T Q x \\ &= x^T U D U^T x \\ &= y^T D y \\ &= \lambda_1 y_1^2 + \lambda_2 y_2^2 + \lambda_3 y_3^2 \end{aligned}$$

Thus, using the change of variables described by $y = U^T x$ *we have eliminated the cross-product terms. This is quite general procedure, it can be used for quadratic forms of arbitrary number* n *of real variables.*

It is also a very important procedure, because we often need to know if some quadratic form is positive for all $x \neq 0$. *In this example, we see that the quadratic form* $q(x)$ *is positive for all* $x \neq 0$ *if and only if all eigenvalues of* Q *are positive. This is so because real symmetric matrices have* n *(in this case* $n = 3$*) orthonormal eigenvectors, so* $\det(U) \neq 0$, *and therefore to any* $x \neq 0$ *corresponds exactly one* $y = U^T x$ $(\neq 0)$. *Therefore,* $q(x)$ *is positive for all* $x \neq 0$ *if and only if it is positive for all* $y \neq 0$ *and this is true if and only if* $\lambda_1, \lambda_2, \lambda_3 > 0$. □

Example C.5.2 *Here are a few "real" examples to introduce and motivate some new terminology:*

a) $q(x) = x_1^2 + x_2^2 + x_3^2$ *is positive for any* $x \neq 0$. *It is called positive definite.*

b) $q(x) = x_1^2 + x_3^2$ *is not positive for all* $x \neq 0$. *For example* $q(0, 1, 0) = 0$. *But it is nonnegative for all* $x \neq 0$. *It is called nonnegative definite or positive semi-definite.*

c) $q(x) = x_1^2 - x_3^2$ *can be positive, negative or zero when* $x \neq 0$. *It is called indefinite.* □

Now we define the terminology used in the previous example:

Definition C.5.2 *Quadratic form* $x^T Q x$ *is definite if*

$$x^T Q x > 0 \quad (\forall x \neq 0) \qquad \textit{(positive definite)}$$

or

$$x^T Q x < 0 \quad (\forall x \neq 0) \qquad \textit{(negative definite)}$$

Definition C.5.3 *Quadratic form* $x^T Q x$ *is semi-definite if*

$$x^T Q x \geq 0 \quad (\forall x \neq 0) \qquad \textit{(positive semi-definite)}$$

or

$$x^T Q x \leq 0 \quad (\forall x \neq 0) \qquad \textit{(negative semi-definite)}$$

Definition C.5.4 *Quadratic form* $x^T Q x$ *is indefinite if it is not semi-definite.*

The matrix (in)definiteness is defined via (in)definiteness of the corresponding quadratic forms.

Definition C.5.5 *A real square matrix* Q *is positive (negative) definite if* $x^T Q x$ *is positive (negative) definite. We write* $Q > 0$ $(Q < 0)$.

Definition C.5.6 *A real square matrix* Q *is positive (negative) semi-definite if* $x^T Q x$ *is positive (negative) semi-definite. We write* $Q \geq 0$ $(Q \leq 0)$.

Definition C.5.7 *A real square matrix* Q *is indefinite if* $x^T Q x$ *is indefinite.*

Example C.5.3 *Consider a differentiable function of three variables $f(x_1, x_2, x_3)$. A necessary condition for $f(x)$, ($x = [x_1\ x_2\ x_3]^T$) to have an extremum at $x = P = [p_1\ p_2\ p_3]^T$ is that all three partial derivatives of $f(x)$ at $x = P$ be zero:*

$$\left.\frac{\partial f}{\partial x_i}\right|_{x=P} = 0, \quad (i = 1, 2, 3) \tag{C.13}$$

To see that this condition is not sufficient, consider $f(x) = x_1x_2x_3$.

A sufficient condition for $f(x)$ to have a minimum at $x = P$ can be found as follows: Taylor expansion of $f(x)$ around $x = P$ is

$$f(x) = f(P) + \sum_{i=1}^{3} \frac{\partial f}{\partial x_i}(P)(x_i - p_i) + \frac{1}{2!}\sum_{i=1}^{3}\sum_{j=1}^{3} \frac{\partial^2 f}{\partial x_i \partial x_j}(P)(x_i - p_i)(x_j - p_j) + \ldots$$

If (C.13) *is true, then*

$$\sum_{i=1}^{3}\sum_{j=1}^{3} \frac{\partial f}{\partial x_i}(P)(x_i - p_i)(x_j - p_j) > 0 \quad (\forall x \neq P)$$

guarantees that $f(P)$ is indeed a minimum. Note that this condition is not necessary.

With $u_k = x_k - p_k$ ($k = 1, 2, 3$) we can write this condition in the following form

$$u^T H u > 0 \quad (\forall u \neq 0) \tag{C.14}$$

where

$$H = \begin{bmatrix} h_{11}(P) & h_{12}(P) & h_{13}(P) \\ h_{21}(P) & h_{22}(P) & h_{23}(P) \\ h_{31}(P) & h_{32}(P) & h_{33}(P) \end{bmatrix} \quad \text{and} \quad h_{ij}(P) = \frac{\partial^2 f}{\partial x_i \partial x_j}(P), \quad (i, j = 1, 2, 3)$$

H is often called a Hessian matrix of $f(x)$.

In other words, if (C.13) *is true and $H > 0$ (read: H is positive definite), then $f(x)$ has a minimum at $x = P$. Similarly, if* (C.13) *is true and $H < 0$ (read: H is negative definite), then $f(x)$ has a maximum at $x = P$. If H is semi-definite, further investigation is needed. If H is indefinite, $f(P)$ is not an extremal point.*

For "well-behaved" functions $\partial^2 f/\partial x_i \partial x_j = \partial^2 f/\partial x_j \partial x_i$, hence H is symmetric, and we can use the eigenvalue test presented in Example C.5.1 *to test H.*

Just like Example C.5.1, *this Example is easy to generalize to functions of n variables. Note that we don't need the eigenvectors of H, only its eigenvalues.* □

The matrices introduced in the Examples C.4.3 and C.4.4, viz. Gram and correlation matrices, are positive semi-definite. This is true even in the complex case, when these matrices are Hermitian. To prove this more general statement, we consider the Hermitian form, the complex generalization of quadratic forms.

Example C.5.4 *Let A be a complex $m \times n$ ($m \geq n$) matrix and $G(A) = A^H A$ its Gram matrix. Then $G(A)$ is $n \times n$. Consider the Hermitian form $z^H G(A) z$ for all complex vectors $z \neq 0$:*

$$z^H G(A) z = z^H A^H A\, z = w^H w = \sum_{k=1}^{n} |w_k|^2 \geq 0$$

Thus, even when complex, Gram matrix is positive semi-definite, $G(A) \geq 0$. Note that $G(A)$ is singular if and only if A is not of full rank, i.e., if $\rho(A) < \min(m,n) = n$ (cf. Problem C.8.11). Thus, $G(A)$ is positive definite if and only if A has a full rank, and $m \geq n$.

Example C.5.5 *If R is a correlation matrix of a complex wide-sense stationary signal, then $R = E[uu^H]$. Consider the Hermitian form $z^H R z$ for all complex vectors $z \neq 0$:*

$$z^H R z = z^H E[uu^H] z = E[z^H u u^H z]$$

Note that $(z^H u)^H = u^H z$ are scalars, therefore

$$z^H R z = E\left[|z^H u|^2\right] \geq 0, \quad i.e., \ R \geq 0$$

Most often however, $R > 0$, because R is singular if and only if the signal $u(t)$ is a sum of $k \leq m$ sinusoids, where m is the length of vector u (cf. Problem C.8.12).

Sometimes we have to work with matrices which are not symmetric. In such cases we can not apply the eigenvalue test of Example C.5.1 directly[7].

First, observe that every matrix can be represented as a sum of two matrices, one symmetric and the other skew-symmetric[8]:

$$A = \frac{1}{2}(A + A^T) + \frac{1}{2}(A - A^T) = A_s + A_{ss}$$

Second, observe that quadratic form corresponding to a skew-symmetric matrix is zero. To see that, use the facts that $x^T A_{ss} x$ is a scalar and that A_{ss} is skew-symmetric,

$$x^T A_{ss} x = (x^T A_{ss} x)^T = -x^T A_{ss} x$$

therefore $x^T A_{ss} x = 0$.

If we now consider a quadratic form corresponding to A:

$$\begin{aligned} x^T A x &= x^T (A_s + A_{ss}) x \\ &= x^T A_s x \end{aligned}$$

we see that we can reduce the problem to analyzing the eigenvalues of the symmetric part of matrix A. This proves the following theorem:

Theorem C.5.1 *A real matrix A is positive (negative) definite if and only if its symmetric part*

$$A_s = \frac{1}{2}(A + A^T)$$

is positive (negative) definite.

[7] Even if a matrix is diagonalizable, for the eigenvalue test to work, we need the eigenvectors to be mutually orthogonal, which is the case only for normal matrices (see Problem C.8.9.)

[8] Matrix B is skew-symmetric if $B^T = -B$.

Similarly, we can prove the following theorem:

Theorem C.5.2 *A real matrix A is positive (negative) semi-definite if and only if its symmetric part*

$$A_s = \frac{1}{2}(A + A^T)$$

is positive (negative) semi-definite.

Let us formalize the method of the Example C.5.1, commonly known as the Rayleigh-Ritz theorem:

Theorem C.5.3 *A real symmetric matrix A is positive (negative) definite if and only if all of its eigenvalues are positive (negative).*

Proof. Proof of this theorem is essentially the same as the derivation in the Example C.5.1. □

Similarly, the following theorem is true:

Theorem C.5.4 *A real symmetric matrix A is positive (negative) semi-definite if and only if all of its eigenvalues are positive or zero (negative or zero).*

The following tests are due to Sylvester. After stating them in the following two theorems, we shall prove only the first of them, because ideas involved are the same. Similarly to the eigenvalue test, they work for real symmetric matrices only. If a matrix is not symmetric, the test should be applied to its symmetric part.

Theorem C.5.5 *A real symmetric matrix $A = [a_{ij}]_{n \times n}$ is positive definite if and only if all of its leading principal minors are positive, that is*

$$\Delta_1 = a_{11} > 0, \quad \Delta_2 = \begin{vmatrix} a_{11} & a_{12} \\ a_{21} & a_{22} \end{vmatrix} > 0$$

$$\Delta_3 = \begin{vmatrix} a_{11} & a_{12} & a_{13} \\ a_{21} & a_{22} & a_{23} \\ a_{31} & a_{32} & a_{33} \end{vmatrix} > 0, \quad \ldots, \quad \Delta_n = \det(A) > 0$$

Theorem C.5.6 *A real symmetric matrix $A = [a_{ij}]_{n \times n}$ is positive semi-definite if and only if all of its principal minors (not only the leading principal minors) are nonnegative, that is for all $i, j, k, \ldots$*

$$a_{ij} \geq 0, \quad \begin{vmatrix} a_{ii} & a_{ij} \\ a_{ji} & a_{jj} \end{vmatrix} \geq 0, \quad \begin{vmatrix} a_{ii} & a_{ij} & a_{ik} \\ a_{ji} & a_{jj} & a_{jk} \\ a_{ki} & a_{kj} & a_{kk} \end{vmatrix} \geq 0, \quad \ldots, \quad \det(A) \geq 0$$

The negative (semi-) definiteness of A is tested as the positive (semi-) definiteness of $-A$.

Proof. We use the induction on the size of A.

- $a_{11}x_1^2 > 0 \Leftrightarrow a_{11} > 0$.
- Suppose that a matrix of size $n-1$ is positive definite if and only if $\Delta_1 > 0, \Delta_2 > 0, \dots, \Delta_{n-1} > 0$.
- Consider A, an $n \times n$ matrix.

  The "only if" part. If

$$q_n(x_1, x_2, \dots, x_n) = [x_1\ x_2\ \dots\ x_n] \begin{bmatrix} a_{11} & a_{12} & \dots & a_{1n} \\ a_{21} & a_{22} & \dots & a_{2n} \\ & & \dots & \\ a_{n1} & a_{n2} & \dots & a_{nn} \end{bmatrix} \begin{bmatrix} x_1 \\ x_2 \\ \dots \\ x_n \end{bmatrix}$$

  is positive definite, so is $q_{n-1}(x_1, x_2, \dots, x_{n-1}) = q_n(x_1, x_2, \dots, x_{n-1}, 0)$. Thus, by the inductive conjecture, if A is positive definite, then $\Delta_1 > 0, \Delta_2 > 0, \dots, \Delta_{n-1} > 0$, and the only remaining thing to prove in this part is that if A is positive definite, then $\Delta_n > 0$. But

$$\Delta_n = \det(A) = \lambda_1 \lambda_2 \dots \lambda_n > 0$$

  because according to Theorem C.5.3, if A is positive definite, each of its eigenvalues is positive.

  The "if" part. Let $\Delta_1 > 0, \Delta_2 > 0, \dots, \Delta_n > 0$. Quadratic form $q_n(x_1, x_2, \dots, x_n)$ can be written as

$$x^T A x = \sum_{i=1}^{n-1} \sum_{j=1}^{n-1} a_{ij} x_i x_j + 2 \sum_{i=1}^{n} a_{in} x_i x_n + a_{nn} x_n^2$$

  The term $\sum\sum a_{ij}x_ix_j$ can be diagonalized and written as $\sum b_i y_i^2$. Since $\Delta_1 > 0, \Delta_2 > 0, \dots, \Delta_{n-1} > 0$, according to the inductive conjecture, this term is positive definite, so we can write $b_i = c_i^2 > 0$ $(i = 1, 2, \dots, n-1)$. Thus, for some coefficients c_i and d_i

$$\begin{aligned} x^T A x &= \sum_{i=1}^{n-1} c_i^2 y_i^2 + 2 \sum_{i=1}^{n} d_{in} y_i y_n + a_{nn} y_n^2 \\ &= \sum_{i=1}^{n-1} (c_i y_i + \frac{d_i}{c_i} y_n)^2 - \sum_{i=1}^{n-1} \left(\frac{d_i}{c_i}\right)^2 y_n^2 + a_{nn} y_n^2 \\ &= \sum_{i=1}^{n-1} z_i^2 + \alpha z_n^2 \ = \ z^T \mathrm{diag}(1, \dots, 1, \alpha)\, z \end{aligned}$$

where the z_i's are the linear combinations of the x_i's, and

$$\alpha = a_{nn} - \sum_{i=1}^{n-1} \left(\frac{d_i}{c_i}\right)^2$$

It can be seen that we can write $z = Px$, where P is some nonsingular matrix. Therefore

$$x^T A x = x^T P^T \mathrm{diag}(1, \ldots, 1, \alpha) P x$$

Since $P^T \mathrm{diag}(1, \ldots, 1, \alpha) P$ is symmetric, we see that

$$A = P^T \mathrm{diag}(1, \ldots, 1, \alpha) P$$

and finally

$$\det(A) = \alpha\,(\det(P))^2 \quad \Rightarrow \quad \mathrm{sgn}(\alpha) = \mathrm{sgn}(\Delta_n) \quad \Rightarrow \quad \alpha > 0 \quad \Rightarrow$$

$$\Rightarrow \quad z^T \mathrm{diag}(1, \ldots, 1, \alpha)\, z > 0 \quad (\forall z \neq 0) \quad \Rightarrow \quad A \text{ is positive definite}$$

This concludes the proof of Theorem C.5.5. □

Let us now consider the conditions under which we can extract a "square root" of a real matrix A, i.e., write $A = B^T B$.

Since $(B^T B)^T = B^T B$, A must be symmetric. If that is so, A is diagonalizable via some orthogonal matrix U (by Theorem C.4.3), and its eigenvalues are real:

$$A = U^{-1} D U = U^T D U$$

Now the question is when can we extract the "square root" of the diagonal matrix D, and the answer is: only when its diagonal elements, which are at the same time the eigenvalues of A, are nonnegative. Therefore, the necessary conditions are that A is real symmetric and positive semi-definite. These conditions are also sufficient. Thus, we proved the following theorem:

Theorem C.5.7 *Extracting the square root of a matrix A, i.e., writing it in the form $A = B^T B$ is possible if and only if A is real symmetric and positive semi-definite.*

Note that this decomposition is not unique, because if B is a square root of A, so is VB, for any orthogonal matrix V. Similar theorem holds for Hermitian matrices.

Quadratic forms are often used to express a cost function in the optimization problems. It is therefore important to know how to differentiate them with respect to vector x, or with respect to some of its components. Since every quadratic form can be written in terms of some symmetric matrix, in the following we assume that $Q^T = Q$.

First consider the differentiation with respect to the m-th component of x:

$$\begin{aligned} \frac{\partial(x^T Q x)}{\partial x_m} &= \frac{\partial}{\partial x_m}\left(\sum_{k=1}^{n} q_{kk}x_k^2 + 2\sum_{i<j} q_{ij}x_i x_j\right) \\ &= 2\sum_{i=1}^{n} q_{mi}x_i \;=\; 2q_{(m)}x \end{aligned}$$

where $q_{(m)}$ denotes the m-th row of Q.

If we define the differentiation with respect to a vector as

$$\frac{\partial(x^T Q x)}{\partial x} = \mathrm{grad}(x^T Q x) = \begin{bmatrix} \partial/\partial x_1 \\ \partial/\partial x_2 \\ \vdots \\ \partial/\partial x_n \end{bmatrix} (x^T Q x)$$

we see that

$$\frac{\partial(x^T Q x)}{\partial x} = 2Qx$$

Since Q is real-symmetric, the Hessian matrix of $q(x) = x^T Q x$ is $H = 2Q$. Recall that in the calculus of functions of more than one variable, the Hessian matrix takes the role the second derivative has in the "standard" calculus, just like the gradient vector takes the role of the first derivative.

C.6 Some special matrices

Several classes of special matrices appear frequently in applied and pure mathematics. We already encountered some such classes: diagonal and Jordan matrices in Section C.2, companion matrices in Section C.3, real symmetric, Hermitian, normal, Gram, and some other special matrices in Section C.4, and definite and semi-definite matrices in Section C.5. In this Section we shall define more classes of special matrices: Hankel, Toeplitz, Vandermonde, and Hurwitz.

Hankel matrices. Matrix A is said to be a Hankel matrix if the elements along its anti-diagonals are equal.

Example C.6.1 In linear control systems we use an $n \times n$ Hankel matrix made up of Markov parameters $h_1, h_2, \ldots, h_{2n-1}$

$$\mathcal{M} = \begin{bmatrix} h_1 & h_2 & \cdots & h_n \\ h_2 & h_3 & \cdots & h_{n+1} \\ \vdots & \vdots & \ddots & \vdots \\ h_n & h_{n+1} & \cdots & h_{2n-1} \end{bmatrix}$$

If the elements of a Hankel matrix are real, then it is real symmetric, and therefore its eigenvalues are real.

Toeplitz matrices. Matrix A is said to be Toeplitz if the elements along its diagonals are equal.

Example C.6.2 A discrete-time convolution describes the relation between the input and the output of a discrete-time system:

$$y[k] = \sum_{i=1}^{n} f_{k-i} x[i] \qquad (k = 1, 2, \ldots, n)$$

This relation can be written using matrix notation

$$y = Fx$$

where

$$x = [\ x[1]\ \ x[2]\ \ \ldots\ \ x[n]\ ]' \qquad y = [\ y[1]\ \ y[2]\ \ \ldots\ \ y[n]\ ]'$$

and

$$F = \begin{bmatrix} f_0 & f_{-1} & \cdots & f_{-(n-1)} \\ f_1 & f_0 & \cdots & f_{-(n-2)} \\ \vdots & \vdots & \ddots & \vdots \\ f_{n-1} & f_{n-2} & \cdots & f_0 \end{bmatrix}$$

is a Toeplitz matrix.

This special structure of Toeplitz matrices is used to speed-up their inversion. An example of such algorithms is the well known Levinson algorithm which originated in signal processing (see [19] and [46]).

Vandermonde matrices. Matrix V is said to be a Vandermonde matrix if it has the following form:

$$V_n = \begin{bmatrix} 1 & a_1 & \dots & a_1^{n-1} \\ 1 & a_2 & \dots & a_2^{n-1} \\ \vdots & \vdots & \ddots & \vdots \\ 1 & a_n & \dots & a_n^{n-1} \end{bmatrix}$$

In Appendix B.4 (see Example B.4.2) we show that the determinant of a Vandermonde matrix is

$$\det(V_n) = \prod_{1 \leq i < j \leq n} (a_j - a_i)$$

For a useful generalization of the Vandermonde matrices and determinants see Problem 3.2.7.

Hurwitz matrices. Matrix A is said to be Hurwitz if all of its eigenvalues have strictly negative real parts, i.e., if they all lie in the left complex half-plane.

Such matrices are also called ***stability*** matrices, because of their role in the theory of continuous-time linear control systems. In the realm of discrete-time systems, such an important role is played by matrices with eigenvalues inside the unit circle of the complex plane. Apparently they do not have a special name, except perhaps ***discrete stability*** matrices.

The celebrated Lyapunov stability criterion states that A is Hurwitz if and only if for any given positive definite symmetric matrix Q there exists a positive definite symmetric matrix P such that

$$A'P + PA = -Q$$

This equation is known as the Lyapunov equation.

There is an analogous criterion for discrete stability matrices in which the Lyapunov equation is replaced by the discrete-time Lyapunov equation:

$$A'PA - P = -Q$$

Proofs of these criteria are given in Section 2.2 of this book. The Routh, Hurwitz, and Jury criteria of Sections 1.1 and 1.2 can also be used. How about the Sylvester definiteness criterion from Section C.5 of this appendix?

C.7 Rank, pseudoinverses, SVD, and norms

In this Section we shall consider several loosely related topics, which in some applications come together quite nicely.

Rank

This Subsection is a brief survey of some important properties of the rank of a matrix. No proofs are given.

Definition C.7.1 *If A is an $m \times n$ complex matrix, its rank, $\rho(A)$, is the size of A's largest nonsingular submatrix. The rank of a null-matrix is 0.*

Theorem C.7.1 *Let A be an $m \times n$ matrix. If $r = \rho(A)$, then A has exactly r linearly independent columns and exactly r linearly independent rows.*

Theorem C.7.2 (Frobenius inequality) *If P, Q, and R are rectangular matrices such that the product PQR is well defined, then*

$$\rho(PQ) + \rho(QR) \leq \rho(Q) + \rho(PQR)$$

Corollary C.7.1 (Sylvester's law) *Let A be $m \times n$ and let B be $n \times p$. Then*

$$\rho(A) + \rho(B) - n \leq \rho(AB) \leq \min(\rho(A), \rho(B))$$

Corollary C.7.2 *If P and Q are nonsingular, and C is $m \times n$, then*

$$\rho(PCQ) = \rho(C)$$

Pseudoinverses

If the number of *independent* equations is greater than the number of unknowns, the system is overdetermined, and the solution to that system does not exist. On the other hand, if there are more unknowns than the *independent* equations, the system is underdetermined, and there are infinitely many solutions to that system.

If there are dependent equations, situation becomes more complicated, but reduces to the above.

Example C.7.1 *Consider the following two systems:*

$$\begin{bmatrix} 1 & 0 \\ 0 & 1 \\ 1 & 1 \end{bmatrix} \begin{bmatrix} x \\ y \end{bmatrix} = \begin{bmatrix} 3 \\ 4 \\ 6 \end{bmatrix}$$ *does not have any solutions.*

$$\begin{bmatrix} 1 & 1 & 0 \\ 1 & 0 & 1 \end{bmatrix} \begin{bmatrix} x \\ y \\ z \end{bmatrix} = \begin{bmatrix} 5 \\ 6 \end{bmatrix} \quad \textit{has infinitely many solutions}$$

□

In this Subsection we shall see what we can do in such cases. If the system is overdetermined, we can obtain some approximation to the solution. If the system is underdetermined, we can put some additional constraints on the solution, so that only one out of infinitely many solutions is selected.

When the system is overdetermined, we are often interested in an approximate solution which is in some sense optimal. Let the overdetermined system be

$$Ax = b \tag{C.15}$$

where A is $m \times n$, $m > n$, and $\rho(A) = \min(m, n) = n$, i.e., A has a full rank[9].

Often we define the optimal approximate solution of (C.15) to be the vector $x = x_0$ which minimizes the Euclidean length of the error vector $e = Ax - b$

$$||e|| = \sqrt{e^T e} = \sqrt{(Ax - b)^T (Ax - b)}$$

or equivalently its square

$$||e||^2 = e^T e = (Ax - b)^T (Ax - b)$$

In order to determine x_0, let us form the cost function

$$J(x) = e^T e = (Ax - b)^T (Ax - b) = x^T A^T A x - 2b^T A x + b^T b$$

and minimize it

$$\frac{\partial J}{\partial x} = 2A^T A x - (2b^T A)^T = 0 \quad \Rightarrow \quad x_0 = (A^T A)^{-1} A^T b$$

Note that we could take the inverse of $A^T A$, because we assumed A to have a full rank and that $m > n$ (cf. Problem C.8.11).

This is a minimum of $J(x)$ because the Hessian matrix of $J(x)$ is $2A^T A$, a positive definite matrix. (Recall that the Gram matrices are always positive semi-definite, and add the fact that $A^T A$ is nonsingular.)

---

[9]Later, in a Subsection about the singular value decomposition (SVD), we discuss the more general case when $\rho(A) \leq \min(m, n)$.

Let the system be underdetermined, i.e., let it be given by

$$Ax = b \tag{C.16}$$

where A is $m \times n$, $m < n$, and $\rho(A) = \min(m, n) = m$, i.e., A is of full rank[10].

Among the infinitely many solutions of this system we often wish to find the solution closest to the origin in Euclidean distance sense. This time the cost function is

$$J(x) = \|x\|^2$$

and we have to minimize it over vectors x that satisfy the Equation (C.16). This is a typical setup for the application of the Lagrange's method of multipliers.

Let $L(x, \lambda) = J(x) + \lambda^T(Ax - b)$, then

$$\left. \begin{array}{l} \partial L/\partial x = 2x + A^T\lambda = 0 \\ \partial L/\partial \lambda = Ax - b = 0 \end{array} \right\} \quad \Rightarrow \quad x_0 = A^T(AA^T)^{-1}b$$

In this case, the Hessian matrix of $J(x)$ is $2I$, a positive definite matrix, so we are sure we minimized $J(x)$.

Matrices $(A^TA)^{-1}A^T$ (for $m > n$) and $A^T(AA^T)^{-1}$ (for $m < n$) are called the left and the right pseudoinverses of A, respectively.

Singular value decomposition

In this Subsection we prove the singular value decomposition theorem, and see some of its consequences.

Theorem C.7.3 *Let A be a complex $m \times n$ matrix with rank $\rho(A) = r \leq \min(m, n)$. Then A can be written as*

$$A = USV^H, \quad S = \begin{bmatrix} \Sigma_r & 0 \\ 0 & 0 \end{bmatrix}$$

where U and V are some unitary matrices, $\Sigma_r = \operatorname{diag}(\sigma_1, \ldots, \sigma_r)$, and $\sigma_1 \geq \ldots \geq \sigma_r > 0$ are positive real numbers, the positive singular values of A. If there are any zeros on the main diagonal of S, they are also singular values of A.

Proof. The sizes of these matrices are illustrated in Figure C.2. Since the Gram matrix A^HA is Hermitian and positive semi-definite (Examples C.4.3 and C.5.4), its eigenvalues are real and nonnegative (Theorems C.4.1 and C.5.3). According to Theorem C.4.2 we can find a unitary matrix V such that

$$V^HA^HAV = \begin{bmatrix} \Sigma_r^2 & 0 \\ 0 & 0 \end{bmatrix}$$

[10] Again, the more general case when $\rho(A) \leq \min(m, n)$ will be considered in a Subsection about SVD.

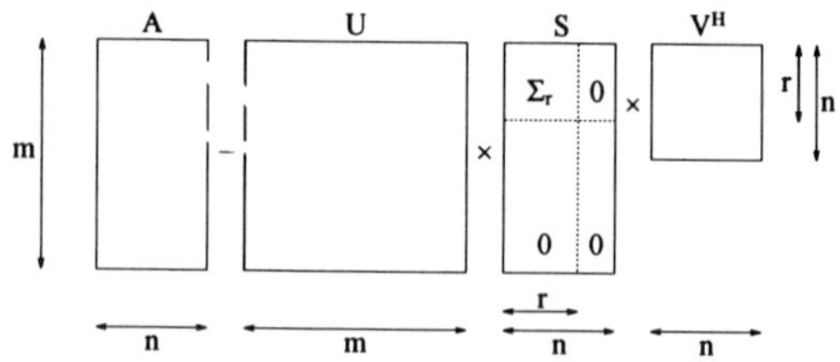

Figure C.2: Illustration of the singular value decomposition (SVD) for the case when $m > n$. In order to reduce the amount of computer memory used in computations, matrix S is often taken to be $k \times k$, where $k = \min(m, n) \geq r = \rho(A)$. If this "economy" representation of S is used, then the dimensions of U and V must be adjusted too.

where $\Sigma_r = \text{diag}(\sigma_1, \ldots, \sigma_r)$, and $\sigma_1 \geq \ldots \geq \sigma_r > 0$. Obviously, $\sigma_1^2, \ldots, \sigma_r^2$ are the non-zero eigenvalues of $A^H A$. Also, according to Corollary C.7.2, $r = \rho(A^H A)$.

Thus

$$A^H A = V \begin{bmatrix} \Sigma_r^2 & 0 \\ 0 & 0 \end{bmatrix} V^H$$

which implies (cf. Theorem C.5.7 and a comment after it) that A is of the form

$$A = U \begin{bmatrix} \Sigma_r & 0 \\ 0 & 0 \end{bmatrix} V^H$$

where U is some unitary matrix. Again, from Corollary C.7.2, $r = \rho(A)$. □

We could prove this theorem by looking at the outer product AA^H instead of the Gram matrix $A^H A$ (also called the inner product). To see that, note that AA^H is the Gram matrix of A^H, hence it is Hermitian and positive semi-definite. Thus, the following theorem is true:

Theorem C.7.4 *Let A be a complex $m \times n$ matrix with rank $\rho(A) \leq \min(m, n)$. Then the non-zero (i.e., positive) singular values of A are the square roots of the non-zero eigenvalues of the Gram matrix $A^H A$ (also called the inner product), and also of the outer product AA^H. In addition to that, if there are r non-zero singular values of A, then*

$$r = \rho(A) = \rho(A^H A) = \rho(AA^H)$$

In the following we shall see a connection between the inverse of a nonsingular matrix, the two pseudoinverses (left and right), and the general pseudoinverse defined via SVD.

Definition C.7.2 *Let the singular value decomposition of an $m \times n$ matrix A be given by*

$$A = USV^H, \quad S = \begin{bmatrix} \Sigma_r & 0 \\ 0 & 0 \end{bmatrix}$$

where U and V are the corresponding unitary matrices, $\Sigma_r = \text{diag}(\sigma_1, \ldots, \sigma_r)$, and $\sigma_1 \geq \ldots \geq \sigma_r > 0$ are the non-zero singular values of A.

Then the general pseudoinverse of A is defined as

$$A^{\#} = V \begin{bmatrix} \Sigma_r^{-1} & 0 \\ 0 & 0 \end{bmatrix} U^H$$

Example C.7.2 *Suppose A is an $n \times n$ nonsingular matrix. Obviously $\rho(A) = n$. Therefore, the SVD of A is*

$$A = USV^H, \quad S = \Sigma_n$$

By definition

$$A^{\#} = V\Sigma_n^{-1}U^H$$

Since U and V are unitary matrices, we have

$$AA^{\#} = U\Sigma_n V^H V\Sigma_n^{-1}U^H = I$$

and

$$A^{\#}A = V\Sigma_n^{-1}U^H U\Sigma_n V^H = I$$

Therefore, for nonsingular matrices the inverse and the general pseudoinverse are the same. □

Example C.7.3 *Consider a full rank $m \times n$ matrix A, that is a matrix for which $\rho(A) = \min(m, n)$.*

- *If $m > n$, matrix A corresponds to an overdetermined system of linear equations, and since $r = \rho(A) = \min(m, n) = n$, the SVD of A is*

  $$A = USV^H, \quad S = \begin{bmatrix} \Sigma_n \\ 0 \end{bmatrix}$$

  Since

  $$(A^H A)^{-1} A^H = (VS^H U^H USV^H)^{-1} VS^H U^H = V\Sigma_n^{-2} [\Sigma_n \ 0] U^H$$

  we have

  $$A^{\#} = V \begin{bmatrix} \Sigma_n^{-1} & 0 \end{bmatrix} U^H = (A^H A)^{-1} A^H$$

  the left pseudoinverse of A.

- *If $m < n$, then matrix A corresponds to an underdetermined system of linear equations. In that case $r = \rho(A) = \min(m,n) = m$, and the SVD of A is*

$$A = USV^H, \quad S = \begin{bmatrix} \Sigma_n & 0 \end{bmatrix}$$

Since

$$A^H(AA^H)^{-1} = VS^HU^H(USV^HVS^HU^H)^{-1} = V\,[\Sigma_n \; 0]\,\Sigma_n^{-2}U^H$$

we have

$$A^{\#} = V \begin{bmatrix} \Sigma_n^{-1} & 0 \end{bmatrix} U^H = A^H(AA^H)^{-1}$$

the right pseudoinverse of A. □

Recall that we derived the left and right pseudoinverses by minimizing the Euclidean norms of vectors $e = Ax - b$ and x, respectively. In the previous example we saw that they are special cases of the general pseudoinverse. It can be shown that the general pseudoinverse minimizes these norms even in cases when A is not of full rank, i.e., when $\rho(A) < \min(m,n)$.

For much more about interpretation and applications of SVD, see [46].

Norms

If we consider a linear transformation

$$y = Ax$$

we often need to know what is the maximum "amplification" done by it. In other words, we want to know what is the value of

$$\max_{x\neq 0} \frac{||y||}{||x||} = \max_{x\neq 0} \frac{||Ax||}{||x||}$$

where $||z||$ denotes the Euclidean length (Euclidean norm) of a vector $z \in C^n$

$$||z|| = \sqrt{z^Hz} = \sqrt{|z_1|^2 + \ldots + |z_n|^2}$$

In this Subsection, we show that this maximum is in fact the largest of the singular values of A. We shall also mention other often used vector and matrix norms.

In order to distinguish it from the other norms to be mentioned later, when talking about the Euclidean norm, we use a subscript 2. As we shall see later, Euclidean norm is a special case of p-norms, when $p = 2$.

Let us begin with a few definitions:

Definition C.7.3 *The Euclidean norm of a vector $z \in C^n$ is*

$$||z||_2 = \sqrt{z^Hz} = \sqrt{|z_1|^2 + \ldots + |z_n|^2}$$

Definition C.7.4 *The induced Euclidean norm of a matrix is*

$$\|A\|_2 = \max_{x \neq 0} \frac{\|Ax\|_2}{\|x\|_2}$$

Theorem C.7.5 *For any complex matrix A of size $n \times n$*

$$\|A\|_2 = \overline{\sigma}(A)$$

where $\overline{\sigma}(A)$ is the largest singular value of A.

Proof. Consider

$$\|A\|_2^2 = \left(\max_{x \neq 0} \frac{\|Ax\|_2}{\|x\|_2} \right)^2 = \max_{\|x\|_2 = 1} \|Ax\|_2^2 = \max_{\|x\|_2=1} [x^H A^H A x]$$

Since $A^H A$ is Hermitian and positive semi-definite, all of its eigenvalues are real and nonnegative:

$$\lambda_1 \geq \ldots \geq \lambda_n \geq 0$$

and its eigenvectors can be chosen to form an orthonormal basis :

$$r_1, \ldots, r_n, \quad r_i^H r_j = \delta_{ij}$$

Therefore, any x such that $\|x\|_2 = 1$ can be represented as

$$x = \alpha_1 r_1 + \ldots + \alpha_n r_n, \quad \text{with} \quad \|\alpha\|_2 = 1$$

Using this decomposition, we see that for any x such that $\|x\|_2 = 1$

$$\begin{aligned} x^H A^H A x &= x^H (A^H A x) \\ &= x^H (A^H A (\alpha_1 r_1 + \ldots + \alpha_n r_n)) \\ &= x^H (\alpha_1 \lambda_1 r_1 + \ldots + \alpha_n \lambda_n r_n) \\ &= (\alpha_1 r_1 + \ldots + \alpha_n r_n)^H (\alpha_1 \lambda_1 r_1 + \ldots + \alpha_n \lambda_n r_n) \\ &= |\alpha_1|^2 \lambda_1 + \ldots + |\alpha_n|^2 \lambda_n \end{aligned}$$

Therefore

$$\max_{\|x\|_2=1} [x^H A^H A x] = \max_{\|\alpha\|_2=1} (|\alpha_1|^2 \lambda_1 + \ldots + |\alpha_n|^2 \lambda_n)$$

Since all λ_i's are nonnegative, and λ_1 is the largest among them, and since $\|\alpha\|_2 = 1$, so that $0 \leq |\alpha_i|^2 \leq 1$, we have

$$\max_{\|x\|_2=1} [x^H A^H A x] = \lambda_1 = \lambda_{max}(A^H A) = (\overline{\sigma}(A))^2$$

Therefore $||A||_2 = \overline{\sigma}(A)$. This value is achieved when $\alpha_1 = 1$, while all other α_i's are zero, so that $x = r_1$, the normalized eigenvector of $A^H A$ corresponding to its largest eigenvalue. □

Similarly, we can show that the minimum "amplification" equals the smallest singular value of A.

Theorem C.7.6 *For any complex matrix A of size $n \times n$*

$$\min_{x \neq 0} \frac{||Ax||_2}{||x||_2} = \underline{\sigma}(A)$$

where $\underline{\sigma}(A)$ is the minimum singular value of A.

Proof. Like in the proof of the previous theorem, we use the decomposition

$$x = \alpha_1 r_1 + \ldots + \alpha_n r_n$$

to find that

$$\begin{aligned}
\left(\min_{x \neq 0} \frac{||Ax||_2}{||x||_2}\right)^2 &= \min_{||x||_2=1} ||Ax||_2^2 \\
&= \min_{||x||_2=1} [x^H A^H A x] \\
&= \min_{||\alpha||_2=1} (|\alpha_1|^2 \lambda_1 + \ldots + |\alpha_n|^2 \lambda_n) \\
&= \lambda_n = \lambda_{min}(A^H A) = (\underline{\sigma}(A))^2
\end{aligned}$$

This minimum is achieved when x is the normalized eigenvector of $A^H A$ corresponding to its smallest eigenvalue. □

Theorem C.7.7 *For any complex matrix A of size $n \times n$*

$$\underline{\sigma}(A) \leq |\lambda(A)| \leq \overline{\sigma}(A)$$

for all eigenvalues of A.

Proof. Earlier we proved that for any non-zero vector x

$$\underline{\sigma}(A) \leq \frac{||Ax||_2}{||x||_2} \leq \overline{\sigma}(A)$$

Since for all $i = 1, 2, \ldots, n$

$$\frac{||Ar_i||_2}{||r_i||_2} = ||Ar_i||_2 = ||\lambda_i r_i||_2 = |\lambda_i| \cdot ||r_i||_2 = |\lambda_i|$$

the magnitudes of all eigenvalues of A are bounded by $\underline{\sigma}(A)$ and $\overline{\sigma}(A)$. □

Theorem C.7.8 *For any nonsingular complex matrix A*

$$\overline{\sigma}(A^{-1}) = \frac{1}{\underline{\sigma}(A)}$$

Proof. If we put $x = Ay$,

$$\begin{aligned}
\overline{\sigma}(A^{-1}) &= \max_{x\neq 0} \frac{\|A^{-1}x\|_2}{\|x\|_2} \\
&= \max_{y\neq 0} \frac{\|y\|_2}{\|Ay\|_2} \\
&= 1/\left(\min_{y\neq 0} \frac{\|Ay\|_2}{\|y\|_2}\right) \quad = \quad \frac{1}{\underline{\sigma}(A)}
\end{aligned}$$

□

In different applications, different vector and matrix norms are used, but they are all in a way equivalent, because they must satisfy the axioms of a norm.

Definition C.7.5 *Any real function $\|z\|$ of a vector $z \in C^n$ satisfying the following three axioms*

1. $\|z\| > 0 \quad (\forall z \neq 0),$ *and* $\|0\| = 0.$
2. $\|\alpha \cdot z\| = |\alpha| \cdot \|z\| \quad (\forall \alpha, z).$
3. $\|x + y\| \leq \|x\| + \|y\| \quad (\forall x, y)$ *(triangle inequality).*

is called a vector norm.

Definition C.7.6 *The induced matrix norm is defined as*

$$\|A\| = \sup_{x\neq 0} \frac{\|Ax\|}{\|x\|}$$

where $\|z\|$ denotes some vector norm.

Example C.7.4 *The p-norms are defined as*

$$\|z\|_p = (|z_1|^p + \ldots + |z_n|^p)^{1/p} \qquad (p > 0)$$

Important special cases are:

- $p = 1 \quad \Rightarrow \quad \|z\|_1 = |z_1| + \ldots + |z_n|.$
- $p = 2 \quad \Rightarrow \quad \|z\|_2 = \sqrt{|z_1|^2 + \ldots + |z_n|^2}.$ *This is the Euclidean norm.*
- $p = \infty \quad \Rightarrow \quad \|z\|_\infty = \max(|z_1|, \ldots, |z_n|).$

The corresponding induced p-norms are

$$||A||_p = \sup_{x \neq 0} \frac{||Ax||_p}{||x||_p}$$

For special cases when $p = 1, 2, \infty$ *it can be shown (see Problem* C.8.13, *Theorem* C.7.5, *and Problem* C.8.14, *respectively) that* sup *can be substituted by* max, *and that:*

$$||A||_1 = \max_j \sum_{i=1}^{n} |a_{ij}|$$

$$||A||_2 = \overline{\sigma}(A)$$

$$||A||_\infty = \max_i \sum_{j=1}^{n} |a_{ij}|$$

□

At the end, let us just mention two other matrix norms:

- Frobenius norm

$$||A||_F = \sum_{i=1}^{n} \sum_{j=1}^{n} |a_{ij}|^2 = \text{tr}(A^H A)$$

- The smallest singular value $\underline{\sigma}(A)$.

C.8 Problems

In this Section we present a few problems and illustrations of the advanced matrix results presented in this Appendix.

Problem C.8.1 Let $E_k(A)$ $(k = 1, 2, \ldots, n)$ denote the sum of all order-k principal minors of A. Then

$$\det(\lambda I - A) = \lambda^n - E_1(A)\lambda^{n-1} + \ldots + (-1)^n E_n(A)$$

Hint: Proof of this useful property can be found in [36, page 21]. Note that $E_1(A) = a_{11} + a_{22} + \ldots + a_{nn} = \text{tr}(A)$, and $E_n(A) = \det(A)$. Also note that there are $\binom{n}{k}$ order-k principal minors of A.

Problem C.8.2 Apply the result of Problem C.8.1 to

$$A = \begin{bmatrix} 1 & 0 & 0 & 0 \\ 0 & 2 & 0 & 0 \\ 0 & 0 & 3 & 0 \\ 0 & 0 & 0 & 4 \end{bmatrix}$$

Solution: The characteristic polynomial of A is

$$\det(\lambda I - A) = \lambda^4 - 10\lambda^3 + 35\lambda^2 - 50\lambda + 24$$

We can check:

$$1 + 2 + 3 + 4 = 10$$

$$\begin{vmatrix} 1 & 0 \\ 0 & 2 \end{vmatrix} + \begin{vmatrix} 1 & 0 \\ 0 & 3 \end{vmatrix} + \begin{vmatrix} 1 & 0 \\ 0 & 4 \end{vmatrix} + \begin{vmatrix} 2 & 0 \\ 0 & 3 \end{vmatrix} + \begin{vmatrix} 2 & 0 \\ 0 & 4 \end{vmatrix} + \begin{vmatrix} 3 & 0 \\ 0 & 4 \end{vmatrix} = 35$$

$$\begin{vmatrix} 1 & 0 & 0 \\ 0 & 2 & 0 \\ 0 & 0 & 3 \end{vmatrix} + \begin{vmatrix} 1 & 0 & 0 \\ 0 & 2 & 0 \\ 0 & 0 & 4 \end{vmatrix} + \begin{vmatrix} 1 & 0 & 0 \\ 0 & 3 & 0 \\ 0 & 0 & 4 \end{vmatrix} + \begin{vmatrix} 2 & 0 & 0 \\ 0 & 3 & 0 \\ 0 & 0 & 4 \end{vmatrix} = 50$$

$$\begin{vmatrix} 1 & 0 & 0 & 0 \\ 0 & 2 & 0 & 0 \\ 0 & 0 & 3 & 0 \\ 0 & 0 & 0 & 4 \end{vmatrix} = 24$$

Problem C.8.3 Matrices of the form

$$D = ab^H$$

where a and b are vectors of size n, are called the dyads. Use the result of Problem C.8.1 to show that the eigenvalues of D are $\lambda_1 = a^H b$ and $\lambda_i = 0$ $(i = 2, \ldots, n-1)$. What are the corresponding eigenvectors?

Problem C.8.4 Prove that if λ_1 is a multiplicity r eigenvalue of an $n \times n$ matrix A, then $\nu(\lambda_1 I - A)$, the number of linearly independent eigenvectors corresponding to λ_1, satisfies

$$1 \leq \nu(\lambda_I - A) \leq r$$

Thus, for example, if the eigenvalues of A are $1, 1, 1, 7, 7$, at most three linearly independent eigenvectors corresponding to the triple eigenvalue 1 can exist.

Solution: Since $\det(\lambda_1 I - A) = 0$, there is at least one eigenvector corresponding to λ_1.

On the other hand, $w = 0$ is a multiplicity r root of

$$\det(A - (\lambda_1 + w)I) = 0$$

hence this equation must have w^r as a factor:

$$(-1)^n(w^n - \alpha_1 w^{n-1} + \ldots + (-1)^{n-r}\alpha_{n-r}w^r) = 0$$

with $\alpha_{n-r} \neq 0$.

Since $\alpha_{n-r} \neq 0$ is a sum of all principal minors of order $n-r$ of the matrix $A - \lambda_1 I$, we see that at least one of them is $\neq 0$. Therefore

$$\rho(\lambda_1 I - A) \geq n - r, \qquad \text{i.e.,} \qquad \nu(\lambda_1 I - A) \leq r$$

Problem C.8.5 Prove that $\text{tr}(AB) = \text{tr}(BA)$. Use this fact and the fact that every square matrix is similar to some Jordan matrix to give an alternative proof for Formula (C.5):

$$\lambda_1 + \lambda_2 + \ldots + \lambda_n = \text{tr}(A)$$

Solution: Let $C = AB$ and $D = BA$. Then $c_{ii} = \sum_k a_{ik}b_{ki}$ and $d_{ii} = \sum_k b_{ik}a_{ki}$, hence

$$\text{tr}(AB) = \sum_{i=1}^{m} c_{ii} = \sum_{i=1}^{m}\sum_{k=1}^{n} a_{ik}b_{ki} = \sum_{k=1}^{n}\sum_{i=1}^{m} a_{ik}b_{ki} = \sum_{i=1}^{n}\sum_{k=1}^{m} b_{ik}a_{ki} = \text{tr}(BA)$$

Since for any A we can write $A = SJS^{-1}$, where J is in Jordan form, we can write

$$\text{tr}(A) = \text{tr}(SJS^{-1}) = \text{tr}(JS^{-1}S) = \text{tr}(J) = \lambda_1 + \lambda_2 + \ldots + \lambda_n$$

Problem C.8.6 Prove that if A is normal, than $B = U^{-1}AU$, where U is a unitary matrix, is also normal. Also prove that if B is both upper triangular and normal then it is diagonal. Use these results to generalize Theorem C.4.2 to normal matrices.

Problem C.8.7 Square matrix A is skew-Hermitian if $A^H = -A$.

a) Prove that the skew-Hermitian matrices are normal.

b) Prove that the eigenvalues of the skew-Hermitian matrices are imaginary.

Problem C.8.8 Prove that the unitary matrices are normal. Also, prove that all eigenvalues of the unitary matrices lie on the unit circle in a complex plane.

Problem C.8.9 Prove that the set of eigenvectors of a matrix forms an orthonormal basis if and only if the matrix is normal.

Problem C.8.10 Generalize Theorems C.5.3, C.5.4, and C.5.7 to Hermitian matrices.

Problem C.8.11 Let A be $m \times n$ $(m \geq n)$. Show that the Gram matrix $G(A)$ is singular if and only if A is not of full rank. Try a direct proof. Try also a proof using the Theorem C.7.4.

Problem C.8.12 Show that the correlation matrix $R = E[u^H u]$ is singular if and only if the signal $u(t)$ is a sum of $k \leq m$ sinusoids, where m is the length of vector u.

Problem C.8.13 Show that

$$\|A\|_1 = \max_j \sum_{i=1}^{n} |a_{ij}|$$

Solution: If $y = Ax$ then $y_i = \sum_{j=1}^{n} a_{ij}x_j$, and

$$\|Ax\|_1 = \|y\|_1 = \sum_{i=1}^{m} |y_i| = \sum_{i=1}^{m} \left| \sum_{j=1}^{n} a_{ij}x_j \right|$$

Therefore

$$\begin{aligned}
\|A\|_1 &= \sup_{x \neq 0} \frac{\|Ax\|_1}{\|x\|_1} = \sup_{\|x\|_1 = 1} \|Ax\|_1 \\
\|A\|_1 &= \sup_{\|x\|_1 = 1} \sum_{i=1}^{m} \left| \sum_{j=1}^{n} a_{ij}x_j \right| \leq \sup_{\|x\|_1 = 1} \sum_{i=1}^{m} \sum_{j=1}^{n} |a_{ij}|\,|x_j| \\
\|A\|_1 &\leq \underbrace{\left(\sup_{\|x\|_1 = 1} \sum_{j=1}^{n} |x_j| \right)}_{1} \left(\max_j \sum_{i=1}^{m} |a_{ij}| \right) = \max_j \sum_{i=1}^{m} |a_{ij}|
\end{aligned}$$

Since this upper bound is actually achieved for $x = e^{(k)}$, where k is such that $\sum_{i=1}^{m} |a_{ik}| = \max_j \sum_{i=1}^{m} |a_{ij}|$, we can write

$$\|A\|_1 = \sup_{x \neq 0} \frac{\|Ax\|_1}{\|x\|_1} = \max_{x \neq 0} \frac{\|Ax\|_1}{\|x\|_1} = \max_j \sum_{i=1}^{m} |a_{ij}|$$

Problem C.8.14 Show that

$$\|A\|_\infty = \max_i \sum_{j=1}^{n} |a_{ij}|.$$

Hint: This upper bound is achieved for $x = [1\ 1\ \ldots\ 1]^T$, so we can write

$$\|A\|_\infty = \sup_{x \neq 0} \frac{\|Ax\|_\infty}{\|x\|_\infty} = \max_{x \neq 0} \frac{\|Ax\|_\infty}{\|x\|_\infty} = \max_i \sum_{j=1}^{m} |a_{ij}|$$

Problem C.8.15 Prove that

$$\|A\|_2 \leq \|A\|_1 \cdot \|A\|_\infty$$

Solution:

$$\|A\|_2^2 = (\overline{\sigma}(A))^2 = \lambda_{max}(A^H A)$$

For any square matrix B we can write

$$|\lambda(B)| \leq \max_{x \neq 0} \frac{\|Bx\|_1}{\|x\|_1} = \max_j \sum_{i=1}^m |b_{ij}|$$

Therefore, for $B = A^H A$, when $b_{ij} = \sum_{k=1}^m a_{ki} a_{kj}$, we have

$$\|A\|_2^2 \leq \max_j \sum_{i=1}^m |b_{ij}| = \max_j \sum_{i=1}^m \left| \sum_{k=1}^m a_{ki} a_{kj} \right|$$

Finally we see that

$$\|A\|_2^2 \leq \left(\max_i \sum_{k=1}^m |a_{ki}| \right) \left(\max_k \sum_{j=1}^m |a_{kj}| \right)$$

Problem C.8.16 Prove that

$$\overline{\sigma}(A) - 1 \leq \overline{\sigma}(I + A) \leq \overline{\sigma}(A) + 1$$

Solution: The right-hand side inequality is a simple consequence of the triangle inequality:

$$\overline{\sigma}(I + A) \leq \overline{\sigma}(A) + \overline{\sigma}(I) \leq \overline{\sigma}(A) + 1$$

The left-hand side inequality is also a consequence of the triangle inequality:

$$\begin{aligned} \overline{\sigma}(A) - 1 &= \overline{\sigma}(I + A - I) - 1 \\ &\leq \overline{\sigma}(I + A) + \overline{\sigma}(-I) - 1 = \overline{\sigma}(I + A) \end{aligned}$$

Problem C.8.17 Prove the following property:

$$\underline{\sigma}(I + A) \geq 1 - \overline{\sigma}(A)$$

Solution: Since the smallest singular value of a matrix is a norm, we can use the triangle inequality:

$$1 = \underline{\sigma}(I) = \underline{\sigma}(I + A - A) \leq \underline{\sigma}(I + A) + \underline{\sigma}(A) \leq \underline{\sigma}(I + A) + \overline{\sigma}(A)$$

Problem C.8.18 Prove the Cauchy-Schwarz-Buniakowski inequality:

$$\left(\sum_{i=1}^n x_i y_i \right)^2 \leq \left(\sum_{i=1}^n x_i^2 \right) \left(\sum_{i=1}^n y_i^2 \right)$$

which can also be written as

$$|x^T y| \leq \|x\|_2 \|y\|_2$$

Solution: For any scalar a and vectors x and y

$$||ax + y||_2^2 \geq 0$$

i.e.,

$$||x||_2^2 a^2 + 2x^T y a + ||y||_2^2 \geq 0$$

therefore for any a the discriminant of this quadratic trinomial must be ≤ 0:

$$(2x^T y)^2 - 4||x||_2^2 ||y||_2^2 \leq 0$$

i.e.,

$$|x^T y| \leq ||x||_2 ||y||_2$$

The equality is satisfied if and only if $ax + y = 0$, i.e., if x and y are linearly dependent.

Bibliography

[1] S. Barnett and D. D. Šiljak, "Routh's Algorithm, A Centennial Survey," *SIAM Review,* Vol. 19, pp. 472–489, April 1977.

[2] R. E. Bellman, *Dynamic Programming,* Princeton University Press, 1957.

[3] S. Bennett, *A History of Control Engineering 1800-1930,* Peter Peregrinus, 1978.

[4] S. Bennett, *A History of Control Engineering 1930-1955,* Peter Peregrinus, 1993.

[5] H. S. Black, "Inventing the negative feedback amplifier," *IEEE Spectrum,* Vol. 14, pp. 54–60, December 1977.

[6] H. W. Bode, "Relations between attenuation and phase in feedback amplifier design," *Bell System Technical Journal,* Vol. 19, pp. 421–454, July 1940.

[7] H. W. Bode, *Network Analysis and Feedback Amplifier Design,* Van Nostrand, 1945.

[8] R. G. Brown and P. Y. C. Hwang, *Introduction to Random Signals and Applied Kalman Filtering,* Second Edition, John Wiley and Sons, 1992.

[9] A. E. Bryson, Jr., *Control of Spacecraft and Aircraft,* Princeton University Press, 1994.

[10] W. B. Cannon, *The Wisdom of the Body,* Norton, 1936.

[11] D. Cvetković, *Teorija grafova,* Naučna knjiga, 1981.

[12] P. Dorato, "Control history from 1960," *Proceedings of the 13th Triennial IFAC World Congress in San Francisco, USA,* pp. 129–134, 1996.

[13] P. Dorato, C. Abdallah, and V. Cernoe, *Linear-Quadratic Control: An Introduction,* Prentice Hall, 1995.

[14] W. R. Evans, "Control system synthesis by root locus method," *AIEE Trans.,* Vol. 69, pp. 1–4, 1950.

[15] H. Eves, *Elementary Matrix Theory,* Dover, 1980.

[16] J. N. Franklin, *Matrix Theory,* Prentice Hall, 1968.

[17] R. A. Gabel and R. A. Roberts, *Signals and Linear Systems,* Second Edition, John Wiley and Sons, 1980.

[18] M. Green and D. J. N. Limebeer, *Linear Robust Control,* Prentice Hall, 1995.

[19] S. Haykin, *Adaptive Filter Theory,* Second Edition, Prentice Hall, 1991.

[20] J. J. Hopfield, "Neurons with graded response have collective computational properties like those of two-state neurons," *Proceedings of the National Academy of Sciences,* Vol. 81, pp. 3088–3092, 1984.

[21] T. C. Hsia, *System Identification,* Lexington Books, 1977.

[22] T. Kailath, *Linear Systems,* Prentice-Hall, 1980.

[23] R. E. Kalman and R. W. Koepcke, "Optimal synthesis of linear sampling control systems using generalized performance indexes," *Transactions of ASME,* Vol. 80, pp. 1820–1826, 1958.

[24] R. E. Kalman, "On the general theory of control systems," *Proceedings of the First IFAC Congress in Moscow,* Vol. 1, pp. 481–492, Butterworth, 1960.

[25] R. E. Kalman, "Contributions to the theory of optimal control," *Bol. Soc. Mat. Mexicana,* Vol. 5, pp. 102–119, 1960.

[26] R. E. Kalman, "A new approach to linear filtering and prediction problems," *Transactions of ASME, Serial D, Journal of Basic Engineering,* Vol. 82, pp. 35–45, 1960.

[27] R. E. Kalman and J. E. Bertram, "Control system analysis and design via the 'Second Method' of Lyapunov. I. Continuous-time systems," *Transactions of ASME, Serial D, Journal of Basic Engineering,* Vol. 82, pp. 371–393, 1960.

[28] R. E. Kalman and R. S. Bucy, "New results in linear filtering and prediction theory," *Transactions of ASME, Serial D, Journal of Basic Engineering,* Vol. 83, pp. 95–108, 1961.

[29] R. E. Kalman, P. Falb, and M. A. Arbib, *Topics in Mathematical System Theory,* McGraw-Hill, 1969.

[30] E. W. Kamen and B. S. Heck, *Fundamentals of Signals and Systems Using* MATLAB, Prentice Hall, 1997.

[31] S. M. Kay, *Fundamentals of Statistical Signal Processing: Estimation Theory,* Prentice Hall, 1993.

[32] K. H. Khalil, *Nonlinear Systems,* Macmillan, 1992.

[33] P. V. Kokotović, Editor, *Foundations of Adaptive Control,* Springer, 1991.

[34] F. L. Lewis, *Applied Optimal Control and Estimation,* Prentice-Hall, 1992.

[35] L. Ljung, *System Identification: Theory for the User,* Prentice-Hall, 1987.

[36] M. Marcus and H. Minc, *A Survey of Matrix Theory and Matrix Inequalities,* Dover, 1992.

[37] O. Mayr, *The Origins of Feedback Control,* MIT Press, 1970.

[38] D. S. Mitrinović i D. Ž. Djoković, *Polinomi i matrice,* Gradjevinska knjiga, 1986.

[39] H. Nyquist, "Regeneration theory," *Bell System Technical Journal,* Vol. 11, pp. 126–147, January 1932.

[40] K. Ogata, *Discrete-Time Control Systems,* Prentice-Hall, 1987.

[41] K. Ogata, *Modern Control Engineering,* Prentice-Hall, 1990.

[42] K. Ogata, *Designing Linear Control Systems with* MATLAB, Prentice-Hall, 1994.

[43] A. V. Oppenheim and R. W. Schafer, *Discrete-Time Signal Processing,* Prentice Hall, 1989.

[44] L. Padulo and M. A. Arbib, *System Theory,* W. B. Saunders, 1974.

[45] L. S. Pontryagin, V. G. Boltyansky, R. V. Gamkrelidze, and E. F. Mishchenko, *The Mathematical Theory of Optimal Processes,* Wiley, 1962.

[46] W. H. Press, S. A. Teukolsky, W. T. Vetterling, and B. P. Flannery, *Numerical Recipes in C: The Art of Scientific Computing,* Second Edition, Cambridge University Press, 1992.

[47] J. R. Ragazzini and L. A. Zadeh, "The Analysis of Sampled-Data Systems," *AIEE Trans.,* Vol. 71, pp. 225–234, November 1952.

[48] L. L. Scharf, *Statistical Signal Processing,* Addison-Wesley, 1991.

[49] M. R. Schroeder, *Number Theory in Science and Communication,* Second Enlarged Edition, Springer-Verlag, 1990.

[50] C. E. Shannon, "Communications in the presence of noise," *Proc. IRE,* Vol. 37, pp. 10–21, Jan. 1949.

[51] C. E. Shannon and W. Weaver, *The Mathematical Theory of Communication,* University of Illinois Press, 1949.

[52] T. B. Sheridan and W. R. Ferrell, *Man-Machine Systems: Information, Control, and Decision Models of Human Performance,* MIT Press, 1974.

[53] H. A. Simon, "The architecture of complexity," *Proceedings of the American Philosophical Society,* Vol. 106, p. 467, 1962.

[54] H. W. Sorenson, *Parameter Estimation,* Marcel Dekker, 1980.

[55] H. W. Sorenson, Editor, *Kalman Filtering: Theory and Application,* IEEE Press, 1985.

[56] J. Stillwell, *Mathematics and its History,* Springer, 1989.

[57] M. Stojić, *Kontinualni sistemi automatskog upravljanja,* Naučna knjiga, 1988.

[58] M. Stojić, *Digitalni sistemi upravljanja,* Nauka, 1990.

[59] R. F. Strangel, *Optimal Control and Estimation,* Dover, 1994.

[60] D. D. Šiljak, *Nonlinear Systems,* Wiley, 1969.

[61] D. D. Šiljak, "Alexandr Michailovich Liapunov (1857–1918)," *Journal of Dynamic Systems, Measurement, and Control,* pp. 121–122, June 1976.

[62] M. Vidyasagar, *Nonlinear Systems Analysis,* Prentice Hall, 1993.

[63] N. Wiener, *Extrapolation, Interpolation, and Smoothing of Stationary Time Series,* MIT Press, 1966.

[64] C. R. Wylie and L. C. Barrett, *Advanced Engineering Mathematics,* McGraw-Hill, 1982.

[65] D. Younger, "A Simple Derivation of Mason's Gain Formula," *Proceedings of the IEEE,* Vol. 51, No. 7, pp. 1043–1044, July 1963.

Index